Environmental Isotope in Groundwater

Edited by Junxia Li, Xianjun Xie, Yijun Yang

图书在版编目(CIP)数据

Environmental Isotope in Groundwater(同位素水文学)/李俊霞等编著.—武汉:中国地质大学出版社,2022.12

ISBN 978-7-5625-5460-8

Ⅰ.①同…　Ⅱ.①李…　Ⅲ.①同位素-水文学-研究生-教材-英文　Ⅳ.①P33

中国版本图书馆 CIP 数据核字(2022)第 233000 号

Environmental Isotope in Groundwater　　李俊霞　**等编著**

责任编辑:王凤林　　责任校对:何澍语

出版发行:中国地质大学出版社(武汉市洪山区鲁磨路 388 号)　　邮编:430074

电　话:(027)67883511　　传　真:(027)67883580　　E-mail:cbb@cug.edu.cn

经　销:全国新华书店　　http://cugp.cug.edu.cn

开本:787 毫米×1092 毫米　1/16　　字数:566 千字　　印张:22.25

版次:2022 年 12 月第 1 版　　印次:2022 年 12 月第 1 次印刷

印刷:湖北睿智印务有限公司

ISBN 978-7-5625-5460-8　　定价:78.00 元

Contents

Chapter 1 Hydrogeochemical Processes in the Groundwater System

This chapter is concerned solely with groundwater and the application of isotopes in the groundwater system. Some preliminary information on the groundwater chemistry and processes must be introduced to set the proper stage for things to come.

1.1 Groundwater resource

Globally, groundwater represents the largest volume of freshwater readily available to society and ecosystems. Groundwater is the source of drinking water for many people around the world, especially in arid regions, rural areas, and increasingly in urban and suburban environments. It is the most abundant, available source of freshwater and most extracted raw material on earth, representing about 97% of nonfrozen fresh water with withdrawal rates near 982km^3/yr. Worldwide, groundwater accounts for approximately 35% of all water withdrawals by human populations. Groundwater supplies an estimated 38% ~ 42% of the global water used for irrigation, approximately 36% of the water resources needed for households, and roughly 27% of the water needed for industry and manufacturing. Excessive groundwater abstraction, where withdrawals exceed recharge over time, can have many negative consequences, and about 1.7 billion people live in areas where groundwater resources are under threat. Exploitation of limited groundwater can stress aquifers used for water supply, produce ground subsidence, increase saline water intrusion in coastal regions, contribute to sea-level rise, and reduce water supply to groundwater-dependent ecosystems surrounding springs, rivers, estuaries, and wetlands (Gleeson et al., 2016; Wada et al., 2010).

Groundwater availability is highly heterogeneous and varies substantially among different geological and geomorphological terrains. This skewed distribution of groundwater resources, subjected to substantial stress for sustaining a large groundwater-dependent population of more than 5 billion people, has resulted in increasing water crises for both society and ecosystems supported by groundwater, as well as public health and politics. In addition, studies suggest that recent changes in climatic patterns may intensify the problem in some regions. The unaccounted loss of groundwater along with processes such as outflow to oceans, evapotranspiration, and seepage to deeper levels or pipeline leakages, add up to ~50% of the total volume of groundwater that leaves basins annually. Out of the residual groundwater resources that are accounted for 60% is stored in porous, alluvial deposits of large fluvial systems, for example, the Amazon, Nile, Indus-Ganges-Brahmaputra, Yellow, Tigris-Euphrates, Mekong, Murray-Darling, which aerially constitutes only ~40% of the continental landmass.

Groundwater, as a natural resource, is sourced in underground geological units, also known as aquifers. The upper 2km of the Earth's crust contains about 23×10^6 km^3 of groundwater, of which only $(\sim 0.1-5.0) \times 10^6$ km^3 is modern groundwater that is recharged within the last 50 years (Gleeson et al., 2016). The global-scale distribution of the depth of the groundwater table is highly heterogeneous, depending on the dynamic equilibrium between geology/geomorphology of the aquifers, hydrological conditions leading to recharge, regional to local-scale groundwater flow systems, and human impacts in terms of groundwater abstraction. In a natural setting the groundwater level can elucidate patterns of vegetation distribution and wetland extents. Shallow groundwater is believed to influence ~30% of the global continental area, including contributing ~15% to the surface water bodies, for example, perennial or seasonal rivers, and much of the remainder contributing to the plant root zone, through the capillary fringe (Fan et al., 2013).

Presently, >80% of domestic water supplies for many populous countries are met by groundwater (e.g., South Asian countries). However, identification of the presence of natural contaminants, along with delineation of human-sourced contaminants, has resulted in a growing concern about the availability of safe groundwater in many regions globally, with water quality emerging as another facet of groundwater scarcity. While there is a general consensus that the extent and processes related to natural pollutions are more pervasive and nonpoint source, the extent and effect of other emerging and unidentified groundwater contaminants, mostly sourced from agriculture, human sewage, industrial, and medical waste (e.g., nitrate, pesticides, radiogens, antibiotics) are yet to be accounted for. Intensive agriculture is associated with high inputs of chemical and synthetic pesticides that leach into groundwater systems. More acute, but barely reported and discussed, groundwater pollution can be linked to improper sanitation, which has resulted in widespread concerns about public health.

1.2 Groundwater cycling

Schematic presentations of the hydrologic cycle such as Figure 1.1 often lump the ocean, atmosphere, and land areas into single components. Yet another presentation of the hydrologic cycle is one that portrays the various moisture inputs and outputs on a basin scale. This is shown in Figure 1.1, where precipitation is taken as input and evaporation and transpiration (referred to as evapotranspiration) along with stream runoff are outputs.

The stream runoff component, referred to as overland flow, can be augmented by interflow, a process that operates below the surface but above the zone where rocks are saturated with water, and by base flow, a direct component of discharge to streams from the saturated portion of the system. Infiltration of water into the subsurface is the ultimate source of interflow and recharge to the groundwater. When the precipitation leaches into the groundwater, the water would experience complex water-sediment/rock interactions, which involves several general hydrogeochemical processes, including evapotranspiration, dissolution and precipitation, oxidization and reduction, and adsorption and desorption. These processes shape the groundwater chemistry, not only for the general anion and cation compositions, but also for the groundwater pollution.

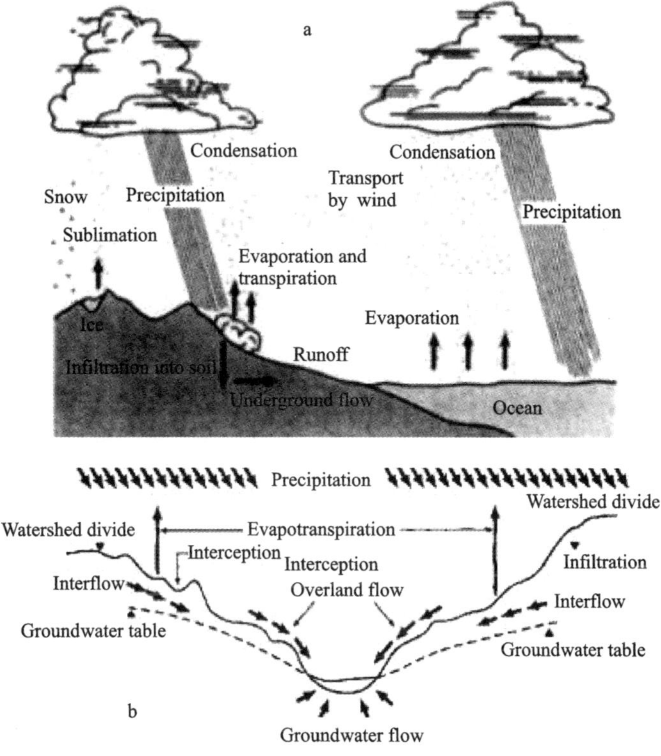

Figure 1.1 Schematic representation of the hydrologic cycle (a) and the basin hydrologic cycle (b).

1.3 Evapotranspiration

Water molecules are continually being exchanged between liquid and atmospheric water vapor. If the number passing to the vapor state exceeds the number joining the liquid, the result is evaporation. The vapor pressure of the liquid is directly proportional to the temperature. Evaporation will proceed until the air becomes saturated with moisture. The absolute humidity of a given air mass is the number of grams of water per cubic meter of air. When an air mass is cooled and the saturation humidity value drops, condensation occurs as the air mass can no longer hold all of its humidity. If the absolute humidity remains constant, the relative humidity will rise. When it reaches 100%, any further cooling will result in condensation. The dew point for an air mass is the temperature at which condensation will begin. Evaporation of water takes place from free-water surfaces—lakes, reservoirs, puddles, dew droplets, for example. The rate depends on factors such as the water temperature and the temperature and absolute humidity of the layer of air just above the free-water surface. Solar radiation is the driving energy force behind evaporation, as it warms both the water and the air. The rate of evaporation is also related to the wind, especially over land. The wind carries vapor away from the free-water surface and keeps absolute humidity low. By disturbing the water surface, the wind may also increase the rate of molecular diffusion from it.

Free-water evaporation is only part of the mechanism for mass transfer of water to the atmosphere. Growing plants are continuously pumping water from the ground into the atmosphere through a process called transpiration. Water is drawn into a plant rootlet from the soil moisture owing to osmotic pressure, whereupon it moves through the plant to the leaves. The process of transpiration accounts for most of the vapor losses from a land dominated drainage basin. The amount of transpiration is a function of the density and size of the vegetation. Transpiration is also limited by available soil water.

Under field conditions it is not possible to separate evaporation from transpiration totally. Indeed, we are generally concerned with the total water loss, or evapotranspiration, from a basin. Whether the loss is due to free-water evaporation, plant transpiration, or soil moisture evaporation is of little importance. The term potential evapotranspiration was introduced by Thornthwaite (1944) as equal to "the water loss, which will occur if at no time there is a deficiency of water in the soil for the use of vegetation." Thornthwaite recognized an upper limit to the amount of water an ecosystem will lose by evapotranspiration. The majority of the water loss due to evapotranspiration takes place during the summer months, with little or no loss during the winter. Because there is often not sufficient water available from soil water, the term actual evapotranspiration is used to describe the amount of evapotranspiration that occurs under field conditions.

Evapotranspiration is the major use of water in all but extremely humid, cool climates. If evapotranspiration were reduced, then runoff or ground-water infiltration or both could increase, as would the available water supply. Studies have shown that basin runoff from a forested waterhed has increased following the timbering of the forest (Hibbert, 1967). The increase is greatest during the first year, when there is little reforestation. As the forest regrows, the runoff again decreases. Cutting of forests to increase runoff may also result in increased erosion from the uplands and concurrent sedimentation in the lowlands. Conversion of one plant cover to another can also affect the evapotranspiration rate.

1.4 Dissolution and precipitation

1.4.1 Silicate weathering

Weathering of silicate minerals is a slow process. In the global cycle of CO_2, the weathering of silicate minerals acts as an important CO_2 sink. Furthermore, silicate weathering is the most important pH-buffering mechanism in sediments without carbonate minerals. Because the rate of silicate dissolution is slow, aquifers in silicate rock are vulnerable towards acidification.

Traditionally, weathering of detrital silicate minerals has been studied in soils. Soils may be exposed to chemical weathering over thousands of years and the variation in mineralogical composition which develops as a function of depthand time displays even very slow degradation and transformation processes of minerals.

Taking the granodiorite as an example, the parent granodiorite rock consists of plagioclase (32%), quartz (28%), K-feldspar (20%), biotite (~13%), muscovite (~7%) and small amounts of hornblende (<2%). The sequential disappearance of different silicate minerals reflects the overall kinetic control on the distribution of primary silicate minerals during weathering. This was recognized as early as 1938 by Goldich who used field observations to compile the weathering sequence shown in Figure 1.2. Olivine and Ca-plagioclase are listed as the most easily weatherable minerals, and quartz as the mineral most resistant to weathering. The minerals that formed at the highest temperature in the Earth (olivine) are the most unstable under weathering conditions at the surface of the Earth.

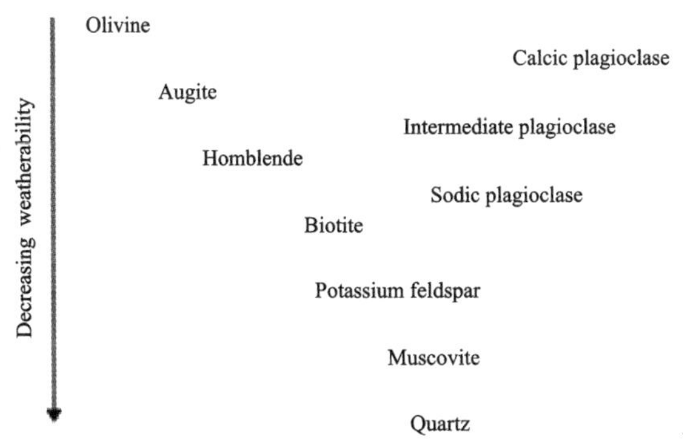

Figure 1.2 The Goldich weathering sequence based on observations of the sequence of their disappearance in soils.

The second important observation is the formation of secondary minerals like clays, here kaolinite, and Fe-oxides during the weathering process. These are the insoluble remnants which form during incongruent dissolution of primary silicate minerals. Incongruent dissolution strictly means that the ratio of elements appearing in the solution differs from the ratio in the dissolving mineral. In silicate weathering studies the term incongruent dissolution is commonly extended to include the effect of secondary precipitates. Weathering reactions for some common primary minerals are listed follow, where the clay mineral kaolinite is used as an example of a weathering product:

$$2Na(AlSi_3)O_8,(Albite)+2H^+ +9H_2O \rightarrow Al_2Si_2O_5(OH)_4(Kaolinite)+2Na^+ +4H_4SiO_4$$

$$Ca(Al_2Si_2)O_8(Anorthite)+2H^+ +H_2O \rightarrow Al_2Si_2O_5(OH)_4(Kaolinite)+Ca^{2+}$$

$$2K(AlSi_3)O_8(Microcline)+2H^+ +9H_2O \rightarrow Al_2Si_2O_5(OH)_4(Kaolinite)+2K^+ +4H_4SiO_4$$

$$(Mg_{0.7}CaAl_{0.3})(Al_{0.3}Si_{1.7})O_6(Augite)+3.4H^+ +1.1H_2O \rightarrow$$

$$0.3Al_2Si_2O_5(OH)_4(Kaolinite)+2Ca^+ +0.7Mg^{2+} +1.1H_4SiO_4$$

$$2K(Mg_2Fe)(AlSi_3)O_{10}(OH)_2(Biotite)+10H^+ +0.5O_2+7H_2O \rightarrow$$

$$Al_2Si_2O_5(OH)_4(Kaolinite)+2K^+ +4Mg^{2+} +2Fe(OH)_3+4H_4SiO_4$$

The formation of secondary products is due to the insolubility of Al-compounds. The effect of silicate weathering on the water chemistry is primarily the addition of cations and silica. Nearly all silicate weathering reactions consume acid and increase the pH. Under unpolluted conditions, carbonic acid and organic acids are the most important sources of protons. As indicated by the last equations, bicarbonate will be produced during weathering of silicates. Finally, iron that is present in silicate minerals, like biotite or hornblende, may form Fe-oxide as an insoluble weathering product.

Besides kaolinite, weathering of primary silicates results in the formation of clay minerals like montmorillonite and gibbsite, which is described by the following equations:

$$3Na(AlSi_3)O_8(Albite)+Mg^{2+} +4H_2O \rightarrow$$

$$2Na_{0.5}(Al_{1.5}Mg_{0.5})Si_4O_{10}(OH)_2(montmorillonite)+2Na^+ +H_4SiO_4$$

$$Na(AlSi_3)O_8(Albite)+H^+ +7H_2O \rightarrow Al(OH)_3(Gibbsite)+Na^+ +3H_4SiO_4$$

The alteration of albite to montmorillonite consumes no acid, but with kaolinite and gibbsite as weathering products, increasing amounts of protons are consumed. Furthermore, when albite alters to montmorillonite, 89% of the Si is preserved in the weathering product, decreasing to 33% for weathering to kaolinite and to 0 for gibbsite. The sequence of weathering products, going from montmorillonite over kaolinite to gibbsite, reflects increasing intensity of leaching, removing silica and cations from the rock. The hydrological conditions in combination with the rate of mineral weathering determine the nature of the weathering product. Montmorillonite is preferentially formed in relatively dry climates, where the rate of flushing of the soil is low. Its formation is further enhanced when rapidly dissolving material such as volcanic rock is available. Gibbsite, on the other hand, typically forms in tropical areas with intense rainfall and well drained conditions. Under such conditions, gibbsite and other Al-hydroxides may form a thick weathering residue, bauxite, that constitutes the most important Al-ore.

1.4.2 Carbonate dissolution

Limestones and dolomites often form productive aquifers with favorable conditions for groundwater abstraction. The main minerals in these rocks are Ca-and Mg-carbonates which dissolve easily in groundwater and give the water its "hard" character. Usually, the carbonate rock consists of recrystallized biological material with a high porosity and a low permeability. Groundwater flow is then restricted to more permeable fracture zones and karst channels produced by carbonate dissolution. Prolonged dissolution over periods of thousands of years may result in the development of karst landscapes and fantastic features in caves. Apart from the carbonate rocks that consist exclusively of carbonate minerals, sands or sandstones, and marls and clays may contain carbonate minerals as accessory minerals or as cement around the more inert grains. On a world-wide basis, the effect of carbonate dissolution on water compositions is quite conspicuous. The high TDS values are mainly related to increases of Ca^{2+} and HCO_3^-, which predominantly are due to carbonate dissolution.

Carbonate minerals crystallize with either a trigonal or an orthorhombic crystal structure, depending on the ionic radius of the cation. The smaller ions form trigonal minerals, while the larger ones form orthorhombic minerals. The Ca^{2+} ion is intermediate with the trigonal mineral calcite, being the main component of older limestones, and the orthorhombic aragonite, which is common in recent carbonate sediments and the mineral forming pearls and mother of pearl. Rock-forming carbonate minerals are calcite, dolomite and aragonite. The other carbonate minerals, like siderite or rhodochrosite, occur only in small amounts if present at all. Still, they may exercise an important control on dissolved concentrations of Fe^{2+} and Mn^{2+}.

The overall reaction between carbon dioxide and $CaCO_3$ is

$$CO_{2(g)} + H_2O + CaCO_3 \leftrightarrow Ca^{2+} + 2HCO_3^-$$

The detailed presentation on the carbon species and reaction was discussed in Chapter 5.

1.4.3 Saturation states

Generally, the saturation index of mineral is used to evaluate the dissolution/precipitation states of the mineral. The calculation is to compare the solubility product K with the analogue product of the activities derived from water analyses. The latter is often termed the Ion Activity Product (*IAP*). For example, for calcite:

$$K_{calcite} = [Ca^{2+}][CO_3^{2-}] \text{ activities at equailibrium} \quad (1.1)$$

And

$$IAP_{calcite} = [Ca^{2+}][CO_3^{2-}] \text{ activities in water sample} \quad (1.2)$$

Saturation conditions may also be expressed as the ratio between *IAP* and *K*, or the saturation state Ω :

$$\Omega = IAP/K \quad (1.3)$$

Thus for $\Omega = 1$ there is equilibrium, $\Omega > 1$ indicates supersaturation and $\Omega < 1$ undersaturation. For larger deviations from equilibrium, a logarithmic scale can be useful given by the saturation index *SI* :

$$SI = \log(IAP/K) \quad (1.4)$$

For $SI = 0$, there is equilibrium between the mineral and the solution; $SI < 0$ reflects undersaturation, and $SI > 0$ supersaturation.

1.5 Oxidization and reduction

1.5.1 Species

Reduction and oxidation processes exert an important control on the natural concentrations of O_2, Fe^{2+}, SO_4^{2-}, H_2S, CH_4, etc. in groundwater. They also determine the fate of pollutants like nitrate leaching from agricultural fields, contaminants leaching from landfill sites, industrial spills, or heavy metals in acid mine drainage. Redox reactions occur through electron transfer from one atom to another and the order in which they proceed can be predicted from standard equilibrium thermodynamics. However, the electron transfer is often very slow and may only proceed at significant rates when mediated by bacterial catalysis. Redox processes in groundwater typically occur through the addition of an oxidant, like O_2 or NO_3^- to an aquifer containing a reductant. However, the addition of a reductant, such as dissolved organic matter (DOC) that leaches from soils or landfills can also be important.

For chemical reactions in which electrons are transferred from one ion to another (oxidation-reduction reactions, or redox reactions), the oxidation potential of an aqueous solution is called the *Eh*. A transfer of electrons is an electrical current; therefore, a redox equation has an electrical potential. At 25℃ and 1atm pressure, the standard potential, E_0 (in volts), has been measured for many reactions. The sign of the potential is positive if the reaction is oxidizing and negative if it is reducing. The absolute value of E_0 is a measure of the oxidizing or reducing tendency.

The oxidation potential of a reaction is given by the Nernst equation:

$$Eh = E_0 + \frac{RT}{nF}\ln K_{sp} \tag{1.5}$$

Where R is the gas constant, 0.008 314kJ/(mol · K), T is the temperature (kelvins), F is the Faraday constant, 96.484kJ/V, and n is the number of electrons.

Eh expresses the activity of the electrons in solution (i. e., the redox level) in units of volts. We can also express the redox level in terms of electron activity, or *pe*. *Eh* and *pe* are related by the following equation:

$$pe = \frac{F}{2.303RT}Eh \tag{1.6}$$

At a temperature of 25℃,

$$Eh = 0.059pe \tag{1.7}$$

$$pe = 16.9Eh \tag{1.8}$$

Oxidation potential is measured with a specific ion electrode meter. A positive value indicates that the solution is oxidizing; a negative value indicates that it is chemically reducing.

If the pH and *Eh* of an aqueous solution are known, the stability of minerals in contact with the water may be determined. This stability relationship is best represented on an *Eh*-pH diagram. Water, itself, is stable only in a certain part of the *Eh*-pH field. Figure 1.3 shows the framework of aqueous *Eh*-pH fields. Water in nature at near-surface environments is usually between pH 4 and pH 9, although values that are more acidic or more basic can occur. One can also make *pe*-pH diagrams; the only way they differ from an *Eh*-pH diagram is in the scale on the vertical axis. *Eh* is converted to *pe* according to Eq. 1.6. The *Eh*-pH diagram can be used to show the fields of stability for both solid and dissolved ionic species. It has been used very effectively for iron. The *Eh*-pH diagram depends on the concentrations of all ionic species present. For the simple ions and hydroxides of iron, the fields depend upon the molality of the iron in solution. Figure 1.4 shows the stability-field diagram for a 10^{-7}-molal solution of iron. The iron may be either in the Fe^{3+} or Fe^{2+} valence state, depending on its position in the stability field. As the iron concentration increases, the line separating the ferrous and ferric state shifts to the left. This is demonstrated by a dashed line in Figure 1.4, which represents a 1-molal iron concentration. Ferrous iron can exist as Fe^{2+}, $Fe(OH)^{+}$, or $Fe(OH)_2$, depending upon the *Eh* and pH of the solution; ferric iron can be in the forms Fe^{3+}, $Fe(OH)^{2+}$, and $Fe(OH)_3$. The pH at which these species change is also a function of the total amount of iron present. Procedures are available to compute an *Eh*-pH field for iron of any molality (Hem and Cropper, 1959).

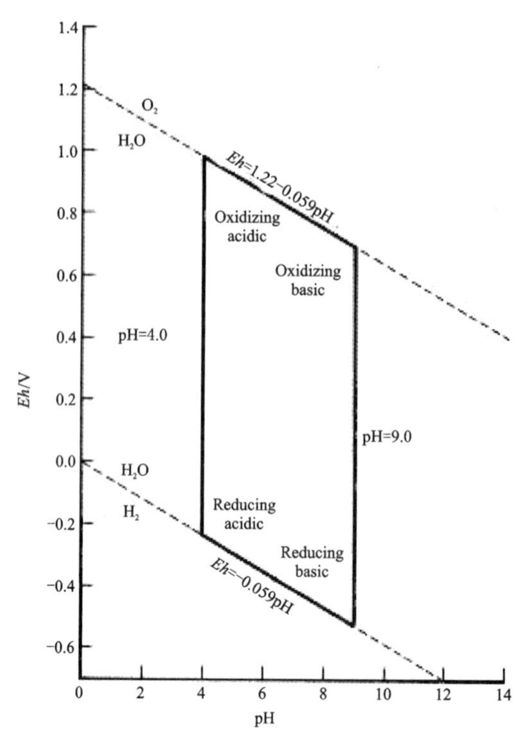

Figure 1. 3 The Eh-pH field where water is a stable component. The usual limits of Eh and pH for near-surface environments are also indicated by solid lines.

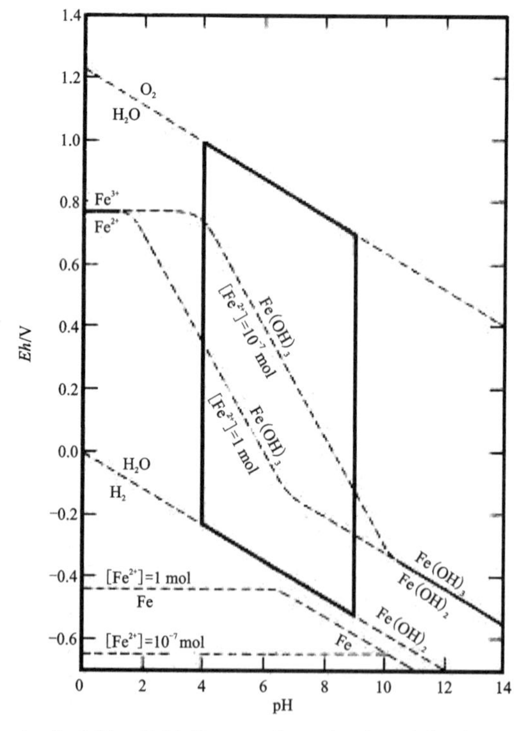

Figure 1. 4 Stability-field diagram for a 10^{-7}-molal solution of iron.

High *Eh* is generally the direct result of dissolved oxygen in the water. For deep ground-water systems, the *Eh* is usually sufficiently low that, for a pH of less than about 8, iron is present as the soluble Fe^{2+} ion. Near a recharge zone, the ground water may have sufficient dissolved oxygen to elevate the *Eh*. As the water travels through the aquifer, the oxygen is chemically reduced by contact with reducing species, and the *Eh* is lowered. The oxygen can react with the small amount of ferrous iron to form ferric hydroxide, $Fe(OH)_3$. Interestingly, the ferric hydroxide thus formed may be colloidal, and can move through the aquifer with the ground water. In the *Eh*-pH range where Fe^{2+} exists, large amounts of dissolved iron can be present.

Natural water contain many ionic species. Again, using iron as an example, an *Eh*-pH diagram can be used to show the stable iron minerals in a mixed aqueous solution with iron, sulfur, and carbonate present. This is done in Figure 1.5; the given concentrations are iron 56μg/L (10^{-6} molar), 96mg/L dissolved sulfur as SO_4^{2-}, and 61mg/L dissolved carbon dioxide as HCO_3^-. Shaded areas indicate *Eh*-pH domains where solid species would be thermodynamically stable. The stable ionic species are also indicated on the figure. For a thorough discussion of *Eh*-pH diagrams, see Drever (1997).

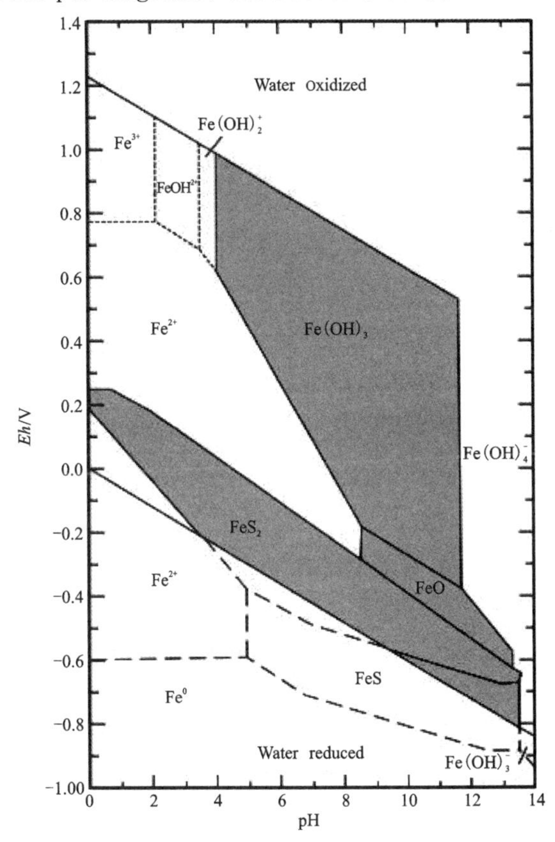

Figure 1.5 Stability fields based on *Eh* and pH for solid and dissolved forms of iron in an aqueous solution of 56μg/L iron, 96mg/L of sulfur as SO_4^{2-}, 61mg/L carbon dioxide as HCO_3^- at 25℃, and 1atm pressure.

1.5.2 Biodegradation (natural organic carbon)

As water contacts reductants in the subsoil, it loses its oxidants in a sequence that follows the *pe* from high to low and the changes in water chemistry can be predicted by the redox diagrams. The sequence of predominant redox half reactions that can be predicted from the redox diagrams is summarized in Figure 1.6 for pH 7. The upper part lists the reduction reactions going from O_2 reduction, NO_3 reduction and reduction of Mn-oxides that occur at a high *pe*, to the reduction of Fe oxides, sulfate reduction and methanogenesis taking place at a lower *pe*. The lower half of Figure 1.6 lists in the same way the oxidation reactions, with the oxidation of organic matter having the lowest *pe*. A reduction reaction will proceed with any oxidation reaction that is located (origin of the arrow) at a lower *pe*. For example, the reduction of sulfate can be combined with the oxidation of organic matter, but not with the oxidation of Fe^{2+}.

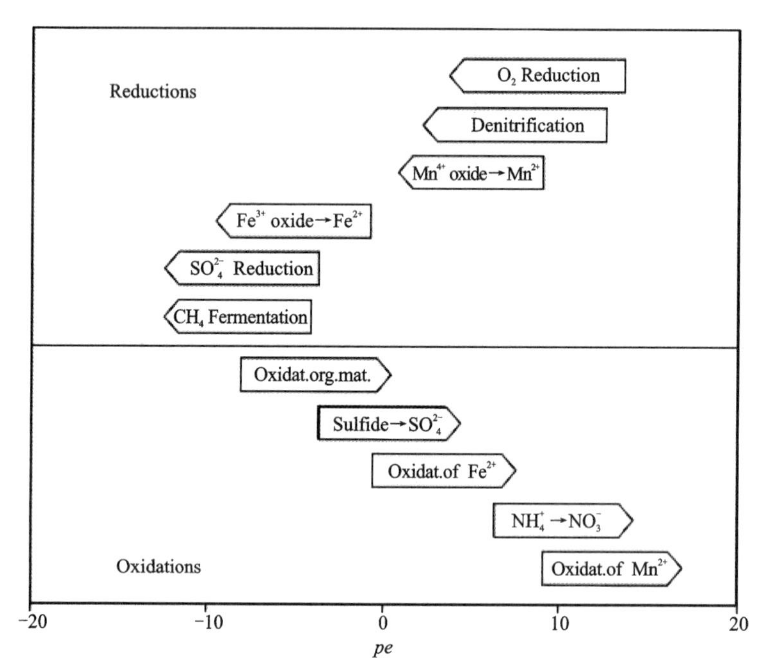

Figure 1.6 Sequences of important redox processes at pH=7 in natural systems (modified and corrected after Stumm and Morgan, 1996).

The degradation of organic matter proceeds via multiple enzymatic reactions involving different organisms and oxidants as well as a number of intermediate compounds. Organic matter oxidation is coupled to the sequential utilization of terminal electron acceptors (TEAs), typically in the order of O_2, NO_3^-, Mn^{4+}, Fe^{3+} and SO_4^{2-} followed by methanogenesis and/or fermentation. Depending on the degradation pathway, organic matter is directly oxidized to CO_2, partly oxidized to intermediate compounds or reduced to

CH_4. Ultimately only a small fraction of the deposited organic carbon escapes degradation. Yet, compilation of field data reveals that the degradation efficiency is not constant and that organic carbon burial rates vary significantly in space and time. Many different factors have been invoked to explain this spatial and temporal variability. They can be conveniently divided into factors that have an indirect influence on organic matter degradation, for instance deposition rate or macro activity and factors that have a direct effect on the degradation process. The latter include, but are not limited to, organic matter composition, for instance non-hydrolyzable substrates that resist fermentative breakdown, electron acceptor availability, community composition, microbial inhibition by specific metabolites, priming or physical protection (Figure 1. 7). The degradation of organic matter is thus a true reaction – transport problem that involves chemical, biological and physical processes. the microbial community exerts an important influence on the degradability of organic matter. The degradation of the deposited material is thermodynamically and/or kinetically controlled by the different abilities of the physiological groups that compete for the common substrate. Transport processes, such as bioturbation, bioirrigation and sedimentation accumulation control the supply of substrates and terminal electron donors and, therefore, exert an additional indirect influence on the degradation process. The relative significance of these controls and the degree to which they influence each other strongly depend on the characteristics of the depositional setting and the timescale of interest.

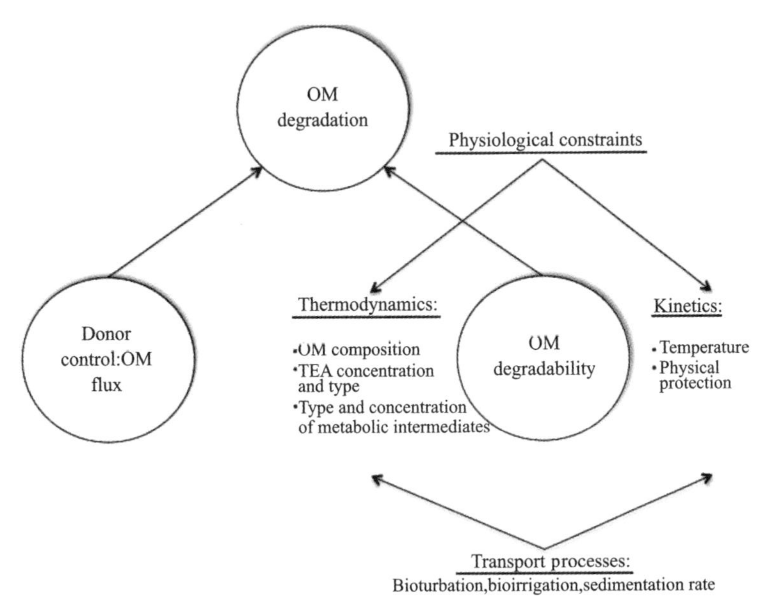

Figure 1. 7 Schematic illustration of the main controls of organic matter degradation in marine sediments (Arndt et al. , 2013).

1.6 Adsorption and desorption

1.6.1 Ion exchange

Under certain conditions, the ions attracted to a solid surface may be exchanged for other ions in aqueous solution. This process is known as ion exchange. Both cation exchange and anion exchange can occur, but in some natural soils cation exchange is the dominant process. The presence of exchange sites is a function of the same general conditions affecting adsorption sites. The ion-exchange process can be conceptualized as the preferential absorption of selective ions with concomitant loss of other ions. Ion-exchange sites are found primarily on clays and soil organic materials, although all soils and sediments have some ion-exchange capacity.

The ion-exchange reactions of different soils must be studied individually in the laboratory. Results are reported in terms of milliequivalents per 100g of soil. A general ordering of cation exchangeability for common ions in ground water is: $Na^+ > K^+ > Mg^{2+} > Ca^{2+}$. The divalent ions are more strongly bonded and tend to replace monovalent ions. However, it is a reversible reaction and, at high activities, the monovalent ions can replace divalent ions. This is the concept behind the home water softener. The divalent Ca^{2+} and Mg^{2+} ions replace the monovalent Na^+ ions on the exchange media. The exchange medium is regenerated when a brine solution with very high Na^+ activity is forced through the softener. The Na^+ replaces the Ca^{2+} and Mg^{2+} at the exchange sites, for instance, seawater intrusion. Ion-exchange capacities of organic colloids and clays can also remove heavy metal cations and thus provide some protection to groundwater supplies, but enough cases of ground-water pollution from heavy metals have been documented to demonstrate that such protection is limited to areas with clay in the soils.

While in principle all solid surfaces in soils and aquifers can act as adsorbers, solid phases with a large specific surface area will adsorb most, and the adsorption capacity therefore depends on the grain size. Solids with a large specific surface area reside in the clay fraction ($<2\mu m$), but coarser grains in a sediment are often coated with organic matter and iron oxyhydroxides. The adsorption capacity is therefore linked to the clay content (fraction $<2\mu m$), clay minerals, organic matter (%C), and oxide or hydroxide content. The current convention is to express the CEC, or Cation Exchange Capacity of a soil in meq/kg, replacing the unit meq/100g. A standard laboratory test is available to determine the cation exchange capacity of soils. A 100g sample of dry soil is mixed with a solution of ammonium acetate to saturate the exchange sites with NH^{4+} ions. The pH of the pore water is adjusted to a value of 7.0. The soil is leached with a strong NaCl solution to replace the NH_4^+ on the exchange sites with Na^+ ions. The sodium content of the leaching solution is then determined and the CEC computed as the difference between the sodium in the original solution and the sodium in the leaching solution at equilibrium. It is reported in milliequivalents per 100g of soil. The CEC is frequently used as an indication of the potential of a soil to attenuate pollutants with exchangeable ions. Table 1. 1 gives the CEC of common soil constituents.

Table 1. 1 Cation exchange capacities of common soil and sediment materials.

	CEC(meq/kg)
Kaolinite	30～150
Halloysite	50～100
Montmorillonite	800～1200
Vermiculite	1000～2000
Glauconite	50～400
Illite	200～500
Chlorite	100～400
Allophane	up to 1000

1. 6. 2 Surface sorption

Sorption is determined experimentally by measuring how much of a solute can be sorbed by a particular sediment, soil, or rock type. Aliquots of the solute in varying concentrations are well mixed with the solid, and the amount of solute removed is determined. The capacity of a solid to remove a solute is a function of the concentration of the solute. The results of the experiment are plotted on a graph called an isotherm, which shows the solute concentration versus the amount sorbed onto the solid. If the sorptive process is rapid compared with the flow velocity, the solute will reach an equilibrium condition with the sorbed phase. This process can be described by an equilibrium sorption isotherm.

Two equations are often employed to describe the relation, the Langmuir and the Freundlich isotherm. The Freundlich isotherm has the form:

$$s_I = K_F \cdot c_I^n \tag{1.9}$$

where s_I is the sorbed concentration (mol/kg, μg/g, etc.), c_I is the solute concentration of chemical I (mol/L, μg/mL, the same chemical mass units as for s), and K_F and n are adjustable coefficients. Usually, n is smaller than 1, so that the increase of sorbed concentration lessens as the solute concentration increases. For fitting of experimental data, the Freundlich equation can be linearized by a log transform to

$$\log s_I = \log K_F + n \log c_I \tag{1.10}$$

which enables the constants to be derived by linear regression.

With the Freundlich equation, sorption extends infinitely as concentrations increase, which is unrealistic since the number of sorption sites can be expected to be limited. Also, it is generally observed that the distribution coefficient, i. e. , the ratio $K'_d = s_I / c_s$, becomes a constant when concentrations are small, but in the Freundlich equation the value of

$$K'_d = K_F \cdot c^{n-1} \tag{1.11}$$

increases indefinitely as concentrations decrease (for $n < 1$). Langmuir used a kinetic approach to derive the equation for gas adsorption on a surface and obtained:

$$s_I = \frac{s_{\max} c_I}{K_L + c_I} \tag{1.12}$$

It can be checked that the equations are identical if $s_{tot} = s_{max}$ and $K_L = 1/K_s$.

In the groundwater, the hydrogeochemical behavior of trace elements is to a large extent controlled by sorption processes. Variable charge solids sorb ions from solution without releasing other ions in equivalent proportion. Their surface charge can be positive or negative depending on the pH and the solution composition. Variable charge solids are important in regulating the mobility of both positively charged heavy metals such as Pb^{2+} and Cd^{2+} , and of oxyanions such as $HAsO_4^{2-}$ and $H_2PO_4^-$. The surfaces of Fe-oxides, clays, other minerals or organics may become protonated or deprotonated and obtain different surface properties as a function of pH. The surface charge of the solid is the sum of structural deficits, unbalanced bonds at the crystal surface and charge generated by the adsorbed ions. At a given pH, the proton charge will compensate all other charge. This pH is called the point of zero charge, PZC, or pH_{PZC}. Table 1. 2 lists PZC's for a number of minerals. Iron oxides have PZC's ranging from 8. 5 to 9. 3 and will at most groundwater pH values be neutral or weakly positive. Birnessite, δ-MnO_2 , has a PZC of 2. 2 and is negatively charged at the pH of most groundwater. Minerals have a general capacity for anion exchange (in the double layer) when the pH is below the PZC, and a cation exchange capacity when the pH is above PZC. The capacity for exchange depends on the difference between the PZC and the pH of the solution. When the pH is higher than the PZC the oxide surface adsorbs Na^+ from solution, when the pH is below the PZC, Cl^- is adsorbed.

Table 1. 2 The Point of Zero Charge, pH_{PZC}, of clays and common soil oxides and hydroxides.

	pH_{PZC}
Kaolinite	4. 6 (Parks)
Montmorillonite	<2. 5 (Parks)
Corundum, α-Al_2O_3	9. 1 (Stumm and Morgan)

Continue

	pH_{PZC}
γ-Al_2O_3	8.5 (Stumm and Morgan)
alpha-$Al(OH)_3$	5.0 (Stumm and Morgan)
Hematite, α-Fe_2O_3	8.5 (Davis and Kent)
Goethite, α-FeOOH	9.3 (Venema et al.)
$Fe(OH)_3$	8.5 (Stumm and Morgan)
Birnessite, δ-MnO_2	2.2 (Davis and Kent)
Rutile, TiO_2	5.8 (Davis and Kent)
Quartz, SiO_2	2.9 (Davis and Kent)
Calcite, $CaCO_3$	9.5 (Parks)
Hydroxyapatite, $Ca_5OH(PO_4)_3$	7.6 (Davis and Kent)

References

ARNDT S, JøRGENSEN B B, LAROWE D E, et al., 2013. Quantifying the degradation of organic matter in marine sediments: A review and synthesis[J]. Earth Sci Rev, 123(0): 53-86.

FAN Y, LI H, MIGUEZ-MACHO G, 2013. Global patterns of groundwater table depth[J]. Science,339 (6122): 940-943.

GLEESON T, BEFUS KM, JASECHKO S, et al., 2016. The global volume and distribution of modern groundwater[J]. Nat Geosci, 9: 161-167.

HIBBERT A R,1967. Forest treatment effects on water yield[M]// Sopper W E, Lull H W Forest hydrology. Oxford, England: Pergamon Press: 527-543.

HEM J D, W H CROPPER,1959. Survey of the ferrousferric chemical equilibria and redox potential[C]. U. S. Geological Survey Water-Supply Paper, 1459-A.

STUMM W, MORGAN J J, 1996. Aquatic chemistry[M]. 3rd ed. New York: Wiley and Sons.

THORNTHWAITE C W, 1944. Report of the committee on transpiration and evaporation[R]. Transactions, American Geophysical Union, 25:687.

WADA Y, VAN BEEK L P, VAN KEMPEN C M, et al., 2010. Global depletion of groundwater resources[J]. Geophys Res Lett, 37 (20): L20402-1—L20402-5.

Chapter 2 Theory and Principle of Isotopes

Isotopes are atoms whose nuclei contain the same number of protons but a different number of neutrons. Isotopes can be divided into two fundamental kinds, stable and unstable (radioactive) species. The number of stable isotopes is about 300; whilst over 1200 unstable ones have been discovered so far. The term "stable" is relative, depending on the detection limits of radioactive decay times. In the range of atomic numbers from ^{1}H to ^{83}Bi, stable nuclides of all masses except 5 and 8 are known. Only 21 elements are pure elements, in the sense that they have only one stable isotope. All other elements are mixtures of at least two isotopes. The relative abundance of different isotopes of an element may vary substantially. In copper, for example, ^{63}Cu accounts for 69% and ^{65}Cu for 31% of all copper nuclei. For the light elements, however, one isotope is predominant, the others being present only in trace amounts. Radioactive isotopes can be classified as being either artificial or natural. Only the latter are of interest in geology, because they are the basis for radiometric dating methods. Radioactive decay processes are spontaneous nuclear reactions and may be characterized by the radiation emitted, i. e. a, b and/or c-emission. Decay processes may also involve electron capture. Radioactive decay is one process that produces variations in isotope abundance. A second cause of differences in isotope abundance is isotope fractionation caused by small chemical and physical differences between the isotopes of an element.

2.1 Definition

2.1.1 Isotope expression

Stable isotope concentrations are measured as a ratio of the rare to the abundant isotope and expressed as the difference in this ratio between the sample and a known reference, taking Oxygen as an example:

$$\delta^{18}O_{sample} = \frac{(^{18}O/^{16}O)_{sample} - (^{18}O/^{16}O)_{reference}}{(^{18}O/^{16}O)_{reference}} \tag{2.1}$$

This normalized difference between the sample and reference is then multiplied by 1000 to express the measurement in permil (‰) units:

$$\delta^{18}O_{sample} = \left[\frac{\left(\frac{^{18}O}{^{16}O}\right)_{sample}}{\left(\frac{^{18}O}{^{16}O}\right)_{V\text{-}SMOW}} - 1 \right] \times 1000‰ \text{ V-SMOW} \tag{2.2}$$

where δ or delta indicates the difference in ratio from the standard, V-SMOW (Vienna Standard Mean Ocean Water) is the standard used in this example, and ‰ is the permil notation.

A δ-value that is positive, say, $\delta^{18}O = +10‰$, signifies that the sample has 10 permil more ^{18}O than the reference, and so is enriched in ^{18}O. Similarly, a sample with $\delta^{18}O = -10‰$ has 10‰ less ^{18}O than V-SMOW, and so is depleted in ^{18}O. The δ-‰ notation for stable isotope data is a derived unit of concentration that comes from the requirement to normalize measurements to a reference material for accurate measurements. However, isotope measurements could be expressed in concentration units similar to those used for solutes. For example, the $^{18}O/^{16}O$ abundance or atomic ratio in V-SMOW is 0.002 005 or 2005×10^{-6} (2005 atoms of ^{18}O per million atoms of ^{16}O or 2009 atoms of ^{18}O per million oxygen atoms (^{16}O plus ^{18}O). Similarly, deuterium in V-SMOW has a concentration 150×10^{-6}. The routinely measured stable environmental isotopes are given in Table 2.1 along with their average natural abundance and the international references used for their measurement.

Table 2.1 Stable environmental isotopes.

Isotope	Ratio	Abundance(%)	Reference(Abundance Ratio)	Common Sample Types
D or ^2H	D/H	0.015	V-SMOW($1.557\ 5 \times 10^{-4}$)	H_2O, CH_4, clays
^{13}C	$^{13}C/^{12}C$	1.11	VPDB($1.123\ 7 \times 10^{-2}$)	DIC, CO_2, $CaCO_3$, CH_4, organic C
^{15}N	$^{15}N/^{14}N$	0.366	AIR N_2(3.677×10^{-3})	NO_3^-, NH_4^+, N_2, N_2O
^{18}O	$^{18}O/^{16}O$	0.204	V-SMOW($2.005\ 2 \times 10^{-3}$)	H_2O, NO_3^-, SO_4^{2-}, O_2, minerals
^{34}S	$^{34}S/^{32}S$	4.21	CDT($4.500\ 5 \times 10^{-2}$)	SO_4^{2-}, H_2S, gypsum, sulfide minerals
^3He	$^3He/^4He$	0.000 138	AIR(1.38×10^{-6})	groundwater, minerals
^6Li	$^6Li/^7Li$	7.6	LSVEC(0.082 15)	water, brines, minerals
^{11}B	$^{11}B/^{10}B$	80.1	NBS 951(4.044)	water, brines, minerals
^{37}Cl	$^{37}Cl/^{35}Cl$	24.23	SMOC(0.324)	water, brines, solvents
^{87}Sr	$^{87}Sr/^{86}Sr$	7.0 and 9.8	Direct measurement	water, brines, minerals

V-SMOW, Vienna Standard Mean Ocean Water; VPDB, Vienna Pee Dee Belemnite, fossil carbonate; CDT, Canon Diablo Troilite, FeS from meteorite; LSVEC, Lithium carbonate standard (Flesch et al., 1973; Coplen, 2011), now also used as a carbonate standard (Coplen et al., 2006); NBS-951, Boric acid standard, also SRM 951, National Bureau of Standards; SMOC, Standard Mean Ocean Chloride.

The stability of a nuclide is compromised by the addition or subtraction of neutrons. For example, the addition of one neutron to ^{13}C produces the radioisotope ^{14}C with a half-life of 5730 years. Adding one neutron to deuterium produces tritium, T, with a half-life of 12.32 years. The radioisotopes are unstable nuclides that decay with time to a more stable configuration by emission of particles and gamma radiation. Although many radionuclides exist, only a few are currently of interest in hydrogeology. These include radionuclides used for dating and tracing (Table 2.2) and those of concern as radio hazards.

Abundance of radioisotopes in environmental samples can be measured and expressed by either their activity as decay eventsper time or, like stable isotopes, by their abundance ratio. The becquerel (Bq) is equal to one disintegration per second. This measurement is then normalized to a given sample size, using units of Bq/L for water or Bq/g for solids.

Table 2.2 The common environmental radioisotopes in groundwater.

Isotope	Half-life (years)	Decay mode	Atmospheric ratio	Activity (Bq)	Common sample types
^{85}Kr	10.76	β^-	$^{85}Kr/Kr=2.7\times10^{-11}$	1.5 Bq/cc_{Kr}	Dissolved Kr
T or ^{3}H	12.32	β^-	$T/H=10^{-17}$	0.12 Bq/L	H_2O, CH_2O, H_2
^{39}Ar	269	β^-	$^{39}Ar/Ar=8.2\times10^{-16}$	1.8×10^{-6} Bq/cc_{Ar}	Dissolved Ar
^{14}C	5730	β^-	$^{14}C/C=1.18\times10^{-12}$	0.226 Bq/gC	DIC, DOC
^{81}Kr	229 000	EC	$^{81}Kr/Kr=5.2\times10^{-13}$	1.34×10^{-6} Bq/cc_{Kr}	Dissolved Kr
^{36}Cl	301 000	β^-	$^{36}Cl/Cl\text{-}10^{-13}$	10^{-6} Bq/L (10×10^{-6} Cl)	Cl^-
^{129}I	15.7×10^{6}	β^-	$^{129}I/I\text{-}10^{-8}$	10^{-4} Bq/L ($\times10^{-6}$ I)	I^-, IO_4^-, CH_3I

2.1.2 Fractionation factor

In 1947, Harold Urey published the theoretical basis for isotope fractionation as an exchange of isotopes between molecules that are participating in a reaction. This established the equilibrium fractionation factor, α, for isotope reactions, which is essentially the equilibrium constant, K:

$$aA_1 + bB_2 \rightarrow aA_2 + bB_1 \tag{2.3}$$

where the subscripts indicate that species A and B contain either the light or heavy isotope 1 or 2, respectively. For this reaction the equilibrium constant K is expressed by:

$$K = \left(\frac{A_2}{A_1}\right)^a / \left(\frac{B_2}{B_1}\right)^b \tag{2.4}$$

K defines the equilibrium condition for mineral solubilities and solute concentrations in geochemical reactions, the fractionation factor, α, establishes the isotope ratios in components of isotope-exchange reactions. Unlike K, which can vary enormously (e. g.

$10^{1.58}$ for dissociation of halite to $10^{-8.4}$ for the dissociation of calcite), isotope fractionation factors are for the most part very close to 1. This is because the partitioning of isotopes between compounds is very minor.

For isotope exchange reactions in geochemistry, the equilibrium constant K is often replaced by the fractionation factor a . The fractionation factor is defined as the ratio of the numbers of any two isotopes in one chemical compound A divided by the corresponding ratio for another chemical compound B :

$$\alpha_{A-B} = \frac{R_A}{R_B} \tag{2.5}$$

Consider the evaporation and condensation of water as an example of a reaction. As the reaction approaches equilibrium, the transfer of reactants to products is matched by the reverse reaction such that some water is evaporating to vapor and some vapor is condensing to water:

$$H_2O_{water} + H_2O_{vapor} \leftrightarrow H_2O_{vapor} + H_2O_{water}$$

Adding isotopes to the reaction, there will be forward and backward exchange of isotopes until there is an equilibrium distribution with either the reactant or the product having more of the heavy isotope. In an equilibrium isotope-exchange reaction, there is no change in the geochemistry of the solution, only an exchange of isotopes:

$$H_2{}^{16}O_{water} + H_2{}^{18}O_{vapor} \leftrightarrow H_2{}^{18}O_{vapor} + H_2{}^{16}O_{water}$$

The thermodynamic reaction constant becomes:

$$K_{w\text{-}v} = \frac{[H_2{}^{18}O]_w[H_2{}^{16}O]_v}{[H_2{}^{16}O]_w[H_2{}^{18}O]_v} = \frac{[{}^{18}O]_w}{[{}^{16}O]_w} \times \frac{[{}^{16}O]_v}{[{}^{18}O]_v} = \frac{\left(\frac{{}^{18}O}{{}^{16}O}\right)_w}{\left(\frac{{}^{18}O}{{}^{18}O}\right)_v} = \frac{R_w}{R_v} = \alpha^{18}O_{w\text{-}v} \tag{2.6}$$

For the two compounds A and B, the δ-values and fractionation factor a are related by:

$$\delta_A - \delta_B = \Delta_{A\text{-}B} \approx 10^3 \ln\alpha_{A\text{-}B} \tag{2.7}$$

2.1.3 Enrichment factor

Isotope geochemistry is based on the variations of isotope concentrations, expressed as δ -‰ values. To compare isotope fractionation with δ -‰ values, we can express the fractionation factor, α, in permil units, using the enrichment factor, ε :

$$\varepsilon = (\alpha - 1) \times 1000 ‰ \tag{2.8}$$

This is the same expression as the δ -value used to express isotope data, and so the enrichment factor can be added and subtracted with the δ -values measured in an equilibrium reaction. Recalling that R is the isotope ratio in a sample or standard, and using the example of fractionation for water evaporating to vapor:

$$\varepsilon = \left(\frac{R_{water}}{R_{vapor}} - 1\right) \times 1000 \tag{2.9}$$

$$\varepsilon_{\text{water-vapor}} = \delta_{\text{water}} - \delta_{\text{vapor}} \tag{2.10}$$

$$\varepsilon_{\text{reactant-product}} = \delta_{\text{reactant}} - \delta_{\text{product}} \tag{2.11}$$

2.2 Isotope fractionation processes

2.2.1 Rayleigh distillation

Isotope fractionation during a geochemical reaction acts to partition isotopes between reservoirs. A good example is a pond of water that is evaporating until it is dry. The fractionation between water and vapor, $\varepsilon^{18}O_{w\text{-}v}$, is 9.3‰. If the original $\delta^{18}O$ of the water is −10‰, then the vapor will initially have $\delta^{18}O=-19.3‰$. This low isotope value for the vapor product means that there is a gradual accumulation of ^{18}O in the pond as evaporation proceeds. This process is a Rayleigh distillation, represented by the general equation:

$$R_f = R_0 f^{\alpha-1} \tag{2.12}$$

where R_f is the isotope ratio of the reactant reservoir after some reaction to some residual fraction, f. R_0 is the initial isotopic ratio of the reactant reservoir when $f=1$, and α is the fractionation factor for the reaction.

The equilibrium fractionation factor is written in the form of the product over the reactant, $\varepsilon_{\text{prod-react}}$, just like the mass action constant, K. A simplification of the Rayleigh equation can be used to allow isotope ratios in δ-‰ notation and the fractionation factor in ε-‰ notation:

$$\delta_{\text{react}} = \delta_{\text{initial react}} + \varepsilon_{\text{prod-react}} \times \ln f \tag{2.13}$$

$$\delta^{18}O_{\text{water}} = \delta^{18}O_{\text{initial water}} + \varepsilon^{18}O_{\text{vapor-water}} \times \ln f \tag{2.14}$$

Plotting this relationship for the residual water in the evaporating pond gives the exponential curve in Figure 2.1. When the residual water has been reduced to a small fraction of its original volume, $\delta^{18}O$ becomes greatly enriched due to the discrimination against ^{18}O during reaction (evaporation). Evaporation is a nonequilibrium process, as it takes place under conditions of low humidity (kinetic reaction).

This extreme depletion (or enrichment, depending on whether ε is positive or negative) is characteristic of Rayleigh distillation reactions. One of the most important examples is the partitioning of ^{18}O and D through the hydrological cycle. This is essentially driven by the Rayleigh distillation of isotopes from vapor masses as they cool and rain out over the continents. This is why tropical rains are enriched in ^{18}O and D, while the rain and snow of cold-climate regions is highly depleted. Rayleigh distillation of isotopes during rain out is the process that partitions isotopes throughout the meteorological cycle and provides a versatile tracer of meteoric water.

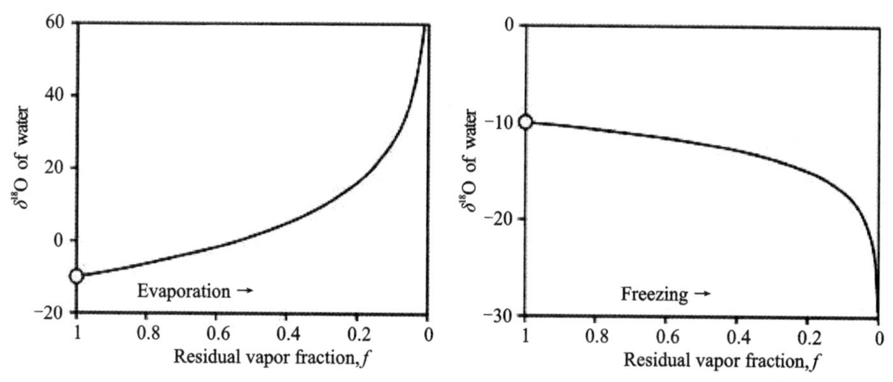

Figure 2. 1 Rayleigh distillation for ^{18}O in water during evaporation (left) and in water during freezing (right). The initial $\delta^{18}O$ of the water is set at $-10‰$ for these examples. The ^{18}O preferentially partitions into the liquid phase during evaporation and so the water becomes enriched as evaporation of the pond proceeds. In contrast, ^{18}O partitions into ice during freezing and so the residual water becomes more depleted as freezing proceeds.

From isotope measurements made during a Rayleigh distillation reaction, it is then possible to calculate the enrichment factor for the reaction. Take the evaporating pond in Figure 2. 1. For example, measurements of $\delta^{18}O$ for the residual water are given in Table 2. 3 for each different amount of residual water (diminishing depth of the pond). Rearranging the Rayleigh distillation equation given earlier to isolate epsilon (ε) allows calculation of the enrichment factor for the pond water:

$$\varepsilon^{18}O_{vapor\text{-}water} = (\delta^{18}O_{water\text{-}f}\delta^{18}O_{water\text{-}initial})/\ln f_{water\text{-}f} \quad (2.15)$$

For a simple Rayleigh distillation, $\varepsilon^{18}O_{vapor\text{-}water}$ can be calculated for any measurement made at a given residual fraction, f. In this case, the measurements give a consistent value for $\varepsilon^{18}O_{vapor\text{-}water}$ of $-16.4‰$, which can also be expressed as $\varepsilon^{18}O_{water\text{-}vapor}$ with a value of $16.4‰$.

Table 2. 3 Isotope values for evaporating pond water at successive residual fractions of water in the pond at 25℃.

f	$\delta^{18}O_{water\ evaporating}$	$\delta^{18}O_{water\ freezing}$
1	−10	−10
0. 6	−1. 7	−11. 6
0. 4	4. 9	−12. 8
0. 2	16. 2	−15. 0
0. 1	27. 5	−17. 1
0. 05	38. 9	−19. 3

2. 2. 2 Temperature effects

Isotope fractionation is a thermodynamic process and so is a function of temperature.

Considerable effort has been made to establish the temperature relationship for isotope fractionation in various reactions, such as evaporation, condensation, freezing, aqueous reactions, and mineral crystallization. These relationships are important in calculations of isotope fractionation at the range of temperatures in natural systems.

At high temperature, α is very close to 1, with a departure from 1 as the temperature of reaction decreases. Fractionation factors are greatest for the low-temperature geochemical reactions in groundwater and surface water.

The temperature dependence of α is determined as its natural logarithm, $\ln\alpha$, and generally represented in equations of the following form:

$$\ln\alpha_{x-y} = aT^{-2} + bT^{-1} + c \tag{2.16}$$

As α is close to 1 for most isotope reactions, $\ln\alpha$ is very close to 0, and so multiplying by 1000 fts with the ‰ convention for δ-values. Providing α is close to 1, the enrichment factor, ε, is then close to 1000 $\ln\alpha$:

$$\varepsilon = 1000(\alpha - 1) \approx 1000\ln\alpha \tag{2.17}$$

The temperature equations for the principal isotope exchange reactions useful in groundwater work are given in Table 2.4.

Table 2.4 Temperature and fractionation for some common isotope-exchange reactions. (T in ℃)

Isotope	Exchange Reaction	Temperature Equation	ε-‰ at 25℃
D	H_2O-H_2O_{vapor}	$\varepsilon D_{water-vapor} = 0.0066T^2 - 1.36T + 106$	79
	H_2O_{vapor}-H_2	$\varepsilon D_{vapor-H_2} = 0.02T^2 - 6.39T + 1395$	2485
	H_2O_{water}-H_2	$\varepsilon D_{water-H_2} = 0.026T^2 - 7.75T + 1502$	2762
	H_2O_{water}-CH_4	$\varepsilon D_{water-CH_4} = -0.0018T^2 + 0.64T + 12$	27
	H_2O_{water}-H_2S	$\varepsilon D_{water-H_2S} = 0.011T^2 - 3.93T + 949$	1358
	Gypsum-H_2O	$\varepsilon D_{water-gypsum} = 0.00008T^2 - 0.028T - 14$	−15
	Illite-H_2O	$\varepsilon D_{water-illite} = -0.0008T^2 + 0.4804T - 66.904$	−55
	Kaolinite-H_2O	$\varepsilon D_{water-kaolinite} = -0.0003T^2 + 0.152T - 35.368$	−32
^{18}O	H_2O_{water}-H_2O_{vapor}	$\varepsilon^{18}O_{water-vapor} = 0.0004T^2 - 0.103T + 11.64$	9.3
	CO_2-H_2O	$\varepsilon^{18}O_{CO_2-water} = 0.0007T^2 - 0.240T + 45.6$	41.0
	Calcite-H_2O	$\varepsilon^{18}O_{CaCO_3-water} = 0.0011T^2 - 0.265T + 34.3$	28.8
	Gypsum-H_2O	$\varepsilon^{18}O_{gypsum-water} = 0.00009T^2 - 0.0304T + 4.72$	4.0
	SO_4^{2-}-H_2O	$\varepsilon^{18}O_{SO_4-water} = 0.0011T^2 - 0.275T + 34.5$	28.7

Continue

Isotope	Exchange Reaction	Temperature Equation	ε -‰ at 25℃
^{18}O	Illite-H_2O	$\varepsilon^{18}O_{illite\text{-}water}=0.0004T^2+0.19T+27.86$	33.1
	Kaolinite-H_2O	$\varepsilon^{18}O_{kaolinite\text{-}water}=0.0004T^2+0.201T+29.12$	34.4
	$SiO_{2(amorph)}$-H_2O	$\varepsilon^{18}O_{SiO_2\text{-}water}=0.0014T^2-0.336T+42.8$	35.9
^{13}C	H_2CO_3-$CO_{2(g)}$	$\varepsilon^{13}C_{H_2CO_3\text{-}CO_{2(g)}}=-0.000014T^2+0.0049T-1.18$	−1.1
	HCO_3^--$CO_{2(g)}$	$\varepsilon^{13}C_{HCO_3\text{-}CO_{2(g)}}=0.00032T^2-0.124T+10.87$	8.0
	CO_3^{2-}-$CO_{2(g)}$	$\varepsilon^{13}C_{CO_3\text{-}CO_{2(g)}}=0.00033T^2-0.083T+8.25$	6.4
	$CaCO_3$-HCO_3^-	$\varepsilon^{13}C_{CaCO_3\text{-}HCO_{3(g)}}=-0.0002T^2+0.056T-0.39$	0.9
	$CaCO_3$-CO_2	$\varepsilon^{13}C_{H_2CO_3\text{-}CO_2}=0.0009T^2-0.184T+14.4$	10.3
	CO_2-CH_4	$\varepsilon^{13}C_{CO_2-CH_4}=0.0015T^2-0.418T+77.7$	70.5
^{34}S	SO_4^{2-}-$H_2S_{(aq)}$	$\varepsilon^{34}S_{SO_4\text{-}H_2S}=0.0019T^2-0.484T+74.2$	65.4
	SO_4^{2-}-$HS^-_{(aq)}$	$\varepsilon^{34}S_{SO_4\text{-}HS}=0.0017T^2-0.493T+83.9$	75.4

From data and references in Clark and Fritz, 1997.

2.2.3 Kinetic effects

Another phenomenon producing fractionations are kinetic isotope effects, which are associated with incomplete and unidirectional processes like evaporation, dissociation reactions, biologically mediated reactions and diffusion. The latter process is of special significance for geological purposes. A kinetic isotope effect also occurs when the rate of a chemical reaction is sensitive to atomic mass at a particular position in one of the reacting species.

Quantitatively, many observed deviations from simple equilibrium processes can be interpreted as consequences of the various isotopic components having different rates of reaction. Isotope measurements taken during unidirectional chemical reactions always show a preferential enrichment of the lighter isotope in the reaction products. The isotope fractionation introduced during the course of a unidirectional reaction may be considered in terms of the ratio of rate constants for the isotopic substances. Thus, for two competing isotopic reactions:

$$k_1 \rightarrow k_2, A_1 \rightarrow B_1, \text{ and } A_2 \rightarrow B_2$$

the ratio of rate constants for the reaction of light and heavy isotope species k_1/k_2, as in the case of equilibrium constants, is expressed in terms of two partition function ratios, one for the two reactant isotopic species, and one for the two isotopic species of the activated complex or transition state A^x :

$$\frac{k_1}{k_2} = \left[\frac{Q^*_{(A_2)}}{Q^*_{(A_1)}} / \frac{Q^*_{(A_2^x)}}{Q^*_{(A_1^x)}}\right]\frac{V_1}{v_2} \tag{2.18}$$

The factor v_1/v_2 in the expression is a mass term ratio for the two isotopic species. The determination of the ratio of rate constants is, therefore, principally the same as the determination of an equilibrium constant, although the calculations are not so precise because of the need for detailed knowledge of the transition state. The term "transition state" refers to the molecular configuration that is most difficult to attain along the path between the reactants and the products. This theory follows the concept that a chemical reaction proceeds from some initial state to a final configuration by a continuous change, and that there is some critical intermediate configuration called the activated species or transition state. There are a small number of activated molecules in equilibrium with the reacting species and the rate of reaction is controlled by the rate of decomposition of these activated species.

Taking the calcium isotope fractionation during the calcite precipitation for example (Tang et al., 2008). In the laboratory, a conditional experiment was set up to simulate the process of calcite precipitation. And the calcium isotope ratios of calcite, aqueous Ca, and $CaCl_2 \cdot 2H_2O$ were measured during the process. The isotope values of Ca are reported as $\delta^{44/40}Ca$ (‰) values relative to the NIST standard SRM915a, where $\delta^{44/40}Ca = [(^{44/40}Ca)_{sample}/(^{44/40}Ca)_{SRM915a} - 1] \times 1000$. The samples are normalized to the mean $^{44/40}Ca$ of four SRM915a analyses, run on the same turret. The calcium isotope fractionation between calcite and solution is expressed as

$$\Delta^{44/40}Ca_{calcite\text{-}aq} = \delta^{44/40}Ca_{calcite} - \delta^{44/40}Ca_{aq} \tag{2.19}$$

At a constant precipitation rate absolute $\Delta^{44/40}Ca_{calcite\text{-}aq}$ values generally decrease with increasing temperature, especially at relatively high precipitation rates (Figure 2.2). Accordingly, a negative temperature-dependence of Ca isotope fractionation is observed from our spontaneous calcite formation experiments from 5℃ to 40℃, i.e., elevated temperature results in less Ca isotope fractionation.

As shown in Figure 2.2, absolute $\Delta^{44/40}Ca_{calcite\text{-}aq}$ values are positively related to the precipitation rates, log$R$, at 5 and 25℃. For experiments done at 40℃ this relationship is rather weak, which indicates that the influence of precipitation rate on Ca isotope fractionation is relatively weak at higher temperature. Discounting these two experiments a significant positive correlation between absolute $\Delta^{44/40}Ca_{calcite\text{-}aq}$ and log$R$ values is obtained at 40℃. The rate-related Ca isotope fractionation can be followed by the individual equations:

$$\Delta^{44/40}Ca_{calcite\text{-}aq} = (-0.57 \pm 0.15) \cdot \lg R + 0.52 \pm 0.37 \quad R^2 = 0.91, p = 1.82 \times 10^{-5}, n = 10 \tag{2.20}$$

$$\Delta^{44/40}Ca_{calcite\text{-}aq} = (-0.43 \pm 0.11) \cdot \lg R + 0.40 \pm 0.38 \quad R^2 = 0.86, p = 4.18 \times 10^{-6}, n = 13 \tag{2.21}$$

$$\Delta^{44/40}Ca_{calcite\text{-}aq} = (-0.17 \pm 0.11) \cdot \lg R - 0.21 \pm 0.33 \quad R^2 = 0.44, p = 6.99 \times 10^{-3}, n = 15 \tag{2.22}$$

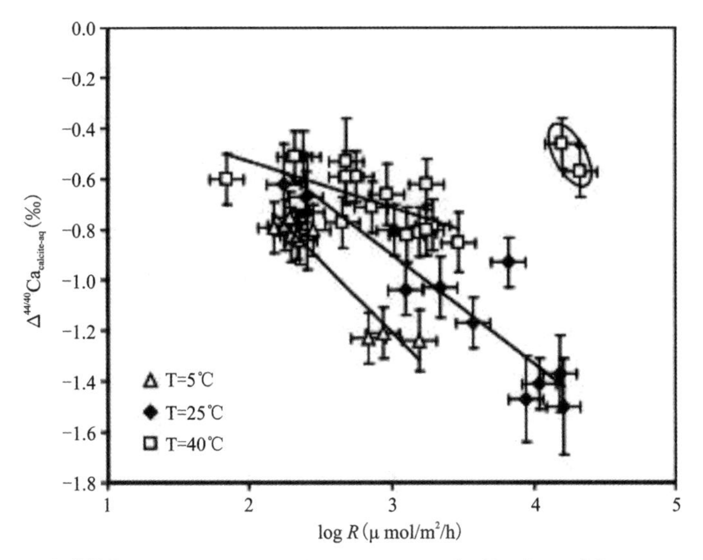

Figure 2. 2 $\Delta^{44/40}Ca_{calcite-aq}$ versus precipitation rate, log$R$, for calcite grown from the stirred $CaCl_2$-$SrCl_2$-NH_4Cl or CaC_2-NH_4Cl solutions at 5℃, 25℃, and 40℃ ($I \approx$ 0. 035M). Experiment Nos. 39 and 40 with abnormal $\Delta^{44/40}Ca_{calcite-aq}$ values are circled.

at 5℃, 25℃, and 40℃, respectively. The respective slopes of the regression lines decrease with increasing temperature. The obtained rate-dependence of Ca isotope fractionation is consistent with the results of Fantle and DePaolo (2007).

2. 2. 4 Diffusive fractionation

Diffusion is an additional process that can fractionate isotopes and in theory can be related to the relative molecular velocity of a heavy isotope species relative to the light isotope species. It is an important process in aquitards where groundwater advection is minimal and transport is dominated by molecular diffusion in the porewater.

The simplest case is the diffusion of gas in a vacuum, where steady-state fractionation equals the ratio of the velocities of the two isotopes, which resolves to the root of the mass ratio, which is Graham's law:

$$\alpha_{m'-m}^{*} = \frac{v^*}{v} = \frac{\sqrt{\frac{kT}{2\pi m^*}}}{\sqrt{\frac{kT}{2\pi m}}} = \sqrt{\frac{m}{m^*}} \tag{2.23}$$

where is v, molecular velocity (cm/s); k, Boltzmann constant (1. 380 658$\times10^{-23}$ J/K, gas constant per molecule); m, molecular mass; and T, absolute temperature(K).

Derivatives of Graham's law for diffusion through a medium such as air or water include the molecular mass of the medium. However, in groundwater, diffusive fractionation is limited to aquitards, where low permeability brings solute transport into the domain of diffusion. Here, it is concentration gradients that determine the rate of diffusion together with the diffusion coefficient for the isotope species in water and through the porous network. Fractionation of isotopes by diffusion through aquitards is examined by the rate of diffusion of the individual isotope species (e.g., $H_2{}^{18}O$, HDO, or $H^{13}CO_3^-$), each with its own coefficient of diffusion. The decoupling of isotopes in this way does not allow us to represent fractionation in terms of α and ε values as shown earlier. It is a mass dependent process like physical - chemical fractionation, but can be considered more precisely as diffusive mixing.

2.3 Radioisotopes

The addition or loss of neutrons from the stable nuclides produces increasingly unstable radioisotopes, which decay to a more stable configuration, principally by emission of alpha (α) or beta (β) particles, sometimes accompanied by a photon as electromagnetic gamma, γ, radiation, from the nucleus. Radioactivity is measured as decay events per time. The original unit is the curie, Ci, which is equal to the radioactivity of 1 g of ^{226}Ra—the isotope of radium that Marie Curie isolated from the ^{238}U decay series. The Becquerel(Bq) is now used, defined as one nuclear disintegration per second ($1Ci=3.7\times10^{10}$ Bq).

The most common decay mode is beta decay, β, which is the emission of an electron from the nucleus, converting a neutron to a proton with a translocation to a higher atomic number and lower neutron number. The result is an isobaric transformation to a nuclide of a different element, but with the same atomic mass. Similarly, decay by electron capture transforms a proton to a neutron, producing a more stable nuclide with lower atomic number but the same atomic mass. Radioisotopes of the heavy actinides, including U and Th, decay largely by emission of an alpha particle, α, which is a ^4He nucleus of 2n and 2p. Many of the actinides, such as ^{238}U and ^{239}Pu, can also decay by spontaneous fission into two mid-weight nuclides. The electromagnetic radiation that accompanies decay is high-energy gamma, γ, radiation from the nucleus. The intensity and wavelength of γ-radiation is characteristic for each radioisotope, and can be used in γ-counters for their identification.

Radioactivity is measured in Becquerels as the number of decay events per second. Samples with a greater number of atoms of a given radioisotope will naturally have greater radioactivity. Through time, the number of atoms decreases and so does its radioactivity. This is the decay rate that is equal to the change in the number of atoms of the radioisotope per second, $\partial n/\partial t$, and so the rate of decay is proportional to the number of atoms. Each radioisotope has its unique rate of decay, represented by the decay constant, λ, which is the constant of this proportionality.

$$-\frac{\partial n}{\partial t} = \lambda n \text{(decay/time) radioactive decay} \tag{2.24}$$

or

$$-\frac{\partial n}{n} = \lambda \partial t \tag{2.25}$$

Integrating gives the general form of the radioactive decay equation that determines the number of atoms after decay from an initial number, no, over a given period, t:

$$n_t = n_0 e^{-\lambda t} \tag{2.26}$$

Taking the natural log of this exponential equation and inverting gives

$$\ln\left(\frac{n_0}{n_t}\right) = \lambda t \tag{2.27}$$

Solving the decay equation for the period of time required to decay the number of atoms, n, to half ($n_0/n_t=2$), establishes the half-life, $T_{½}$, for a given radioisotope (Figure 2.3):

$$\ln\left(\frac{n_0}{n_t}\right) = \lambda T_{½} \tag{2.28}$$

$$\ln\left(\frac{n_0}{n_t}\right) = \lambda T_{½} \tag{2.29}$$

$$\lambda = \frac{\ln 2}{T_{½}} (\text{time}^{-1}) \tag{2.30}$$

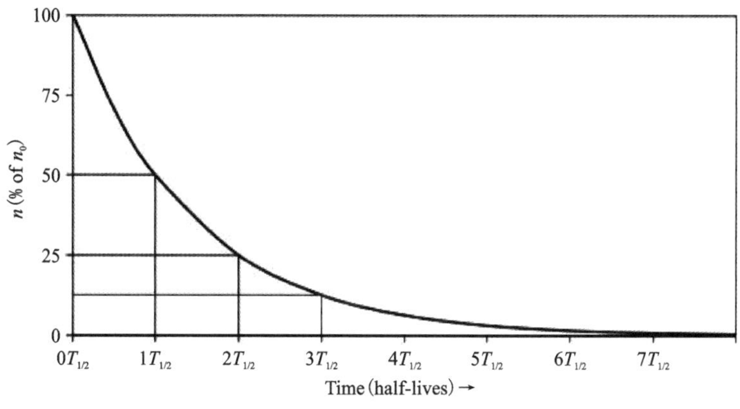

Figure 2.3 Radioactive decay and half-life, $T_{½}$.

Geochronology is based on the decay equation, which can be expressed as loss of the parent radioisotope from some initial number of atoms, n_{P_0} :

$$n_{P_t} = n_{P_0} e^{-\lambda t} \tag{2.31}$$

For the gain of the daughter isotope, n_D, this decay equation becomes:

$$n_{D_t} = n_{P_0}(1 - e^{-\lambda t}) \tag{2.32}$$

The exponential functions of decay and ingrowth are shown in Figure 2.4.

Groundwater dating with radioisotopes relies on the decay of atmospherically derived nuclides that accompany recharge or on the ingrowth of radiogenic nuclides produced in the subsurface. Those that are atmospherically derived are cosmogenic nuclides, produced by

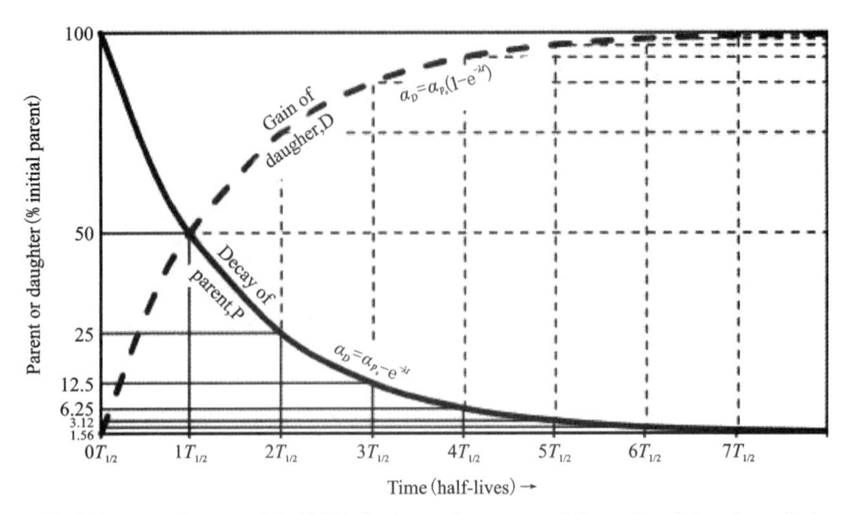

Figure 2. 4　Radioisotope decay and half-life for loss-of-parent and for gain-of-daughter dating schemes.

high-energy particles, mainly protons originating from supernovae elsewhere in the galaxy that constantly bombard our planet. Collisions with the atmosphere generate showers of secondary, lower-energy neutrons that further interact with atmospheric gases to produce a range of cosmogenic radioisotopes that reach the surface as dry fallout or with precipitation wet fallout (Figure 2. 5). The cosmogenic neutron flux can also activate nuclides such as ^{35}Cl on the earth's surface, which is epigenic production. These atmospheric or epigenic radioisotopes are then incorporated into groundwater during recharge where subsequent decay from their initial value can be used as a chronometer.

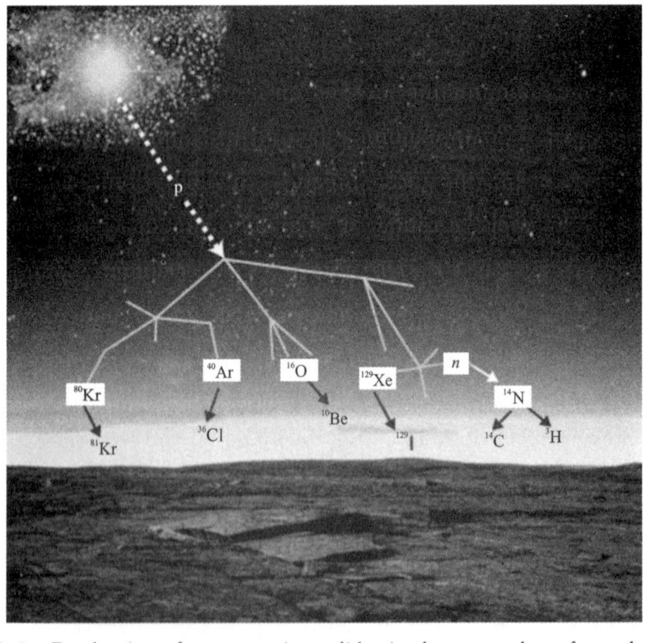

Figure 2. 5　Production of cosmogenic nuclides in the atmosphere from the shower of secondary neutrons produced by impact of high energy particles (mainly protons) originating from supernovae in the galaxy.

Geogenic nuclides are those produced in the subsurface by the decay and spontaneous fission of uranium and other actinides present in all rocks. The range of radiogenic nuclides produced in the subsurface, which are incorporated by groundwater and can be used for chronology. The release of high energy α particles during the decay schemes for uranium and thorium isotopes (^{238}U, ^{235}U, and ^{232}Th) generates a secondary neutron flux through impact on light nuclei such as O, F, Na, Al, Si, and K. This secondary neutron flux impacts on other target nuclides to produce a range of geogenic nuclides through neutron-activation reactions (uptake of neutron by the target nuclide) or spallation reactions (loss of mass as protons, neutrons, or both). Two examples useful in groundwater dating are as follows:

The third class of geogenic nuclide production is by radioactive decay. Considering

$$^{35}Cl + n \rightarrow {}^{36}Cl \quad T_{1/2}(^{36}Cl) = 301\ 000 \text{ years}$$

$$^{6}Li + n \rightarrow {}^{3}He + {}^{4}He$$

the more than billion-year half-lives of these parent radionuclides, the production of the stable daughter nuclides on the scale of millions of years can be calculated as linear ingrowth rates:

$$^{40}K \rightarrow {}^{40}Ar \qquad T_{1/2}(^{36}K) = 1.28 \times 10^{9} \text{ years} \tag{2.33}$$

$$^{40}Ar \text{ production} = 3.887 \times 10^{-14} \times [K_{\% \text{ in rock}}] cc_{STP}/g/y \tag{2.34}$$

$$^{238}U \rightarrow {}^{234}U + {}^{4}He \rightarrow {}^{206}Pb + 8{}^{4}He \qquad T_{1/2}(^{238}U) = 4.6 \times 10^{9} \text{ years} \tag{2.35}$$

$$^{232}Th_{\alpha} \rightarrow {}^{208}Pb + 6{}^{4}He \qquad T_{1/2}(^{232}Th_{\alpha}) = 14.05B \text{ years}$$

$$^{4}He_{production} = 1.19 \times 10^{-13}[U_{10^{-6}}] + 2.88 \times 10^{-14}[Th_{10^{-6}}] cc_{STP}/g/y \tag{2.36}$$

The groundwater dating using the stable isotope and radioisotopes was further presented in Chapter 12.

References

CLARK I D, FRITZ P, 1997. Environmental Isotopes in Hydrogeology[M]. Boca Raton: FL Lewis Publishers: 328.

FANTLE M, DEPAOLO D J, 2007. Ca isotopes in carbonate sediment and pore fluid from ODP Site 807A: the Ca^{2+} (aq) — calcite equilibrium fractionation factor and calcite recrystallization rates in pleistocene sediments[J]. Geochim. Cosmochim. Acta, 71: 2524-2546.

TANG J, DIETZEL M, BöHM F, et al., 2008. Sr^{2+}/Ca^{2+} and $^{44}Ca/^{40}Ca$ fractionation during inorganic calcite formation: II. Ca isotopes[J]. Geochim. Cosmochim. Acta, 72 (15): 3733-3745.

Chapter 3 Isotopic Analytical Methods

In 1919, British chemist Aston developed a mass spectrometer and achieved high-precision determination of the mass and abundance of various isotopes. The development of isotope testing technology and the precise determination of isotope fractionation in the natural environment expand the depth and breadth of isotopic theory and applied research. Therefore, the isotope testing and analyzing techniques have become a new research direction as well as a scientific research hotspot in the field of environmental geosciences.

In recent years, great progress in environmental isotope research has been made owing to the improvements and breakthroughs in testing and analysis techniques. In this chapter, we focus on isotope testing and analysis, including basic principles of the mass spectrometer and their performance parameters, standard reference materials in the process of spectrometric methods.

We also introduce in detail the advantages and disadvantages of a wide spectrum of mass spectrometers.

3.1 Introduction

3.1.1 Basic Principles of Mass Spectrometry

Mass spectrometric methods are the mosteffective means of measuring isotope abundances to date. A mass spectrometer separates charged atoms and molecules according to their masses and motions in magnetic and/or electrical fields. The motions of ions are usually affected by extraction voltage, which accelerates the ions and focuses them on a beam with the appropriate properties for detection in a mass spectrometer. The extraction voltage depends on the setting mode of the instrument, but it is generally around 1～10kV. The ion extraction efficiency is closely related to the voltage. The higher the extraction voltage, the smaller the solid angle, and the higher the concentration of ions at a specific point in the beamline. The extraction voltage produces an electrostatic field in which electric potential energy is generated and converted to the kinetic energy of the ion. In the end, mass separation is then achieved by applying different electric and magnetic forces to the ion beam. In principle, a mass spectrometer may be divided into four different central constituent parts, including the inlet system, the ion source, the mass analyzer, and the ion detector (Figure 3. 1).

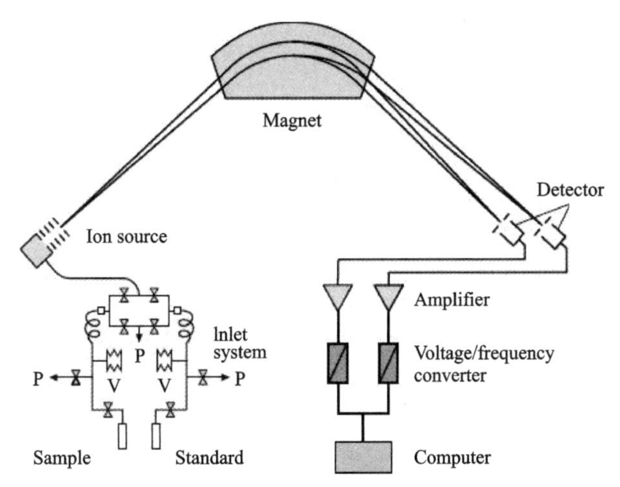

Figure 3. 1 Schematic representation of a gas-source mass spectrometer for stable isotope measurements during the 1960s and 1970s. P denotes pumping system, V denotes a variable volume (Jochen, 2018).

The mass analyzer and the ion detector are common to all kinds of mass spectrometers, the fundamental difference lies in the choice of ion sources. Ion sources vary from product to product in terms of gaseous (inert gases and stable isotopes), solid (thermoelectric ionization) and liquid (inductively coupled plasma mass spectrometry) samples. One advantage of this technique is that the elements are chemically separated, and therefore the elements with higher concentration, resulting in higher ion yield and accuracy, which also helps mass spectrometry eliminate interfering elements, especially atomic interference in the same mass.

During the analysis of the gas in the Noble Gas Mass Spectrometer (NG-MS), the gas is ionized by a filament-based source in the wake of introduction and then separated from the static vacuuming system, during which the gas is gradually consumed. In the analysis of stable isotope gases, the dual-inlet injection system is usually used to switch between the sample gas and the reference gas. The system must be vacuumed dynamically due to these gases are continuously introduced into the mass spectrometer. In the technique of Thermal Ionization Mass Spectrometry (TIMS), the sample was loaded onto the filament, dried and inserted into the Mass spectrometer. The current passes through the filament, causing evaporation and ionization. With regard to MC-ICP-MS, samples in the solution can be directly fed into the plasma for ionization to further determine.

3.1.2 Performance Parameters of Mass Spectrometry

Each type of mass spectrometer has parameters that characterize a variety of performances in which the primary parameters are the ability to describe the mass separation in the instrument and the ability to resolve specific disturbances in the components, which include mass resolution, mass separation and abundance sensitivity. They are referred to as $m/\Delta m$ where m is the mass of the substance to be measured and Δm is the mass difference between the target substance and the interfering substance.

3.1.2.1 Mass resolution

Mass resolution is generally defined as the ability to separate two ion beams at the detector, which differ for a particular quality difference. Peak width is the most common way to define mass resolution. It is based on peak height at 1%, 10%, or 50% of the peak width (m). The advantage of peak width definition is that a simple measurement of peak width at a given level can be recalculated into mass resolution regardless of peak shape (trapezoidal, triangle or Gaussian).

The trapezoidal peak in the sector magnetic mass spectrometer is generated by scanning the ion beam through a resolution slit (outlet or collector slit). The trapezoidal indicates that the image in the entrance or source slit of the extinction conforms to the collector slit. The degree of flat top is proportional to the ratio of the collector slit size to the ion beam width. That is, the wider the slit of the collector relative to the ion beam, the wider the flat top. For a trapezoidal peak, the peak base width is approximately the sum of the extinction slit width and the collector slit width (Figure 3.2). Thus, mass resolution can be improved

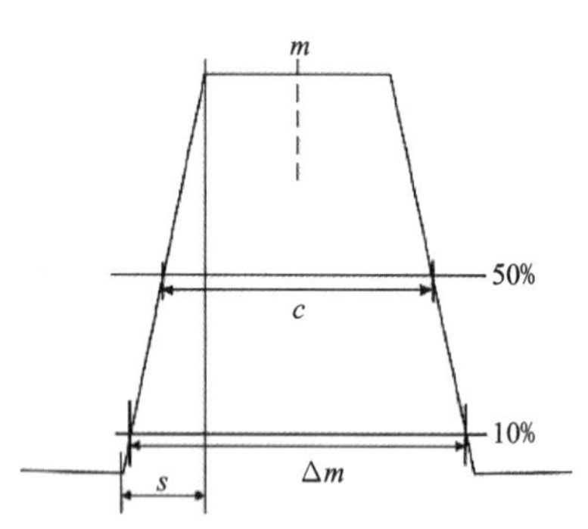

Figure 3.2 Schematic diagram of mass resolution for sector magnetic mass spectrometer. Note: For the trapezoidal peak shape generally produced in the sector magnetic mass spectrometer, the slope of the peak side is the width of the extinction source slit, and 50% of the height is the width of the collector slit. Therefore, the base width is the sum of the width of the collector slit and the amplifier source slit, and the mass resolution of the peak height is 10% ($m/\Delta m$).

by reducing the collector slits, but at the expense of peak flattops, which may eventually affect beam stability. Conversely, reducing the light source slits can improve the flattops, but at the expense of ion beam transmission.

Although the mass resolution is a useful performance parameter, it still has some limitations in practice. In particular, the value of mass resolution cannot completely describe the peak shape. There is no information provided about flat tops and interference produced during analysis can affect the results of the analysis.

3.1.2.2 Mass Separation

The term mass separation is often used interchangeably with mass resolution because it is the ability to separate ion beams by mass. The definition of mass resolution from the International Union of Pure and Applied Chemistry (IUPAC) is based on the mass of the object to be measured divided by the difference between the two masses that can be (fairly) separated. Mass resolution is generally defined as the effective mass resolution of the collector independent of the exit slit. For a ladder peak, mass separation can be defined as the width of the peak side (Figure 3.3) which is actually the extinction width of the source slit, which defines the beam width at the collector as well. To avoid deviation from the peak side, the term Δm can be defined as the width from 10% to 90% of the peak height.

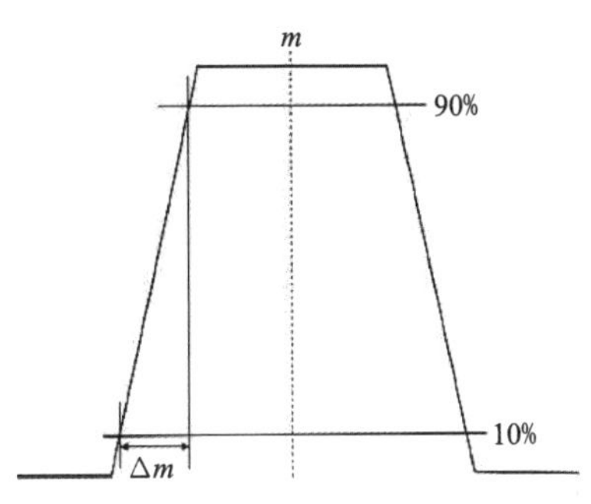

Figure 3.3 Schematic diagram of mass separation for sector magnetic mass spectrometer.
Note: Mass resolution is generally defined as $m/\Delta m$ where Δm is defined as the equivalent mass width between 10% and 90% peak height. The symbol Δm is actually the slit width of the extinction source, ignoring the slit width of the collector.

Mass separation is always higher than mass resolution. Therefore, it is a trend to use mass separation when reporting instrument performance. In particular, when the collector slit is of a fixed width and is analyzed at the peak side, the interference shown at the collection location is minimal. However, while mass resolution can be considered to underestimate the ability to separate masses, mass resolution can also be considered to exaggerate the ability to separate masses at this level.

3.1.2.3 Abundance Sensitivity

The abundance sensitivity is a parameter that represents the tail peak of one atomic mass at another atomic mass (Figure 3.4). The tail peak is generated by ion-optical aberration or gas scattering. At the measured position, the low-intensity peak is more likely to be affected by the peak trailing effect. In this case, Δm is fixed at 1, but the mass can vary, so the relative offset is a function of the mass and therefore the abundance sensitivity must be specified for a particular mass.

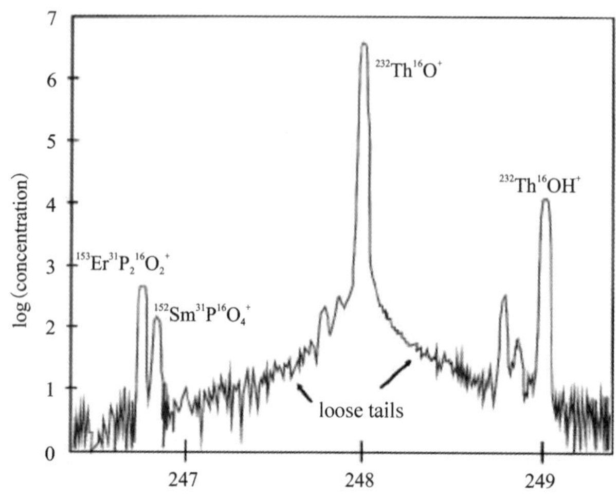

Figure 3.4 Graphical representation of the abundance sensitivity of mass $^{248}ThO^+$ from a single mineral target. Note: Gas scattering and lens aberration make intense peak track to the adjacent peaks. With a quality deviation of 1 amu in this case, peak tailing abundance is around 10^{-6}. Abundance sensitivity is generally defined as 1 amu quality deviation compared with the specified peak. Due to quality differences are not the same for all the elements abundance sensitivity is relevant to mass and must be specified for a particular quality.

3.1.2.4 Count Statistics and Accuracy

The smaller the isotope effect, the higher the required accuracy of the isotope ratio. The limit of the accuracy of the isotope ratio measurement is determined by the poisson meter statistics, where the final accuracy is determined by $1/\sqrt{N}$, where N is the count of the minor isotope. Therefore, the accuracy of 1% requires 10^4 counts and the accuracy of 1‰ requires 10^6 counts, the accuracy of 0.1‰ requires 10^8 counts, and so on. Most mass spectrometers have counting systems, including ion counters for low signals (<1 to 10^5 c/s) and Faraday collectors, whose counting rates can exceed 10^6 c/s (10^{-13} A). In the case of these devices crossing, it is difficult to obtain sufficient precision with Faraday collectors and sufficient accuracy with ion counters.

3.2 Standards

The accuracy of measuring ***absolute*** isotopic abundances is significantly lower than the accuracy of determining ***relative*** differences in isotopic abundances between two samples. However, the determination of the absolute isotope ratio is critical as it forms the basis for calculating the relative difference of the delta value. Comparison of isotopic data from different laboratories requires an internationally accepted set of standards. The widely accepted unit of isotope ratio measurements is the delta value (δ) given in per mil (‰). The δ-value is defined as

$$\delta(‰) = \frac{R_{\text{sample}} - R_{\text{standard}}}{RR_{\text{standard}}} \times 1000 \tag{3.1}$$

where R is the measured isotope ratio. If $\delta_A > \delta_B$, A is preferentially enriched in the rare or "heavy" isotope compared to B. Unfortunately, not all of the δ-values presented in the literature are told by one single universal standard due to several standards of one element in use. The conversion of δ-values calculated by the two standards values can be achieved by the following equation:

$$\delta_{X\text{-}A} = \left[\left(\frac{\delta_{B_{St}-A_{St}}}{10^3}+1\right)\left(\frac{\delta_{X\text{-}B_{St}}}{10^3}+1\right)-1\right]\times 1000 \tag{3.2}$$

where X is the sample, A_{St} and B_{St} are different standards.

For various elements, a convenient *working standard* is applied in each laboratory. Nevertheless, all values, reported in the literature, measured relative to the respective *working standard* are standardized from a universal standard. Take the relationship between the content of an isotope in % and the δ-value in ‰ as an example. Figure 3.5 shows that significant changes in the δ-value only involve little changes in the heavy isotope content (in this case the ^{18}O content). Unfortunately, what kind of reference samples can become the ideal standard used worldwide has not been unified. However, an ideal standard should confirm the following requirements: ①Be homogeneous in composition; ②Be available in relatively large amounts; ③Be easy to handle for chemical preparation and isotopic measurement; ④And have an isotope ratio near the middle of the natural range of variation.

Among the reference samples now used, relatively few enable to meet all of these requirements. For instance, the situation for the SMOW standard is rather confusing. Table 3.1 presents the worldwide standards.

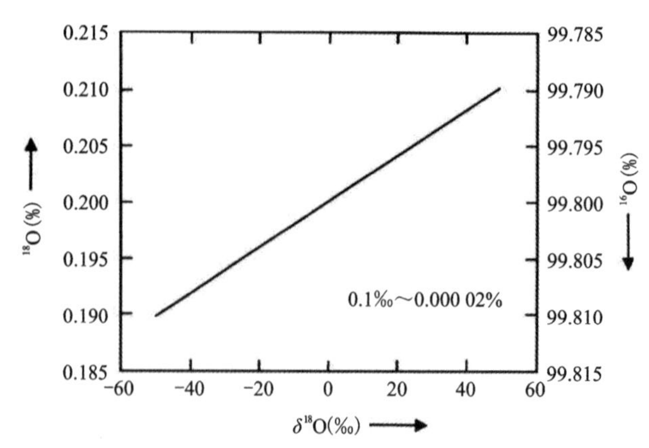

Figure 3. 5　Relationship between ^{18}O (^{16}O) content in percent (%) and $\delta^{18}O$ in per mill (‰) (Jochen, 2018).

Table 3. 1　Worldwide standards in use for the isotopic composition of hydrogen, boron, carbon, nitrogen, oxygen, silicium, sulfur, chlorine and of selected metals (Möller et al., 2012)

Element	Standard 1	Standard 2
H	Standard Mean Ocean Water	C-SMOW
B	Boric acid (NBS)	SRM 951
Cl	Belemnitella americana from the Cretaceous, Peedee formation, South Carolina	V-PDB
N	Air nitrogen	N_2 (atm.)
O	Standard Mean Ocean Water	V-SMOW
Si	Quartz sand	NBS-28
S	Troilite (FeS) from the Canyon Diablo iron meteorite	V-CDT
Cl	Seawater chloride	SMOC
Mg		DSM-3 NIST SRM 980
Ca		NIST SRM 915a
Cr		NIST SRM 979
Fe		IRMM-014
Cu		NIST SRM 976
Zn		JMC3-0749
Mo		NIST 3134
Tl		NIST SRM 997
U		NIST SRM 950a
Ge		NIST SRM 3120a

3.3 Analytical methods

3.3.1 Isotope Radio Mass Spectrometer (IRMS)

Isotope Radio Mass Spectrometer (IRMS) is a specialized mass spectrometer that can measure changes in the abundance of light stable isotopes accurately and accurately (atomic number $Z < 20$, $\Delta A/A \geq 10\%$ ΔA is the mass difference between two isotopes) in natural isotopes. IRMS instruments differ from traditional organic mass spectrometers in that they do not scan for characteristic fragment ions in the mass range to provide structural information about the sample being analyzed. The breakthrough in classical isotope ratio mass spectrometry began in 1948 when Urey introduced the dual inlet ratio mass spectrometer. Since then, the instrument has been further developed and automated to the point where it is now available for commercial use. The isotopes of the following elements are usually measured using IRMS: C (^{13}C, ^{12}C), O (^{16}O, ^{17}O, ^{18}O), H (^{1}H, ^{2}H), N (^{14}N, ^{15}N) and S (^{32}S, ^{33}S, ^{34}S, ^{36}S). Isotopic analysis for Cl, Si and Se are not used much. IRMS consists of an injection system, electron ionization source, ion magnetic sector analyzer, Faraday cup detector and a computer-controlled data acquisition system (Figure 3.6).

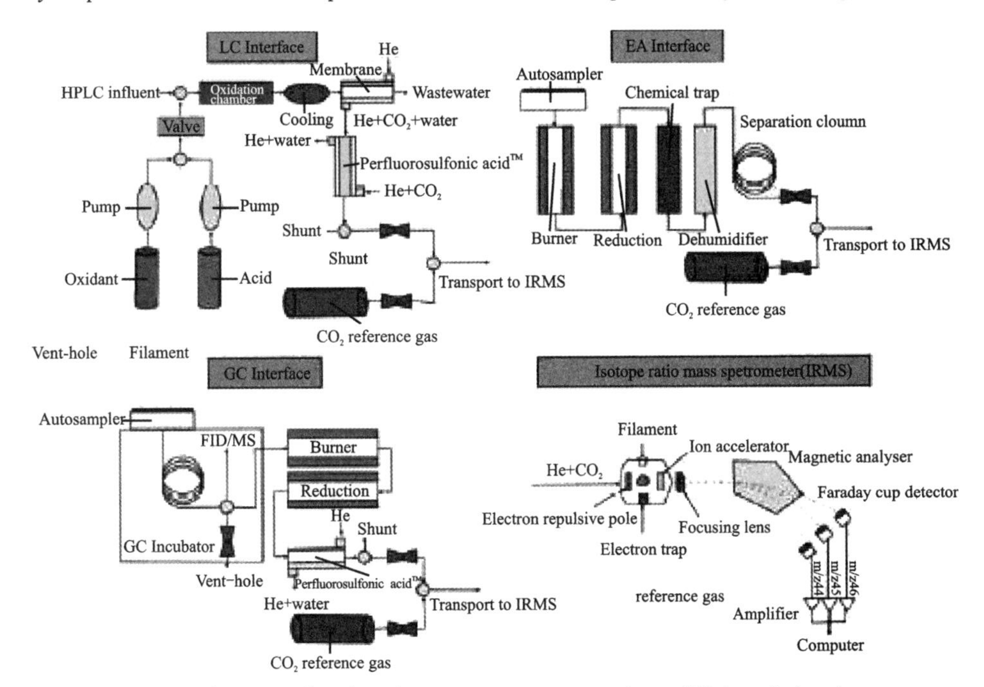

Figure 3.6 Schematic of carbon isotope measurements (e.g. CO_2) and the three most common inlet systems of IRMS. Note: LC denotes liquid chromatography; EA denotes elemental analyzer; GC denotes gas chromatography.

The injection system is designed to process pure gases, primarily CO_2, N_2, H_2, and

SO_2, and can also process other gases, such as O_2, N_2O, CO, CH_3Cl, SF_6, CF_4, and SiF_4. Neutral molecules in the inlet system are introduced into the ion source, ionized by electron collisions, and accelerated to several kilovolts before being separated by magnetic fields and detected by Faraday cups.

IRMS inlet systems include valves, tubes, capillaries, connectors and meters. Homemade inlet systems are usually made of glass, but commercial inlet systems are mostly made of stainless steel. Common IRMS analysis systems include Dual-Inlet-Isotope Radio Mass Spectrometer and Continuous Flow-Isotope Radio Mass Spectrometer. Both of them require the conversion of solid, liquid and gas samples to pure gases before the determination.

3.3.1.1 Dual-Inlet-Isotope Radio Mass Spectrometer (DI-IRMS)

DI-IRMS is equipped with a dual-inlet system that requires off-line preparation of samples for analysis (i. e. conversion to simple gas). The off-line sample preparation process uses a specially designed equipment including vacuum lines, compression pumps, concentrators, reaction furnaces and micro-distillation equipment. This technique is time-consuming, typically requires larger sample sizes, and contamination and isotope fractionation may occur at each stage of the process. Once ready, pure gas enters the IRMS through a variable-volume reservoir called a bellows. The bellows system allows sample and reference gas comparisons to be made under the same circumstances. The bellows are connected to the ion source of the mass spectrometer via capillary inlet lines. A valve system called a "transfer valve" between the capillary and the mass spectrometer is used to transfer the capillary effluent between the ion source of the mass spectrometer and the waste line to keep the capillary flow constant.

3.3.1.2 Continuous Flow-Isotope Radio Mass Spectrometer (CF-IRMS)

The CF-IRMS technique uses helium as the carrier gas, which brings the gas to be analyzed into the ion source of the IRMS. This technology is used to connect the IRMS to a range of automated sample preparation equipment. While dual-channel is the most accurate method for stable isotope ratio measurements, continuous-flow mass spectrometry has several advantages, including online sample preparation, smaller sample volumes, faster analysis, simpler analysis, and more cost-effectiveness, and offers the possibility of interfacing with other preparative techniques, like Elemental Analysis(EA), Gas Chromatography (GC) and Liquid Chromatography (LC).

The main objectives of the ion source system of the IRMS instrument include: ①Enable the linear relationship between ion current intensity and measured ratio; ②No memory-effect between the sample and reference gas subsequently introduced into the mass spectrometer; ③Stagnant sources include welds, copper gaskets (SO_2) and polymer; gas adsorption on polymer gaskets (CO_2, H_2).

In 1947, Nier proposed the simultaneous detection and integration of ion currents of interest using multiple detectors. The advantage of using two independent amplifiers to measure ion current simultaneously is that ion current fluctuations due to temperature changes, electron beam instability, etc. can be completely canceled out. The magnet and accelerating voltage remain the same. No peak hopping is required, which eliminates the corresponding settling time. In addition, each detector channel can be fitted with a high ohmic resistor suitable for the average natural abundance of the ion flux of the target isotope. This static multiple selection principle is still in use, but the collector plate has been replaced by a Faraday cup to minimize false detector currents from secondary electrons.

The Faraday cup (FC) was positioned so that the primary ion current hit each cup of the inlet slit at the same time. Each incoming ion contributes a primary charge, and no stray ions or electrons can enter the cup and no secondary particles formed from impacts on the inner walls of the cup leave the cup. The ionic current is continuously monitored, then amplified, and finally transferred to a computer. The computer integrates the peak areas for each isotope and calculates the corresponding ratio. Take CO_2 as an example, the data consists of three different isotopes: $^{12}C^{16}O_2$, $^{13}C^{16}O_2$, and $^{12}C^{18}O^{16}O$, which have masses of $m/z=44$, 45 and 46, respectively.

3.3.2 Thermal Ionization Mass Spectrometer

Thermal Ionization Mass Spectrometer (TIMS) was developed after the discovery of atoms and the understanding of the behavior of charged particles in electric and magnetic fields. TIMS enables the measurement of isotope ratios with the highest accuracy, precision and sensitivity.

Up to now, TIMS mainly consists of three parts: ① Ion source, the region of ion generation, acceleration and focus;②Analyzer, the region to separate the beam based on the mass/charge ratio;③Collector, the ion beam in the region is measured either continuously (single collector) or simultaneously (multiple selectors). The electronics of these instruments need to be of similar deviation in order for the isotope ratio to be accurate from 0.01% to 0.001%. TIMS operates on a sector magnetic mass spectrometer (Figure 3.7), which accurately measures the isotope ratios of thermally ionized elements, usually by transferring current through a thin metal band or wire in a vacuum environment. In general, TIMS is used for the isotopic analysis of Pb, Sr, U, Fe, Cu, Mg, Os, B and Li.

In addition to the need to develop new methods to improve the efficiency of sample preparation and thus improve the test efficiency, TIMS is expected to further improve the accuracy of isotope analysis, especially in the application of inorganic nutrients in TIMS. The accuracy is mainly affected by the mass-related fractionation effect during the sample

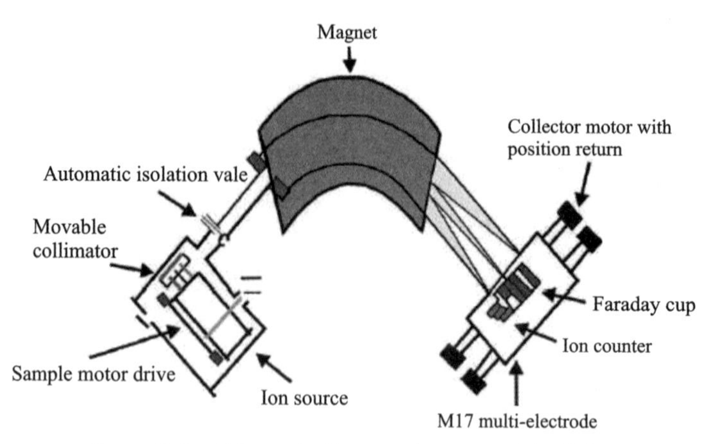

Figure 3. 7 Schematic diagram of commercial TIMS.

evaporation process in the ion source. Lighter isotopes usually evaporate preferentially, resulting in the continuous change of isotope ratios between process and measurement, which directly affects the precision. This effect depends on the relative rather than absolute mass differences of evaporation species. This makes some isotope analysis of inorganic nutrients a special challenge, such as Ca, Mg and Fe.

The advantages of TIMS over other isotope ratio techniques include: ① High accuracy due to the chemical and physical stability of the test environment; ② Lower and more consistent mean mass fractionation; ③ The molecular background and the generation of multi-charged ions are relatively low; ④ Using a single element solution to eliminate the interference; ⑤ Generating ions in a limited energy range avoids the use of energy filters; ⑥ Easy to operate automatically; ⑦ And the ion transfer efficiency from the ion source to the collector is close to 100%.

The disadvantages of TIMS technology are: ① Not all elements are easy to ionize, which limits the application of elements with low ionization potential; ② Ionization is not equivalent to all elements, and generally does not exceed 1%; ③ Mass fractionation changes constantly during analysis; ④ Sample preparation process is cumbersome; ⑤ And accurate mass fractionation correction is limited to elements with three or more isotopes, of which at least two are stable.

3.3.3 Multi-Collector-Inductively Coupled Plasma-Mass Spectrometry (MC-ICP-MS)

Traditionally, TIMS has been the first choice for isotope analysis in terms of accuracy and precision. In recent years, the introduction of multiple collector-ICP-MS (MC-ICP-MS) has enabled the simultaneous determination of multiple elements. In addition to simple and reliable sample injection, large flux and high mass resolution, the technique combines the strength of the ICP technique (high ionization efficiency for nearly all elements) with the high precision of thermal ion source mass-spectrometry (0.001%). Therefore, MC-ICP-MS has been widely used in isotope measurement of various matrix samples.

The MC-ICP-MS is designed to overcome the limitations of other mass spectrometer methods. As shown in Figure 3.8, it is a hybrid mass spectrometer that combines the ICP source with a mass spectrometer equipped with multiple Faraday collectors. This design is intended in order that the isotopic composition of elements with high ionization potentials, which are difficult to analyze with TIMS, can be easily and accurately determined. Meanwhile, the ICP source allows the samples injected flexibly either as a solution or as an aerosol generated by laser ablation. All MC - ICP - MS instruments need Ar as the plasma support gas, in a similar manner to that commonly used in conventional ICP - MS. With a high temperature of Ar plasma from 6000K to 8000K, the ionization rates of elements whose first ionization energy is less than 10eV are over 75%. Thus, almost all elements in the periodic table can be analyzed for isotopes with plasma sources.

The sector magnetic mass analyzer of MC-ICP-MS instrument is similar to that of TIMS instrument, and its ultimate goal is to achieve flat-topped peaks for isotopic measurement of high precision. Considering that all current instruments have a very low mass resolution (about 400), a wide ion source and collector slit are used to effectively reduce the loss. However, it is worth noting that at such quantitative resolution, ions of the same m/z cannot be separated from each other by a mass analyzer. For instance, there is a high resolution of around 2500 required to separate $^{56}Fe^{+}$ from $^{40}Ar^{16}O^{+}$.

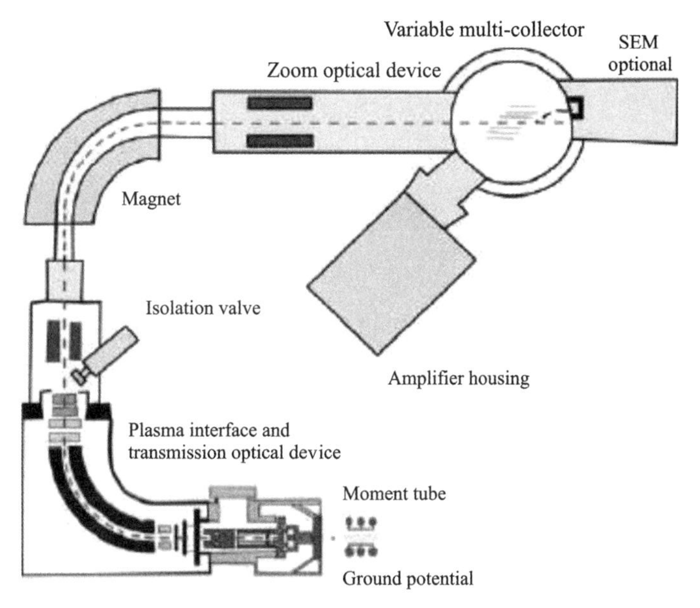

Figure 3.8 Schematic diagram of commercial TIMS.

The detector with multiple Faraday collectors allows simultaneous collection of separated isotopes, which eliminates the impact of noise signals on isotope ratio measurements. This static multiple acquisition technique is particularly important in plasma source mass spectrometry because the ion beam produced by plasma is more unstable than that of TIMS, which is mainly due to the plasma flashing caused by the short-term intense fluctuations. Faraday cups can be adjusted independently to determine the isotopic analysis of various elements with different mass isotopes. Multiple collectors can be configured with as many as 9 Faraday cups and 8 miniature ion counters. The combination of multiple ion collectors is increasingly being applied in mass spectrometry in order to measure precise and accurate isotope ratios for low abundance isotopes. In addition, the dynamic range of isotope ratio measurements can be significantly increased. The measured isotope ratios must be properly corrected for all instrumental deviations, including mass fractionation. Once corrected, these ratios are suitable for plotting in any diagram requiring atomic ratios.

The analytical signals of a single collector are monitored sequentially, whereas the MC-ICP-MS is capable of monitoring the intensity of multiple ions beams simultaneously. Therefore, short-term changes in signal strength affect all isotopes to the same extent, and these factors do not adversely affect the accuracy of isotope ratio measurements. Thus, the isotope ratio precision can reach 0.002% with the latest mass spectrometers, making MC-ICP-MS a strong competitor for TIMS.

In order to ensure the accuracy and precision of the test, it is necessary to optimize the instrument appropriately prior to the test given that MC-ICP-MS is quite complicated in structure. Optimization should achieve the following conditions: ①The peak of all isotopes should be as flat as possible; ②The stability and sensitivity of the analyte should be high and that of the blank sample should be low; ③Faraday detectors are used for all isotopes as possible; ④Daily calibration of Faraday cup efficiencies or gain coefficients; ⑤The ultimate result tends to be static operation, while dynamic operation can provide accurate and precise data; ⑥Optimize the abundance sensitivity appropriately (contribution of adjacent peaks to the strength of a particular analyte's isotope peak) for applications requiring high abundance sensitivity, such as uranium-thorium systems; ⑦Optimize the optimum measurement time to achieve the most accurate results; ⑧Use appropriate cleaning fluid and cleaning time between samples to reduce memory effects.

Compared with other isotope ratio techniques, The advantages of MC-ICP-MS include: ①The ionization efficiency is very high (close to 100%) for most elements, which enables the analysis of most elements in the periodic table, including those elements with high ionization potential which are difficult to be analyzed by TIMS; ② During the analysis process, MC-ICP-MS system is stable, resulting in constant mass fractionation time; ③There is a consistent mass deviation variation over the mass range, which allows the use of adjacent elements to calculate the mass deviation of those elements that do not have more than two stable isotopes; ④The injection system of MC-ICP-MS is flexible and solution can be introduced at atmospheric pressure; ⑤The laser ablation system can also be coupled with the MC-ICP-MS to perform in-situ isotope measurement of solid materials.

The disadvantages of MC-ICP-MS include: ①All components that enter the plasma are ionized (substances with two charges, oxides, etc.), therefore, chemical purification of the sample is required to achieve the highest precision and accuracy; ② Plasma instability limits precision; ③Although the ionization efficiency of plasma is close to 100%, the ion transport efficiency of plasma is lower than TIMS because the ions produced by plasma must be transferred from atmospheric pressure to the high vacuum environment of the mass spectrometer; ④Although adjacent elements can be used to correct the mass deviation, in systems with only two isotopes, such as Yb versus Lu and Ti versus Pb, the mass bias response between them is not the same and must be taken into account.

References

HOEFS J, 2018. Theoretical and Experiment Principles[M]. In J Hoefs (ed.). Springer: Stable Isotope Geochemistry: 27-39.

MöLLER K, SCHOENBERG R, PEDERSEN R B, et al., 2012. Calibration of the new certified reference materials ERM-AE633 and ERM-AE647 for copper and IRMM-3702 for zinc isotope amount ratio determinations[J]. Geostandards and Geoanalytical Research, 36 (2): 177-199.

XIE X J, GAN YQ, LIU Y D, et al., 2019. Principles and applications of environmental isotopes (In Chinese)[M]. Beijing: The Science Press.

Chapter 4 Hydrogen and Oxygen Isotopes

Until 1931, it was assumed that hydrogen consists of only one isotope. Urey et al. (1932) detected the presence of a second stable isotope, which was called deuterium. In addition to these two stable isotopes, there is a third naturally occurring but radioactive isotope, ^{3}H, tritium, with a half-life of approximately 12.5 years. Hydrogen is omnipresent in terrestrial environments, and it is envisaged to play a major role, directly or indirectly, in a wide variety of naturally occurring geological processes(Hoefs, 2018). Rosman and Taylor (1998) gave the following average abundances of the stable hydrogen isotopes:

$^{1}H=99.9885\%$

$^{2}D=0.0115\%$

Oxygen is the most abundant element on earth. It occurs in gaseous, liquid and solid compounds, most of which are thermally stable over large temperature ranges. These facts make oxygen one of the most interesting elements in isotope geochemistry. Because of the higher abundance and the greater mass difference, the $^{18}O/^{16}O$ ratio is normally determined, which may vary in natural samples by about 10% or in absolute numbers from about 1 : 475 to 1 : 525. More recently, with improved analytical techniques, the precise measurement of the $^{17}O/^{16}O$ ratio also became of interest (Hoefs, 2018). Oxygen has three stable isotopes with the following abundances (Rosman and Taylor, 1998) :

$^{16}O=99.757\%$

$^{17}O=0.038\%$

$^{18}O=0.205\%$

As two major elements in nature, hydrogen and oxygen are found throughout the world in both elemental and compound forms. They are not only the important components of various chemical reactions and geological processes in nature, but also the main media of the movement, circulation and energy transmission of various substances in nature. In the earth's crust, the abundance of oxygen is 46.6% and that of hydrogen is 0.14%. Hydrogen is often present in the form of OH^- in various silicate minerals, although its abundance in the atmosphere is small. The volume fraction of hydrogen in the atmosphere is only 0.5×10^{-6}, while oxygen accounts for 21% of the total volume of the atmosphere. Hydrogen and oxygen are also the basic material components of the biosphere and the basis for the survival of all living things. Due to the widespread distribution of hydrogen and oxygen in various substances and their significance in the all kinds of geochemical process, it is vital to study the hydrogen and oxygen isotope composition and their law of variation for exploring the mechanism of various geological processes and solving many geological and hydrogeological problems. Different from other isotopes dissolved in water, they are components of water molecules, so they are of great importance in the study of hydrologic cycle and various hydrologic processes.

4.1 Analytical techniques

The initial method for the determination of stable hydrogen and oxygen isotopes was offline DL-IRMS (Dual channel Isotope Ratio Mass Spectrometer). Then the highly automated Gas Bench-IRMS and TC/EA-IRMS were developed. DL-IRMS is the earliest method for hydrogen and oxygen isotope determination which has the characteristics of high precision and accuracy, but it is time-consuming and labor-consuming. GasBench-IRMS adopts the online test method, which has the advantages of fast operation and high efficiency, but it has some disadvantages such as large sample size and high requirement for temperature stability. TC/EA-Basing on the high temperature conversion principle of carbon reduction, IRMS has gradually realized the simultaneous on-line measurement of hydrogen and oxygen isotopes in trace water, which has the characteristics of convenient, fast and high precision. In recent years, laser water isotope analyzers based on Cavity Enhanced Absorption Spectroscopy (CEAS) have also been widely used to determine hydrogen and oxygen isotopes in water.

4.1.1 Dual channel Isotope Ratio Mass Spectrometer (DL-IRMS)

DL-IRMS hydrogen isotope analysis commonly uses some active metal elements (uranium, zinc, chromium, magnesium) as reducing agents to convert water into hydrogen for mass spectrometry analysis. Chromium (Cr) has good thermal stability and strong reducibility, so it can react quickly with water at high temperature (>800℃). The chemical equation for the reduction reaction to generate H_2 is: $2Cr+3H_2O=Cr_2O_3+3H_2$. The principle of the device for hydrogen production from water samples is as follows: First, the sample preparation system should be at low vacuum and the chromium reactor should be heated to 850℃. After the low vacuum is pumped, the high vacuum should be pumped to reach 1×10^{-3}Pa. Take 1μL water sample by a micro syringe and inject it directly into the chromium reactor for reaction. The sample tube (ST) with activated carbon should be frozen with liquid nitrogen, and then transfer the prepared hydrogen to the sample tube for absorption for 3min. The glass piston between the sample tube and the cold trap should be closed now, and remove the sample tube to send it to the mass spectrometer for testing. The hydrogen isotope composition of the measured water sample was corrected with SMOW or VSMOW standard water sample.

At room temperature (25℃), oxygen in 2mL natural water samples is balanced with standard CO_2 gas (high purity cylinder CO_2) of known isotopic composition by exchange of CO_2-H_2O. After dehydration with refrigerant, CO_2 gas was collected by liquid nitrogen, and then the oxygen isotope composition of equilibrium CO_2 gas is determined by MAT 253 gas mass spectrometer.

4.1.2 Gas Bench-IRMS

The sample tray temperature is set at 28℃ and the column temperature is set at 70℃. Besides, the helium pressure is set at 120kPa. Put 200μL water sample into 12mL flask, subsequently place it in the sample tray with constant temperature after add platinum catalytic rod and tighten the cap. At the same time, fix the injection and blowing needle, and then set the automatic injector working program to fill 2% H_2+He mixture in order to take away the air in the bottle. The time interval between sample inflation and sample analysis is 60min to achieve isotope exchange equilibrium. The ratio of H_2 isotope after isotopic fractionation equilibrium is determined by mass spectrometer.

The sample tray temperature is set at 28℃ and the column temperature is set at 45℃. Besides, the helium pressure is set at 120kPa. Put 200μL water sample and standard water into 12mL flask, subsequently place it in the sample tray with constant temperature. Fix the blowing needle and set the automatic injector working program to fill 0.3% CO_2 + He mixture for 10 minutes in order to take away the air in the bottle. The time interval between sample inflation and sample analysis is 18h to achieve isotope exchange equilibrium. The ratio of CO_2 isotope after isotopic fractionation equilibrium is determined by mass spectrometer.

4.1.3 TC/EA-IRMS

Remove 2mL water sample to the vial and place itin the sample tray of AS 3000 automatic liquid sampler after seal it with a screw hole cap lined with sealing spacer and leave no headspace. Set the working program to order the 0.5μL automatic injection needle to remove 0.2μL water sample from 2mL sample bottle. The water sample is then injected into the high temperature cracking furnace by inserting through the sealing spacer at the injection port of the elemental analyzer. The water vapor formed at high temperature reduces with the glass carbon particles filled in the high temperature cracking furnace at 1400℃. The generated H_2 and CO mixtures were carried by helium carrier gas (flow rate of 100mL/min) and separated by a gas chromatographic column filled with 0.5nm molecular sieve at a column temperature of 90℃. Then, they were sequentially imported into an ion source of a stable isotope mass spectrometer via ConFlo IV to achieve sequential simultaneous determination of hydrogen and oxygen isotopes in a single analysis. The whole test process only takes about 10 minutes.

4.1.4 Cavity Enhanced Absorption Spectroscopy (CEAS)

Laser water stable isotope analysis technology basing on CEAS testing method obtain the concentration of the target gas and calculate its isotope ratio according to the concentration ratio of light and heavy molecules. The method can simultaneously determine $\delta^{2}H$, $\delta^{18}O$ and $\delta^{17}O$ in water samples. The precision is respectively $\delta^{2}H<0.5‰$, $\delta^{18}O<0.1‰$, $\delta^{17}O<0.1‰$.

Taking LWA-45 EP as an example, the analyzer and automatic sampler are turned at first; install injector diaphragm (a diaphragm needs to be replaced every 1～2d after continuous measurement); preheat 3～6h; rinse the injection needle with lubricant (30～50 times), and then rinse the injection needle with deionized water (30～50 times); install injection needle; Check the position of diaphragm inlet. The water sample to be tested is filtered with a 1mL syringe and a PTDE needle cartridge filter with a diameter of 13mm and aperture of 0.45μm. Place the filled sample bottle on the corresponding sample tray. Start running measurement procedure after debugging analyzer parameters.

4.2 Fractionation processes

Isotope fractionation is pronounced when the mass differences between the isotopes of a specific element are large relative to the mass of the element. Therefore, isotope fractionations are especially large for the light elements. All elements that form solid, liquid, and gaseous compounds stable over a wide temperature range are likely to have variations in isotopic composition. Generally, the heavy isotope is concentrated in the solid phase in which it is more tightly bound. Heavier isotopes tend to concentrate in compounds in which they are present in the highest oxidation state and lowest coordination number. Mass balance effects can cause isotope fractionations because modal proportions of substances can change during a chemical reaction. Isotopic variations in biological systems are mostly caused by kinetic effects. During biological reactions (e.g. photosynthesis, bacterial processes), the lighter isotope is very often enriched in the reaction product relative to the substrate (Hoefs, 2018).

4.2.1 Hydrogen isotope

The most effective processes in the generation of hydrogen isotope variations in the terrestrial environment are phase transitions of water between vapor, liquid, and ice through evaporation/precipitation and/or boiling/condensation in the atmosphere, at the Earth's surface, and in the upper part of the crust. Differences in the hydrogen isotope composition arise due to vapor pressure differences of water and, to a smaller degree, to differences in freezing points (Hoefs, 2018). Horita and Wesolowski (1994) have summarized that hydrogen isotope fractionations decrease rapidly with increasing temperatures and become zero at 220～230℃. Above the crossover temperature, water vapor is more enriched in deuterium than liquid water.

Fractionations again approach zero at the critical temperature of water(Figure 4.1). In all processes concerning the evaporation and condensation of water, hydrogen isotopes are fractionated in a similar fashion to those of oxygen isotopes, albeit with a different magnitude, because a corresponding difference in vapor pressures exists between H_2O and HDO in one case and $H_2{}^{16}O$ and $H_2{}^{18}O$ in the other (Hoefs, 2018). Therefore, the hydrogen and oxygen isotope distributions are correlated for meteoric water. Craig (1961a) first defined the generalized relationship:

$$\delta D = 8\,\delta^{18}O + 10 \tag{4.1}$$

which describes the interdependence of H and O isotope ratios in meteoric water on a global scale. This relationship, shown in Figure 4.2, is called in the literature the "Global Meteoric Water Line(GMWL)". Neither the coefficient 8 nor the so-called deuterium excess d of 10 are actually constant in nature. Both may vary due to a superposition of equilibrium and kinetic isotope effects depending on the conditions of evaporation, vapor transport and precipitation and, as a result, offer insight into climatic processes.

D/H fractionations among gases are extraordinarily large. Even in magmatic systems, fractionation factors are sufficiently large to affect the δD-value of dissolved water in melts during degassing of H_2, H_2S or CH_4. The oxidation of H_2 or CH_4 to H_2O and CO_2 may also have an effect on the isotopic composition of water dissolved in melts due to the large fractionation factors (Hoefs, 2018). With respect to mineral-water systems, different experimental studies obtained widely different results for the common hydrous minerals with respect to the absolute magnitude and the temperature dependence of D/H fractionations (Suzuoki and Epstein, 1976; Graham et al., 1980; Vennemann and O'Neil, 1996; Saccocia et al., 2009). Suzuoki and Epstein (1976) first demonstrated the importance of the chemical composition of the octahedral sites in crystal lattices to the mineral H-isotope composition. Subsequently, isotope exchange experiments by Graham et al. (1980, 1984) suggested that the chemical composition of sites other than the octahedral sites can also affect hydrogen isotope compositions. On the basis of theoretical calculations, Driesner (1997) proposed that many of the discrepancies between the experimental studies were due to pressure differences at which the experiments were carried out. Thus, for hydrogen, pressure is a

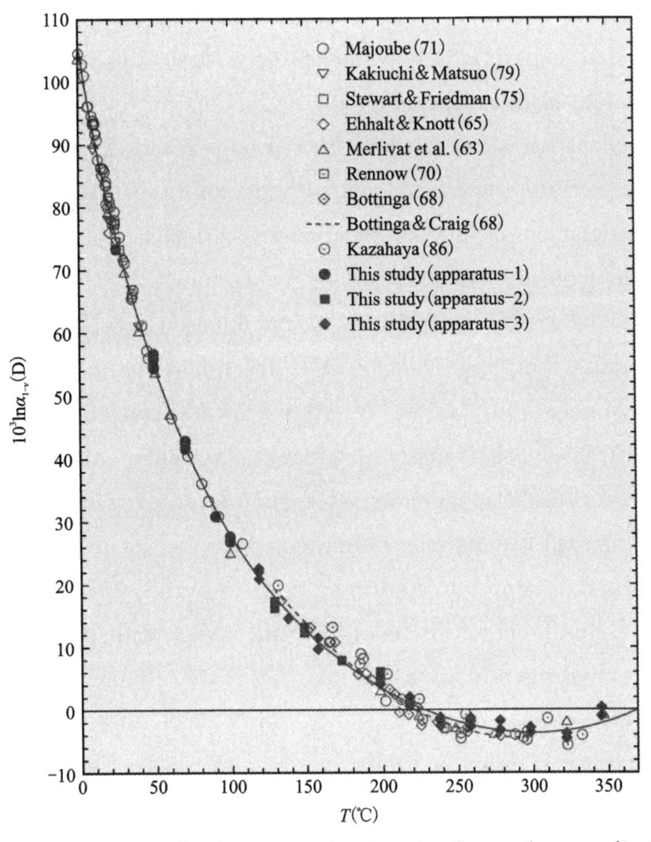

Figure 4.1 Experimentally determined fractionation factors between liquid water and water vapour from 1℃ to 350℃ (after Horita and Wesolowski, 1994).

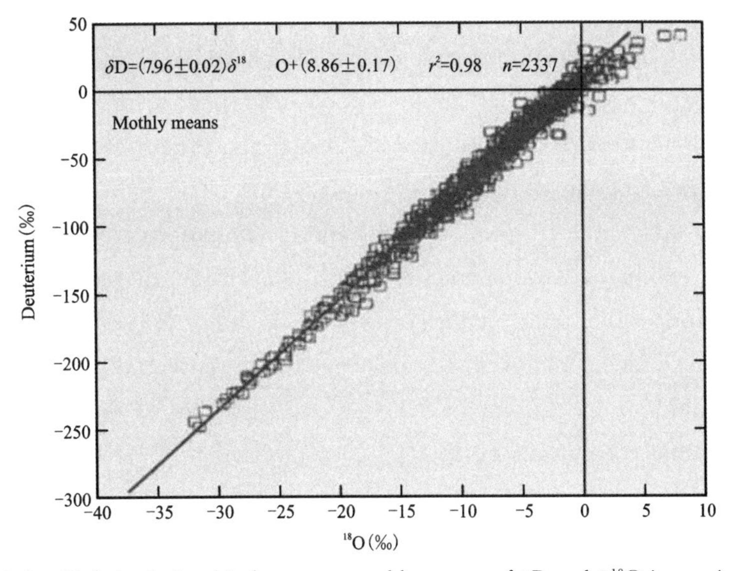

Figure 4.2 Global relationship between monthly means of δD and $\delta^{18}O$ in precipitation, derived for all stations of the IAEA global network. Line indicates the global Meteoric Water Line (GMWL) (after Rozanski et al., 1993).

variable that must be taken into account in fluid-bearing systems. Later, Horita et al. (1999) also presented experimental evidence for a pressure effect between brucite and water. Chacko et al. (1999) developed an alternative method for the experimental determination of hydrogen isotope fractionation factors, which carried out exchange experiments with large single crystals and then analyzed the exchanged rims with the ion probe, and this technique allows the determination of fractionation factors in experiments in which isotopic exchange occurs by a diffusional process. In summary, as discussed by Vennemann and O'Neil (1996), discrepancies between published experimental calibrations in individual mineral-water systems are difficult to resolve, which limits the application of D/H fractionations in mineral-water systems to estimate δD-values of coexisting fluids. As shown by Méheut et al. (2010) first-principles calculations of D/H fractionations may reproduce experimental calculations within a range of about 15‰. These authors also demonstrated that internal fractionations between inner-surface and inner hydroxyl groups may be large and even opposite in sign.

Water is the ultimate source of hydrogen in all naturally organic compounds produced by photosynthesis. Thus D/H ratios in organic matter contain information about climate. During biosynthetic hydrogen conversion of water to organic matter, large H-isotope fractionations with δD-values between −400‰ and +200‰ have been observed (Sachse et al. ,2012). δD-variations in individual compounds within a single plant or organism can be related to differences in biosynthesis. Accurate isotope fractionation factors among organic molecules and water are difficult to be determined, although tremendous progress has been achieved through the introduction of the compound specific hydrogen isotope analysis (Sessions et al. ,1999; Sauer et al. ,2001; Schimmelmann et al. ,2006), which allows the δD analysis of individual biochemical compound. Using a combination of experimental calibration and theoretical calculation Wang et al. (2009a, 2009b) estimated equilibrium factors for various H-positions in molecules such as alkanes, ketones, carboxyl acids and alcoholes. Subsequently, he extended his approach to cyclic compounds (Wang, 2013b). The equilibrium fractionations he obtained are very different to typical biosynthetic fractionations that are between −300‰ and −150‰ due to kinetic isotope fractionations. The biosynthesis of lipids as one of the most common group of organic material involves complex enzymatic reactions in which hydrogen may be added, removed or exchanged, all potentially leading to H-isotope fractionations. Lipids with the smallest D-depletion relative to water are n-alkyl lipids. Isoprenoid lipids show depletions by 200‰~250‰ and phytol and related compounds have the largest D-depletion (Hoefs, 2018).

In salt solutions, isotopic fractionations can occur between water in the "hydration sphere" and free water (Truesdell,1974). The effects of dissolved salts on hydrogen isotope activity ratios in salt solutions can be qualitatively interpreted in terms of interactions between ions and water molecules, which appear to be primarily related to their charge and radius. Hydrogen isotope activity ratios of all salt solutions studied so far are appreciably higher than H-isotope composition ratios (Hoefs, 2018). As shown by Horita et al. (1993), the D/H ratio of water vapor in isotope equilibrium with a solution increase as salt is added to the solution. Magnitudes of the hydrogen isotope effects are in the order $CaCl_2 > MgCl_2 > MgSO_4 > KCl—NaCl > Na_2SO_4$ at the same molality. Isotope effects of this kind are relevant for the proper interpretation of isotope fractionations in aqueous salt solutions and for an understanding of the isotope composition of clay minerals and absorption of water on mineral surfaces (Hoefs, 2018).

4.2.2 Oxygen isotope

Knowledge of the oxygen isotope fractionation between liquid water and water vapor is essential for the interpretation of the isotope composition of different water types. Fractionation factors experimentally determined in the temperature range from 0 to 350℃ have been summarized by Horita and Wesolowski (1994) and are shown in Figure 4.3. Addition of salts to water also affects isotope fractionations. The presence of ionic salts in solution changes the local structure of water around dissolved ions. Taube (1954) first demonstrated that the $^{18}O/^{16}O$ ratio of CO_2 equilibrated with pure H_2O decreased upon the addition of $MgCl_2$, $AlCl_3$ and HCl, remained more or less unchanged for NaCl, and increased upon the addition of $CaCl_2$. To explain this different fractionation behavior, Taube (1954) postulated different isotope effects between the isotopic properties of water in the hydration sphere of the cation and the remaining bulk water. The hydration sphere is highly ordered, whereas the outer layer is poorly ordered. The relative sizes of the two layers are dependent upon the magnitude of the electric field around the dissolved ions. The strength of the interaction between the dissolved ion and water molecules is also dependent upon the atomic mass of the atom to which the ion is bonded. O'Neil and Truesdell (1991) have related the concept of "structure-making" and "structure-breaking" solutes to isotope fractionation: structure makers yield more positive isotope fractionations relative to pure water whereas structure breakers produce negative isotope fractionations. Any solute that results in a positive isotope fractionation is one that causes the solution to be more structured as is the case for the structure of ice, when compared to solutes that lead to less structured forms, in which cation-H_2O bonds are weaker than H_2O-H_2O bonds. the hydration of ions may play a significant role in hydrothermal solutions and volcanic vapors (Driesner and Seward, 2000). Such isotope salt effects may change the oxygen isotope fractionation between water and other phases by several permil.

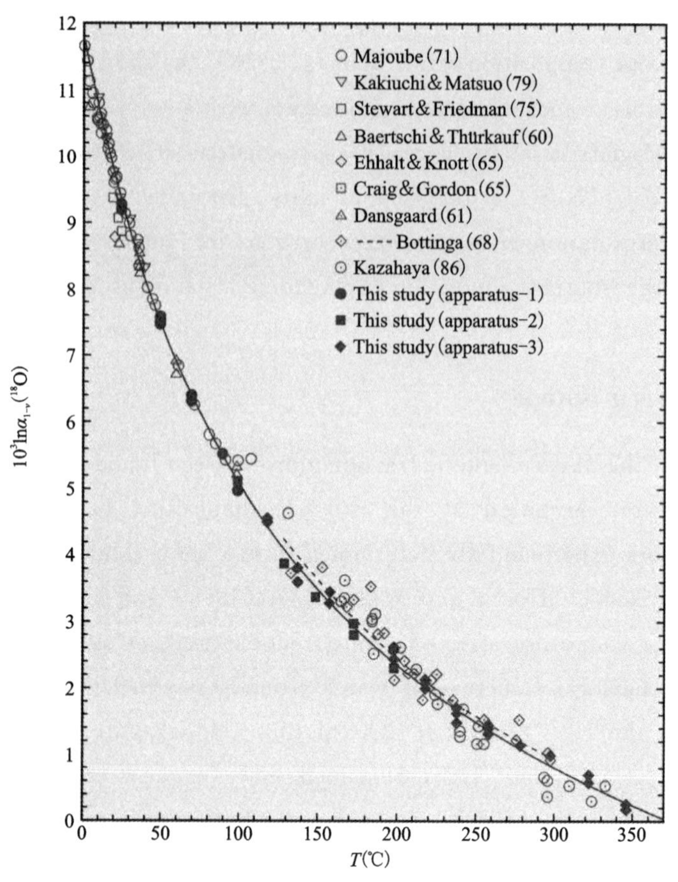

Figure 4.3 Oxygen isotope fractionation factors between liquid water and water vapor in the temperature range 0～350℃ (after Horita and Wesolowski, 1994).

Of equal importance is the oxygen isotope fractionation in the CO_2-H_2O system. Early work concentrated on the oxygen isotope partitioning between gaseous CO_2 and water. In the work by Usdowski et al. (1991), Beck et al. (2005) and Zeebe (2007), it has been demonstrated that the oxygen isotope composition of the individual carbonate species are isotopically different, which is consistent with experimental work of McCrea (1950) and Usdowski and Hoefs (1993). The oxygen isotope fractionation ($1000\ln\alpha$) between aqueous CO_2 and water at 25℃ is 41.6, dropping to 24.7 at high pH values when CO_3^{2-} is the dominant species (Figure 4.4). The pH dependence of the oxygen isotope composition in the carbonate-water system has important implications in the derivation of oxygen isotope temperatures.

The oxygen isotope composition of a rock depends on the ^{18}O contents of the constituent minerals and the mineral proportions. Garlick (1966) and Taylor (1968) arranged coexisting minerals according to their relative tendencies to concentrate ^{18}O. The list given in Table 4.1 has been augmented by data from Kohn et al. (1998a, 1998b, 1998c). This order of

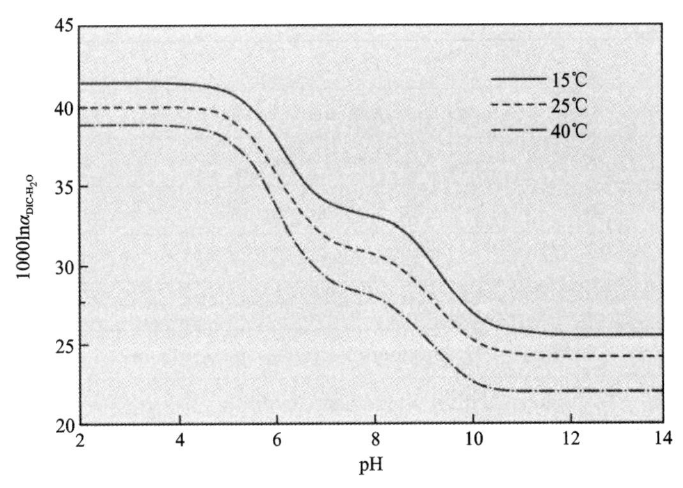

Figure 4.4 Oxygen isotope fractionations between dissolved inorganic carbon (DIC) and water as function of pH and temperatures (after Beck et al., 2005).

decreasing ^{18}O-contents has been explained in terms of the bond-type and strength in the crystal structure. Semi-empirical bond-type calculations have been developed by Garlick (1966) and Savin and Lee (1988) by assuming that oxygen in a chemical bond has similar isotopic behavior regardless of the mineral in which the bond is located, however the accuracy of this approach is not strictly true. Kohn and Valley (1998a, 1998b) determined empirically the effects of cation substitutions in complex minerals such as amphiboles and garnets spanning a large range in chemical compositions. Although isotope effects of cation exchange are generally less than 1‰ at $T > 500$℃, they increase considerably at lower temperatures. On the basis of these systematic tendencies of ^{18}O enrichment found in nature, significant temperature information can be obtained up to temperatures of 1000℃, and even higher, if calibration curves can be worked out for the various mineral pairs. The published literature contains many calibrations of oxygen isotope geothermometers, most are determined by laboratory experiments, although some are based on theoretical calculations. Although much effort has been directed toward the experimental determination of oxygen isotope fractionation factors in mineral—water systems, the use of water as an oxygen isotope exchange medium has some disadvantages (Hoefs, 2018). A more recent summary has been given by Chacko et al. (2001) (Figure 4.5). Many isotopic fractionations between low-temperature minerals and water have been estimated by assuming that their temperature of formation and the isotopic composition of the water in which they formed (for example, ocean water) are well known. This is sometimes the only approach available in cases in which the rates of isotope exchange reactions are slow and in which minerals cannot be synthesized in the laboratory at appropriate temperatures.

Table 4.1 Sequence of minerals in the order (bottom to top) of their increasing tendency to concentrate ^{18}O.

Quartz
Dolomite
K-feldspar, albite
Calcite
Na-rich plagioclase
Ca-rich plagioclase
Muscovite, paragonite, kyanite, glaucophane
Orthopyroxene, biotite
Clinopyroxene, hornblende, garnet, zircon
Olivine
Ilmenite
Magnetite, hematite

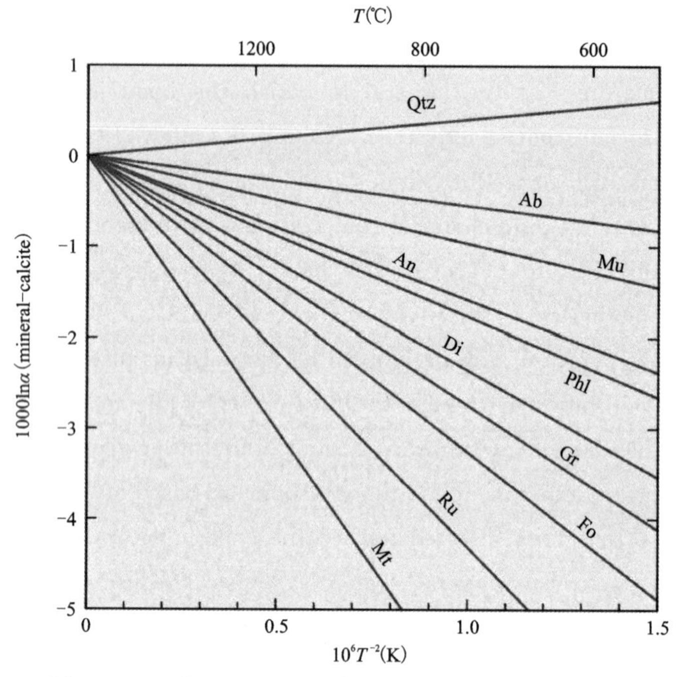

Figure 4.5 Oxygen isotope fractionations between various minerals and calcite (after Chacko et al., 2001).

4.3 Variations of Hydrogen and Oxygen isotopes in groundwater

In temperate and humid climates the isotopic composition of groundwater is similar to that of the precipitation in the area of recharge (Gat, 1971). This is a strong evidence for direct meteoric recharge to an aquifer. The seasonal variation of all meteoric water is strongly attenuated during transit and storage in the ground. The degree of attenuation varies with depth and with surface and bedrock geologic characteristics, but in general deep groundwater show no seasonal variation in δD and δ^{18}O values and have an isotopic composition close to amount-weighted mean annual precipitation values.

The characteristic isotope fingerprint of precipitation provides an effective means for identifying possible groundwater recharge areas and hence subsurface flow paths. For example, in areas close to rivers fed from high altitudes, groundwater represent a mixture of local precipitation and high-altitude low-δ^{18}O water. In suitable cases, quantitative estimates about the fraction of low-^{18}O river water in the groundwater can be carried out as a function of the distance from the river.

The main mechanisms that can cause variations between precipitation and recharged groundwater are (Gat, 1971):

(1) recharge from partially evaporated surface water bodies.

(2) recharge that occurred in past periods of different climate when the isotopic composition of precipitation was different from that at present.

(3) isotope fractionation processes resulting from differential water movement through the soil or the aquifer or due to kinetic or exchange reactions within geologic formations.

In semi-arid or arid regions, evaporative losses before and during recharge shift the isotopic composition of groundwater towards higher *d*-values. Furthermore, transpiration of shallow groundwater through plant leaves may also be an important evaporation process. Detailed studies of soil moisture evaporation have shown that evaporation loss and isotopic enrichment are greatest in the upper part of the soil profile and are most pronounced in unvegetated soils (Welhan, 1987). In some arid regions, groundwater may be classified as paleowater, which were recharged under different meteorological conditions than present in a region today and which imply ages of water of several thousand years. Gat and Issar (1974) have demonstrated that the isotopic composition of such paleowater can be distinguished from more recently recharged groundwater, which have been experienced some evaporation.

4.4 Application

4.4.1 Case study: Influence of irrigation practices on arsenic mobilization: Evidence from isotope composition and Cl/Br ratios in groundwater from Datong Basin, northern China (Xie et al., 2012)

Northern China is one of the most water-stressed regions in the world and groundwater is a major source for both drinking and irrigation purposes. Due to population growth and climate change, the use of groundwater resources has increased dramatically in recent decades. Therefore, in this study, systematic analysis of hydrogeological features of groundwater at Datong Basin was combined with isotope geochemistry to test the applicability of environmental isotopes and Cl/Br ratios in delineating groundwater recharge patterns and to determine the impact of irrigation return flow on arsenic mobilization in the groundwater system.

Datong Basin is a typical Cenozoic basin bounded by Pliocene to Pleistocene NE-SW trending normal faults as part of the Shanxi Rift System (Wang and Shpeyzer, 2000) (Figure 4. 6). Located between the Hengshan Mountain and Hongshou Mountain ranges, the basin is composed of Pliocene to Pleistocene and Holocene unconsolidated sediments with a various thickness from 1500m to 3500m. In the west of the area adjacent to the Hongshou Mountains, the Quaternary aquifer overlies Cambrian and Ordovician limestone and younger sedimentary rocks, mainly Carbonaceous to Permian sandstone and mudstone. The Quaternary sediments are underlain by fractured Archaean metamorphic rocks in the east of the area. The piedmont sediments are mostly alluvial-fluvial gravel and coarse sand, while those of the central basin are lacustrine and alluvial-lacustrine sandy loam, silt, silty clay and clay with high organic contents (Guo and Wang, 2005). Based on palaeosol horizons and fossil assemblages, there are four major stratigraphic units divisions in the Quaternary sediments (Qp—Qh) (Liu et al., 1986). The major types of rocks and sediments in the Datong Basin can be classified into four groups (Xie et al., 2011): ①Archean metamorphic complex (granites and gneiss with greenstone terrain) in the east margin (Heng shan Mountain); ② Cambrian to Ordovician limestone and dolomite with clastic rocks, mainly located in the southwest margin (Hongshou Mountains); ③Carboniferous to Permian coal-bearing clastic rocks in the northwest margin containing varying amounts of inter-bedded sandstone, siltstone and shale; ④ Late Pliocene to Holocene Basin sediments (alluvial and fluvial gravel, sand and silt).

The groundwater mainly occurs within the aquifers, which mainly consist of Middle to Late Pleistocene alluvial-fluvial gravel and sand, with the depth less than 150m below the land surface. In these aquifers, two hydraulically-linked flow systems were discerned. The groundwater table decreased from about 30m to less than 5m from piedmont to the center of the basin. The groundwater movement from recharge areas near the mountain fronts to discharge areas is slow (ranging from 0. 20m to 0. 58m per day) in this area (Xie et al., 2009). In general, two flow regimes can be detected at Datong: ①Flow from mountain front areas to the center of the basin; ② Flow inside the basin along the SW-NE direction of

Sanggan River. Groundwater is recharged by vertically infiltrating meteoric water in the basin and lateral flow of fracture water in bedrock along the basin margins, as well as irrigation return flow (Guo and Wang, 2005). Because of the drying of river channels, no significant contribution of the river to groundwater can be expected. Groundwater is discharged mainly by evapotranspiration and artificial abstraction.

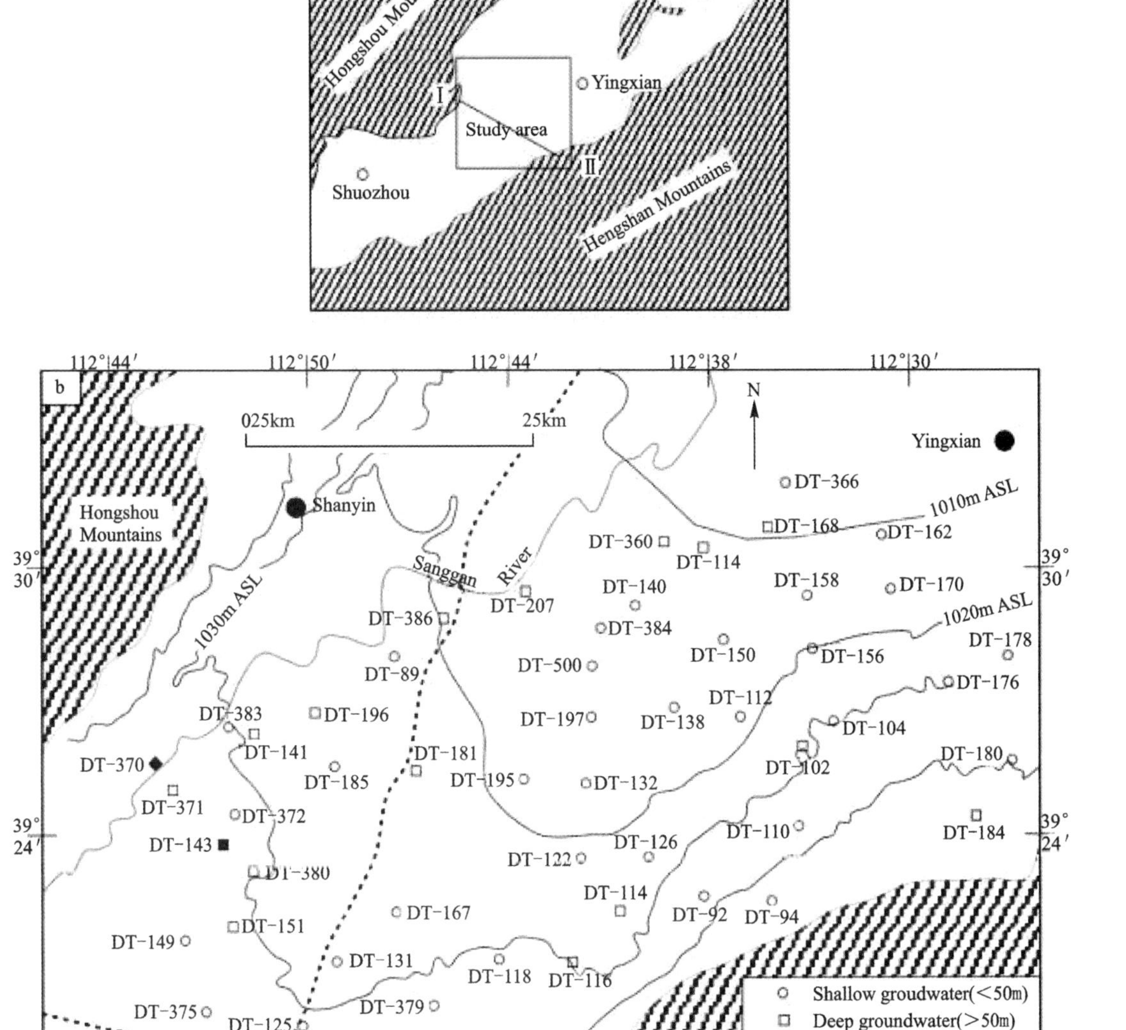

Figure 4. 6 a. Location of the study area in Datong Basin; b. Location of sampling sites.

4.4.1.2 Methods

A total of sixty water samples was collected for isotope and hydrochemical analysis in August 2009 at the Datong Basin. In addition, one surface water sample was collected from the Sanggan River and one spring water sample was taken from mountain front recharge area for hydrochemical analysis. The locations of sampling sites are shown in Figure 4.6. Stable isotopes of oxygen and hydrogen were measured using Finnigan MAT 253 mass spectrometer. $\delta^{18}O$ and $\delta^{2}H$ values were determined relative to internal standards calibrated using V-SMOW and isotopic composition ($\delta^{18}O$ and $\delta^{2}H$) were expressed in standard d notation representing per mille deviations from the V-SMOW standard. Precision for both $\delta^{2}H$ and $\delta^{18}O$ are ±1.0‰ and ±0.1‰.

4.4.1.3 Isotope characteristics in groundwater

Forty-four water samples were analyzed for $\delta^{18}O$ and $\delta^{2}H$. It can be seen that the samples have a quite wide range of $\delta^{18}O$ and $\delta^{2}H$ composition. Low arsenic groundwater samples ($As < 10\mu g/L$) had heavier $\delta^{18}O$ and $\delta^{2}H$ composition than high arsenic groundwater samples ($As > 10\mu g/L$). $\delta^{18}O$ and $\delta^{2}H$ values for low arsenic groundwater samples ranged from −7.9‰ to −11.9‰ (with an average of −10.1‰) and −63‰ to −92‰ (with an average of −75‰), respectively. The $\delta^{18}O$ and $\delta^{2}H$ values of high arsenic groundwater samples ranged respectively between −10.0‰ and −12.7‰ (with an average of −11.9‰) and between −75‰ and −98‰ (with an average of −89‰). Both shallow and deep groundwater samples collected from mountain front area near to Hengshan Mountains contain heavier isotope composition.

Oxygen and hydrogen isotopes can be indicators of groundwater recharge and discharge and subsurface processes (Faure, 1986). $\delta^{18}O$ and $\delta^{2}H$ values of groundwater samples can be compared to the global meteoric water line (GMWL) of Craig (1961) as well as a local meteoric water line. The results of comparison to local precipitation can be used to determine if the groundwater is derived from recent local recharge and various other processes that control fractionation. The low slope is characteristic of water subject to evaporation through a dry surface layer due to non-equilibrium isotope fractionation under the low moisture condition (Allison, 1982; Barnes and Allison, 1983, 1988; Banner et al., 1989; Clark and Fritz, 1997). There has been no record of historic $\delta^{18}O$ and $\delta^{2}H$ values for local precipitation in the study area. An estimate can be made using data for isotopes in precipitation at Baotou station (about 300km north of the study area) with a weighted mean value of −58‰ and −8.3‰ for $\delta^{2}H$ and $\delta^{18}O$ respectively (IAEA/WMO, 2007) and a local meteoric line(LMWL):

$$\delta^{2}H = 6.3\,\delta^{18}O - 5.4 \tag{4.2}$$

In this study, the high and low arsenic groundwater $\delta^{18}O$ and $\delta^{2}H$ values define two regression lines with different slopes (Figure 4.7):

$$\delta^{2}H = 6.3\,\delta^{18}O - 13.5 \text{ (high arsenic groundwater, HAWL).} \quad (4.3)$$

$$\delta^{2}H = 3.6\,\delta^{18}O - 39 \text{ (low arsenic groundwater, LAWL).} \quad (4.4)$$

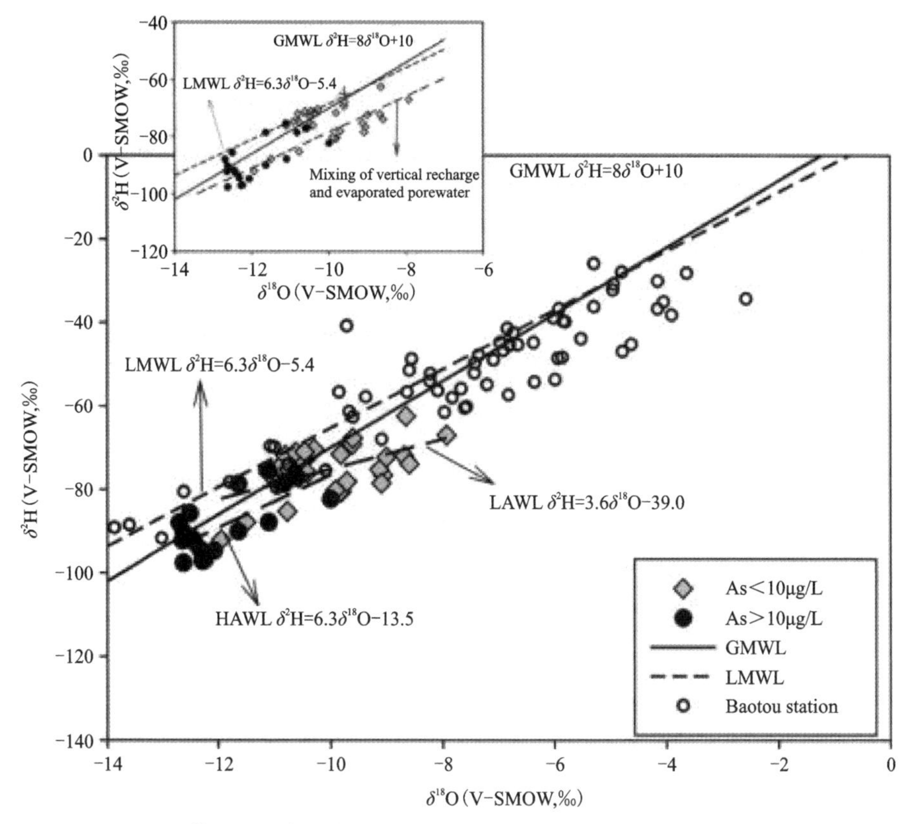

Figure 4.7 $\delta^{18}O$ vs. $\delta^{2}H$ plot of groundwater samples from Datong, as compared with the global meteoric water line (Craig, 1961) and local meteoric water line.

Groundwater $\delta^{18}O$ and $\delta^{2}H$ values plot very close to the global meteoric water line (GMWL) (Craig, 1961) and LMWL (Figure 4.7) indicating a local precipitation origin of groundwater. Samples (DT-184, 92 and 78) collected from mountain areas are plotted on the upper end of local meteoric water line and are therefore recharged by precipitation. However, in general, all Datong water samples fall on the lower end of local meteoric water line. It could be related to the difference in average precipitation between Batou and Datong. The markedly lower slope value of regression line for the low arsenic groundwater than that for LMWL, which has a slope of 6.3, indicates an evaporation effect (Allison, 1982; Barnes and Allison, 1983, 1988; Banner et al., 1989; Clark and Fritz, 1997). Notably, some groundwater samples (including almost all high arsenic groundwater) $\delta^{18}O$ and $\delta^{2}H$ plots are aligned parallel to but shifted apart from the LMWL (Figure 4.7). This could be related to the mixing of direct vertical recharge and evaporated pore water which can lead to a parallel shift away from the LMWL on the $\delta^{18}O$ and $\delta^{2}H$ plot (Clark and Fritz, 1997).

Figure 4.8 shows the spatial distribution of $\delta^{18}O$ values of groundwater. The range of $\delta^{18}O$ in some shallow groundwater ($<$50m) reflects the local variability of the $\delta^{18}O$ of different rainfall events and the extent of local evaporation. However, enrichment of heavier isotope compositions in samples taken from mountain front area may reflect the isotopic character of groundwater recharge via fractures. The relatively homogenized $\delta^{18}O$ values in deep groundwater reflect the effect of mixing of groundwater along the flowpath. It is well documented that stable isotope component of groundwater becomes more homogeneous along groundwater flowpath (Clark and Fritz, 1997; Goller et al., 2005).

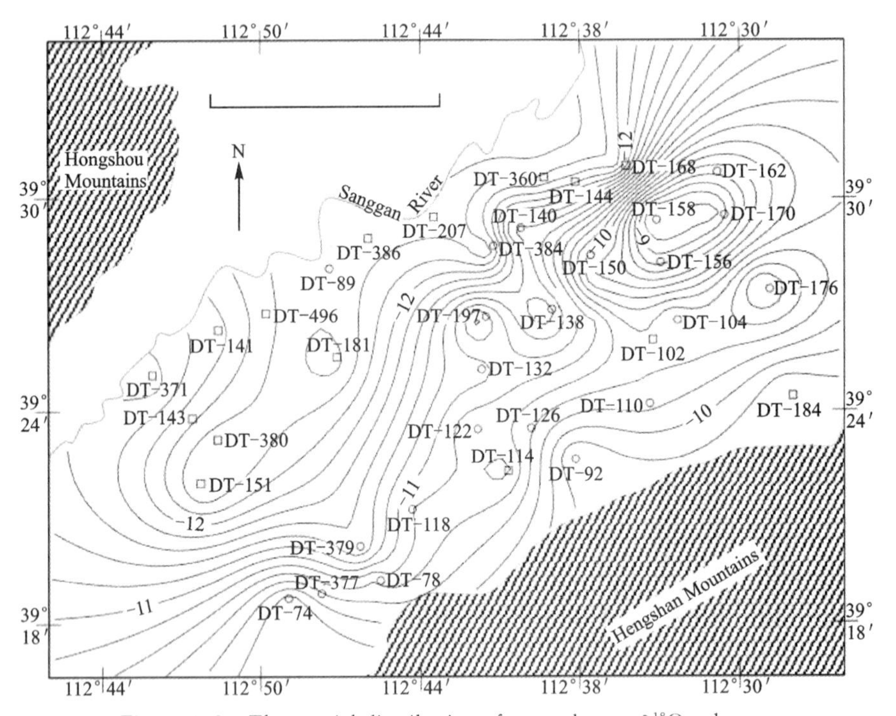

Figure 4.8 The spatial distribution of groundwater $\delta^{18}O$ values.

Intensive extraction of groundwater over decades from wells deeper than 50m for irrigation may have made irrigation return and salt flushing water an important source for recharge of groundwater aquifers at Datong. From the $\delta^{18}O$ vs. Cl concentration plot (Figure 4.9a), three trends can be observed: ① Sharp increase in Cl concentrations with little variation of $\delta^{18}O$ values. As indicated by the $\delta^{18}O$ and $\delta^{2}H$ plot, it can be attributed to vertical mixing of recharge water such as irrigation return or flushing water with pore water. Because of the wide distribution of saline soil, periodical salt flushing has been a common practice in the study area. The dissolution of halite, mixing with irrigation return and flushing water can result in large variation of Cl concentration with no significant change of the $\delta^{18}O$ values. The positive correlation between Na and Cl with Na/Cl ratio close to 1.0

reflects the effect of halite dissolution. ② Drastic increase of $\delta^{18}O$ values with moderate change of Cl concentration. This reflects the effect of evaporation on groundwater because evaporation can modify the $\delta^{18}O$ values of groundwater with moderate change of Cl concentration. ③Change in $\delta^{18}O$ values with no accompanying variation of Cl concentrations. This could be due to mixing of lateral recharge water with deep groundwater because both of them contain low Cl concentrations but different $\delta^{18}O$ values. From Figure 4.9, it can be seen that most of high arsenic groundwater samples are plotted on or close to the leaching line and a few samples contained moderate arsenic concentration falling on evaporation line. This indicates that vertical mixing of salt flushing water and irrigation return with groundwater are major processes controlling the occurrence of high arsenic groundwater. Although evaporation may affect arsenic enrichment in groundwater, the spatial distribution of arsenic concentrations and $\delta^{18}O$ values (Figure 4.9b) indicates that evaporation is not the major factor for enrichment of arsenic in groundwater at Datong.

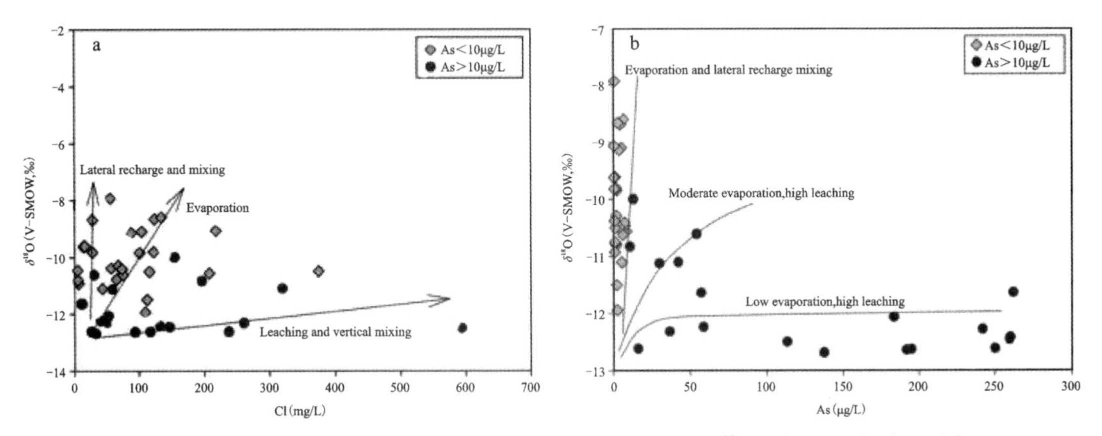

Figure 4.9 a. Relationship between Cl concentrations and $\delta^{18}O$ values in high and low arsenic groundwater, indicating that leaching is a more important process for arsenic mobilization; b. $\delta^{18}O$ and arsenic concentration in high and low arsenic groundwater indicating infiltration of irrigation groundwater controls arsenic mobilization.

Relationship between $\delta^{18}O$ and arsenic concentrations was studied to further reveal the effect of vertical recharge on arsenic concentration in groundwater as documented by plots of $\delta^{18}O$ vs. $\delta^{2}H$ values and $\delta^{18}O$ vs. Cl concentration. The observed non-linear relationship between $\delta^{18}O$ and arsenic concentrations (Figure 4.9b) indicates the effects of evaporation and vertical mixing. From Figure 4.9b, the following three trends can be clearly observed. ① Significant change in arsenic concentration with no or little variation of isotopic composition. The large variation in arsenic concentration with relatively little change in isotopic composition indicates the effect of arsenic mobilization by fast vertical recharge water (irrigation return or salt flushing water). Most of high arsenic groundwater falls on this trend (Figure 4.9b) indicates that vertical fast recharge of salt flushing water and irrigation return govern the mobilization of arsenic. This is consistent with the conclusion inferred from the relationship between $\delta^{18}O$ vs. Cl concentration. ② Moderate change of arsenic concentration with variation of isotopic composition (including a few high arsenic groundwater samples). This trend may reflect the combined effect of vertical recharge and moderate evaporation. Evaporation can cause the enrichment of heavier isotope and arsenic in pore water, while vertical recharge can lead to the enrichment of arsenic in water samples. ③ Large change of $\delta^{18}O$ value with little variation of arsenic concentration, which suggests strong evaporation and weak vertical recharge mixing. Therefore, correlation between $\delta^{18}O$ composition and arsenic concentration (Figure 4. 9b) again confirms that vertical recharge is important in releasing arsenic to the groundwater, as will be discussed later in detail.

4. 4. 2 Case study: Effects of water-sediment interaction and irrigation practices on iodine enrichment in shallow groundwater (Li et al. ,2016)

Iodine is an essential micronutrient for human beings, but excessive intake causes health problems, including goiter, cretinism, thyroid autoimmunity and even thyroid cancers. Cases of iodine enrichment in groundwater have been widely reported in coastal area and in inland basins. In areas affected by waterborne iodine poisoning, groundwater typically provides the dominant source for water supply, therefore understanding the mechanisms of iodine mobilization in the source aquifers is critical both for sustainable water resource management and for effective actions to diminish iodine poisoning.

Datong Basin is a typical iodine-affected area, and its hydrologic and geologic background was well described in the above case study. In this study, the groundwater from the center area of Datong Basin was collected to perform the chemistry and isotope analysis, which was same as to the above study.

The $\delta^{2}H$ and $\delta^{18}O$ values of groundwater ranged from $-90.2‰$ to $-55.6‰$ and from $-12.1‰$ to $-6.5‰$, respectively. All groundwater can be divided into two groups according to the $\delta^{2}H$ and $\delta^{18}O$ values: Group Ⅰ is characterized by enrichment of light isotope with the $\delta^{2}H$ and $\delta^{18}O$ ranging from $-90.2‰$ to $-82.9‰$ and from $-12.1‰$ to $-10.7‰$, respectively; Group Ⅱ, by contrast, has heavier hydrogen and oxygen isotope signatures, changing between $-76.0‰$ and $-55.6‰$ and between $-10.1‰$ and $-6.5‰$, respectively. The distinctive hydrogen and oxygen isotope signatures of these two group samples might reflect the impact of vertical mixing processes in the study area. The two upstream reservoir water samples are characterized by heavier $\delta^{2}H$ and $\delta^{18}O$ values (mean values: $-55.1‰$ and $-6.65‰$, respectively) in comparison with those of the groundwater samples. The rain water collected during the sampling campaign had the $\delta^{2}H$ and $\delta^{18}O$ values ranging from $-87.2‰$ to $-10.31‰$, respectively, which fall within the ranged of groundwater δ^2H and $\delta^{18}O$.

The vertical mixing process using the upstream reservoir water as the main irrigation source can introduce surface water into shallow groundwater, which is further supported by groundwater δ^2H and $\delta^{18}O$ signatures. The plot of δ^2H and $\delta^{18}O$ for all water samples is shown in Figure 4.10 where the global meteoric water line (GMWL) (Craig, 1961) and the local meteoric water line (LMWL) (IAEA/WMO, 2007) were also presented. It can be seen that all groundwater samples from the Datong Basin fell below the GMWL and below the LMWL. The regression line for the Datong groundwater samples is $\delta^2H = 6.018\ \delta^{18}O - 16.35$ ($r^2 = 0.936$) (Li et al., 2016). The precipitation sample from Datong Basin lies below to the three regression lines, which might be related to evaporation.

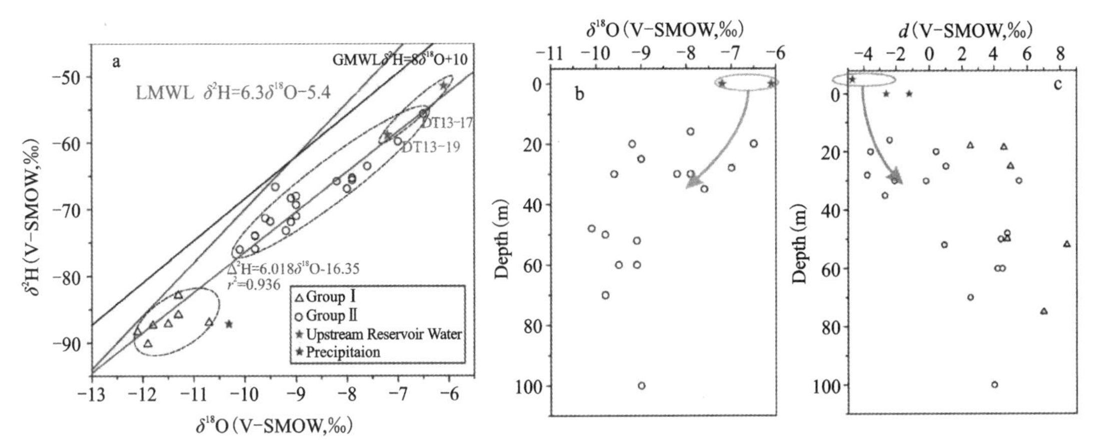

Figure 4.10 The plot of δ^2H and $\delta^{18}O$ of water samples from the Datong Basin(a) with the Global Meteoric Water Line (GMWL; data from Craig, 1961) and Local Meteoric Water Line (LMWL; data from IAEA/WMO, 2007); and the depth profiles of Group Ⅱ $\delta^{18}O$ (b) and δ-excess (c) of groundwater from the Datong Basin. The d-excess of water is defined as $d = \delta^2H - 8\ \delta^{18}O$ (Dansgaard, 1964).

Two groups with distinctive δ^2H and $\delta^{18}O$ features were identified, and Figure 4.10 shows that the Group Ⅱ samples fall along the line between the upstream reservoir water sample and the Group Ⅰ samples, suggesting the Group Ⅱ samples are mixtures. During the irrigation, the upstream reservoir water, as the main irrigation source, is delivered into the shallow groundwater (Li et al., 2016). The depth profile of water $\delta^{18}O$ further suggests that the contribution from reservoir water could explain the higher $\delta^{18}O$ values observed in shallow groundwater (Figure 4.10b). The δ^2H and $\delta^{18}O$ features can therefore be used to construct a balance model to estimate the contributions from the two end-members, as applied in previous studies (Halder et al., 2013; Peng et al., 2012, 2010). The equations in terms of δ value can be expressed as follows:

$$\delta^{18}O_{\text{II}} = \delta^{18}O_{\text{I}} \cdot R_{\text{I}} + \delta^{18}O_{\text{RW}} \cdot R_{\text{RW}} \quad (4.5)$$

$$R_{\mathrm{I}} + R_{\mathrm{RW}} = 1 \tag{4.6}$$

where R_{I} and R_{RW} are the weighted contributions from Group Ⅰ and reservoir water groundwater, respectively. The mean values of $\delta^{18}O$ in the reservoir water and Group Ⅰ samples were $-6.65‰$ ($\delta^{18}O_{RW}$) and $-11.5‰$ ($\delta^{18}O_{I}$), respectively. By this calculation, the contribution of reservoir water to Group Ⅱ approximately varies from 29% to 93%, gradually decreasing with the well depth. The shallow groundwater is heavily influenced by the surface irrigation activities. Notably, the contribution of rainfall was neglected in this two endmembers model. As shown in Figure 4.10c, the lowest δ value of precipitation can act as an end-member leading to the decrease in δ value of shallow groundwater, and therefore, the contribution of reservoir water might be overestimated by the two end-members model (Li et al., 2016).

As shown in Figure 4.10b, the vertical mixing process mainly influences the shallow groundwater with well depths less than 30m, and these groundwater samples have iodine concentrations of 17.4～2180μg/L. From the results of the mixing model, it can be found that the two groundwater samples (DT13-17 and DT13-19), located near the upstream reservoir DT13-28, are dominantly recharged by the upstream reservoir water. The iodine concentrations in these two samples were 143μg/L and 31.1μg/L, respectively, indicating the dilution effect of reservoir water infiltration on groundwater iodine in the shallow aquifer (Li et al., 2016). As reported in our previous studies, (sub)oxidizing condition favors the enrichment of groundwater iodine (Li et al., 2014). The supplement of free oxygen and/or agricultural NO_3, ranking at the top of the electron acceptor ladder, provide more energy to the microorganisms, causing the acceleration of degradation of organic matter and the subsequent release of iodine (Arndt et al., 2013; Rivett et al., 2008). Consequently, the cooccurrence of oxidizing condition and high iodine concentrations were observed in the shallow groundwater.

References

ALLISON G B, 1982. The relationship between ^{18}O and deuterium in water and in sand columns undergoing evaporation[J]. J. Hydrol., 76:1-25.

ARNDT S, et al., 2013. Quantifying the degradation of organic matter in marine sediments: a review and synthesis[J]. Earth Sci. Rev., 123: 53-86.

BANNER J L, WASSERBURG G J, DOBSON P F, et al., 1989. Isotopic and trace-element constraints on the origin and evolution of saline groundwater from central Missouri [J]. Geochim. Cosmochim. Acta, 53:383-398.

BARNES C J, ALLISON G B, 1983. The distribution of deuterium and ^{18}O in dry soils

[J]. J. Hydrol. , 60:141-156.

BARNES C J, ALLISON G B, 1988. Tracing of water movement in the unsaturated zone using stable isotopes of hydrogen and oxygen[J]. J. Hydrol. ,100:143-176.

BECK W C, GROSSMAN E L, MORSE J W, 2005. Experimental studies of oxygen isotope fractionation in the carbonic acid system at 15℃, 25℃, and 40℃[J]. Geochim. Cosmochim. Acta, 69:3493-3503.

CHACKO T, COLE D R, HORITA J, 2001. Equilibrium oxygen, hydrogenand carbon fractionation factors applicable to geologic systems[J]. Rev. Mineral. Geochem. , 43:1-81.

CLARK I D, FRITZ P, 1997. Environmental Isotopes in Hydrogeology[M]. New York:Lewis.

CRAIG H, 1961. Isotope variation in meteoric water[J]. Science, 133:1702-1703.

DANSGAARD W, 1964. Stable isotopes in precipitation[J]. Tell. , 16 (4):436-468.

DRIESNER T, 1997. The effect of pressure on deuterium-hydrogen fractionation in high-temperature water[J]. Science, 277:791-794.

DRIESNER T, SEWARD T M, 2000. Experimental and simulation study of salt effects and pressure/density effects on oxygen and hydrogen stable isotope liquid-vapor fractionation for 4-5 molal aqueous NaCl and KCl solutions to 400℃[J]. Geochim. Cosmochim. Acta, 64:1773-1784.

GARLICK G D, 1966. Oxygen isotope fractionation in igneous rocks[J]. Earth Planet Sci. Lett. , 1:361-368.

GAT J R, 1971. Comments on the stable isotope method in regional groundwater investigation[J]. Water Resour. Res. ,7:980.

GOLLER R, WILCKE, W, LENG M J, et al. , 2005. Tracing water paths through small catchments under a tropical montane rain forest in south Ecuador by an oxygen isotope approach[J]. J. Hydrol. ,308:6780.

GRAHAM C M, SHEPPARD S M F, HEATON T H E, 1980. Experimental hydrogen isotope studies, I. Systematics of hydrogen isotope fractionation in the systems epidote-H_2O, zoisite-H_2O and AlO(OH)-H_2O[J]. Geochim. Cosmochim. Acta, 44:353-364.

GUO H, WANG Y, 2005. Geochemical characteristics of shallow groundwater in Datong Basin, northwestern China[J]. J. Geochem. Explor. , 87:109-120.

HALDER J, DECROUY L, VENNEMANN T W, 2013. Mixing of Rhône River water in Lake Geneva (Switzerland-France) inferred from stable hydrogen and oxygen isotope profiles[J]. J. Hydrol. , 477:152-164.

HOEFS J, 2018. Stable Isotope Geochemistry[M]. Springer International Publishing.

HORITA J, DRIESNER T, COLE D R, 1999. Pressure effect on hydrogen isotope fractionation between brucite and water at elevated temperatures[J]. Science, 286: 1545-1547.

HORITA J, WESOLOWSKI D J, 1994. Liquid-vapor fractionation of oxygen and hydrogen isotopes of water from the freezing to the critical temperature[J]. Geochim. Cosmochim. Acta, 58:3425-3437.

HORITA J, WESOLOWSKI D J, COLE D R, 1993. The activity-composition relationship of oxygen and hydrogen isotopes in aqueous salt solutions. I. Vapor-liquid water equilibration of single salt solutions from 50℃ to 100℃[J]. Geochim. Cosmochim. Acta, 57:2797-2817.

KOHN M J, VALLEY J W, 1998a. Oxygen isotope geochemistry of amphiboles: isotope effects of cation substitutions in minerals[J]. Geochim. Cosmochim. Acta, 62:1947-1958.

KOHN M J, VALLEY J W, 1998b. Effects of cation substitutions in garnet and pyroxene on equilibrium oxygen isotope fractionations[J]. J. Metam. Geol., 16:625-639.

KOHN M J, VALLEY J W, 1998c. Obtaining equilibrium oxygen isotope fractionations from rocks: theory and examples[J]. Contr. Mineral. Petrol., 132:209-224.

LI J, et al., 2014. Iodine mobilization in groundwater system at Datong Basin, China: evidence from hydrochemistry and fluorescence characteristics[J]. Sci. Total Environ., (468-469):738-745.

LI J, WANG Y, XIE X, 2016. Effects of water-sediment interaction and irrigation practices on iodine enrichment in shallow groundwater[J]. Journal of Hydrology, (543): 293-304.

MCCREA J M, 1950. On the isotopic chemistry of carbonates and a paleotemperature scale[J]. J. Chem. Phys., 18:849-857.

MÉHEUT M, LAZZERI M, BALAN E, et al., 2010. First-principles calculation of H/D isotopic fractionation between hydrous minerals and water[J]. Geochim. Cosmochim. Acta, 74:3874-3882.

O'NEIL J R, TRUESDELL A H, 1991. Oxygen isotope fractionation studies of solute-water interactions[J]. In: Stable isotope geochemistry: a tribute to Samuel Epstein[M]. Geochemical Soc. Spec. Publ., 3:17-25.

PENG T-R, et al., 2010. Identification of groundwater sources of a local-scale creep slope: using environmental stable isotopes as tracers[J]. J. Hydrol., 381 (1-2):151-157.

PENG T-R, et al., 2012. Using oxygen, hydrogen, and tritium isotopes to assess pond water's contribution to groundwater and local precipitation in the pediment tableland areas of northwestern Taiwan[J]. J. Hydrol., (450-451):105-116.

RIVETT M O, BUSS S R, MORGAN P, et al., 2008. Nitrate attenuation in groundwater: a review of biogeochemical controlling processes[J]. Water Res., 42 (16): 4215-4232.

ROSMAN J R, TAYLORP D, 1998. Isotopic compositions of the elements (technical report): commission on atomic weights and isotopic abundances[J]. Pure. Appl. Chem., 70:217-235.

SACCOCIA P J, SEEWALD J S, SHANKS W C,2009. Oxygen and hydrogen isotope fractionation in serpentine-water andtalc-water systems from 250℃ to 450℃, 50MPa[J]. Geochim. Cosmochim. Acta,73:6789-6804.

SACHSE D, BILLAULT I, et al. ,2012. Molecular paleohydrology: interpreting the hydrogen-isotopic composition of lipid biomarkers from photosynthesizing organisms[J]. Ann. Rev. Earth Planet Sci. ,40:221-249.

SAVIN S M, LEE M,1988. Isotopic studies of phyllosilicates[J]. Rev. Mineral,19: 189-223.

SCHIAVO M A, HAUSER S, POVINEC P P, 2009. Stable isotopes of water as a tool to study groundwater-seawater interactions in coastal south-eastern Sicily[J]. J. Hydrol. , 364 (1-2):40-49.

SCHIMMELMANN A, SESSIONS A L, MASTALERZ M,2006. Hydrogen isotopic (D/H) composition of organic matter during diagenesis and thermal maturation[J]. Ann. Rev. Earth Planet Sci. , 34:501-533.

SESSIONS A L, BURGOYNE T W, SCHIMMELMANN A, et al. ,1999. Fractionation of hydrogen isotopes in lipid biosynthesis[J]. Org. Geochem. ,30:1193-1200.

SUZUOKI T, EPSTEIN S ,1976. Hydrogen isotope fractionation between OH-bearing minerals and water[J]. Geochim. Cosmochim. Acta,40:1229-1240.

TAUBE H,1954. Use of oxygen isotope effects in the study of hydration ions[J]. J. Phys. Chem. ,58:523.

TRUESDELL A H,1974. Oxygen isotope activities and concentrations in aqueous salt solution at elevated temperatures: Consequences for isotope geochemistry[J]. Earth Planet Sci. Lett. ,23:387-396.

UREY H C, BRICKWEDDE F G, MURPHY G M,1932. A hydrogen isotope of mass 2 and its concentration[J]. Phys. Rev. ,40:1.

USDOWSKI E, HOEFS J, 1993. Oxygen isotope exchange between carbonic acid, bicarbonate, carbonate, and water: a re-examination of the dataof McCrea (1950) and an expression for the overall partitioning of oxygen isotopes between the carbonate species and water[J]. Geochim. Cosmochim. Acta,57:3815-3818.

USDOWSKI E, MICHAELIS J, BöTTCHER M B,et al. ,1991. Factors for the oxygen isotope equilibrium fractionation between aqueous CO_2, carbonic acid, bicarbonate, carbonate, and water[J]. Z. Phys. Chem. ,170:237-249.

VENNEMANN T, O'NEIL J R,1996. Hydrogen isotope exchange reactions between hydrous minerals and hydrogen: I. A new approach for the determination of hydrogen isotope fractionation at moderate temperatures[J]. Geochim. Cosmochim. Acta, 60: 2437-2451.

WANG Y, SESSIONS A L, NIELSEN J R, et al. ,2009a. Equilibrium $^2H/^1H$ fractionations in organic molecules. I. Calibration of ab initio calculations[J]. Geochim. Cosmochim. Acta,73:7060-7086,82-95.

WANG Y, SHVARTSEV S L, SU C, 2009b. Genesis of arsenic/fluoride-enriched soda water: a case study at Datong, northern China[J]. Appl. Geochem., 24 (4):641-649.

WANG Y X, SHPEYZER G, 2000. Hydrogeochemistry of Mineral water from Rrift Systems on the East Asia Continent: Case Studies in Shanxi and Baikal[M]. Beijing: China Environmental Science Press (in Chinese with English abstract).

WANG Y, SESSIONS A L, NIELSEN R J, et al., 2013. Equilibrium $^{2}H/^{1}H$ fractionation in organic molecules. Ⅲ Cyclic ketones and hydrocarbons[J]. Geochim. Cosmochim. Acta,107:82-95.

WELHAN J A, 1987. Stable isotope hydrology. In: Short course in stable isotope geochemistry of low-temperaturefluids[J]. Mineral Assoc. Canada,13:129-161.

WHO, 1993. Guideline for Drinking Water Quality[M]. Geneva:Recommendations.

XIE X, et al., 2011. The sources of geogenic arsenic in aquifers at Datong Basin, northern China: constraints from isotopic and geochemical data[J]. J. Geochem. Explor., 110(2):155-166.

XIE X, WANG Y, SU C, 2012. Influence of irrigation practices on arsenic mobilization: Evidence from isotope composition and Cl/Br ratios in groundwater from Datong Basin, northern China[J]. Journal of Hydrology, (424):37-47.

ZEEBE R E,2007. An expression for the overall oxygen isotope fractionation between the sum of dissolved inorganic carbon and water[J]. Geochem. Geophys. Geosys.: 8.

Chapter 5 Carbon Isotope

Carbon occurs in a wide variety of compounds on Earth, from reduced organic compounds in the biosphere to oxidized inorganic compounds like CO_2 and carbonates. The broad spectrum of carbon-bearing compounds involved in low-and high-temperature geological settings can be assessed on the basis of carbon isotope fractionations.

Carbon has two stable isotopes (Rosman and Taylor, 1998):

$^{12}C = 98.93\%$ (reference mass for atomic weight scale)

$^{13}C = 1.07\%$

The naturally occurring variations in carbon isotope composition are greater than 120‰, neglecting extraterrestrial materials. Heavy carbonates with $\delta^{13}C$-values $> +20$‰ and light methane of < -100‰ have been reported in the literature.

5.1 Carbon species in groundwater

The carbon species in groundwater was mainly controlled by the water-sediment/rock interaction (e.g. carbonate minerals) and the groundwater environment (e.g. pH). When CO_2 dissolves in water, gaseous CO_2 (g) becomes aqueous CO_2 (aq), and some of this associates with water molecules to form carbonic acid, H_2CO_3:

$$CO_2(g) \rightarrow CO_2(aq) \quad (5.1)$$

$$CO_2(g) + H_2O \rightarrow CO_2(aq) + H_2CO_3 \quad K = \frac{[CO_2(aq)][H_2CO_3]}{[P_{CO_2}]} = 10^{-1.5} \quad (5.2)$$

During dissociation, carbonic acid stepwise releases two protons. The concentrations of the dissolved carbonate species therefore depend on the pH of the solution. The constants tabulated are approximate values, adequate for manual calculations.

$$H_2CO_3 \leftrightarrow H^+ + HCO_3^- \quad K_1 = \frac{[H^+][HCO_3^-]}{[H_2CO_3]} = 10^{-6.3} \quad (5.3)$$

$$HCO_3^- \leftrightarrow H^+ + CO_3^{2-} \quad K_1 = \frac{[H^+][CO_3^{2-}]}{[HCO_3^-]} = 10^{-10.3} \quad (5.4)$$

At a pH of 6.3, the activities of HCO_3^- and H_2CO_3 are equal. With pH$>$6.3, HCO_3^- is the predominant species and at pH$<$6.3 there is more H_2CO_3. The same relation for the

CO_3^{2-} / HCO_3^- couple shows that the two species have equal activity at pH=10.3. At a pH > 10.3, CO_3^{2-} becomes the predominant species while HCO_3^- is more abundant at pH<10.3. Figure 5.1 summarizes how the different aqueous carbonate species vary with pH.

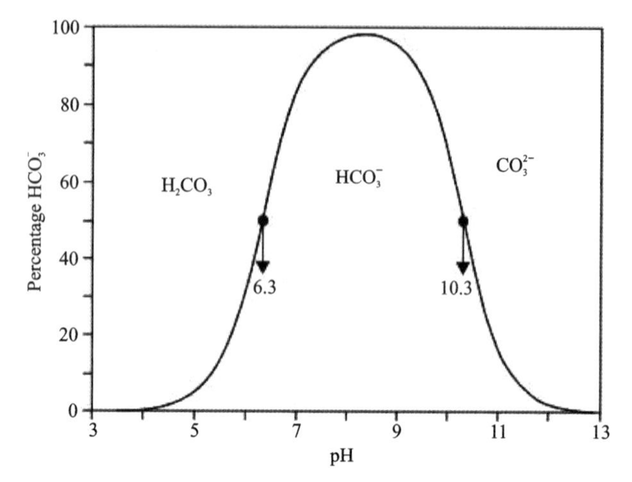

Figure 5.1 Percentage of HCO_3^- of total dissolved carbonate as function of pH.

The distribution of dissolved carbonate species can be calculated for two idealized cases. In the first case the CO_2 pressure is known and constant, a system that is termed open with respect to carbon dioxide gas. In the second case, total inorganic carbon (TIC) is known and constant and the system is called closed with respect to CO_2.

When the gas pressure of CO_2, or the activity $[P_{CO_2}]$, is known, the activities of all dissolved carbonate species can be calculated according to the equations (5.2), (5.3) and (5.4).

The sum of the dissolved carbonate species is given by the mass balance equation:

$$TIC = m_{H_2CO_3} + m_{HCO_3^-} + m_{CO_3^{2-}} \tag{5.5}$$

In a system closed with respect to CO_2, the sum of the dissolved carbonate species is considered constant. The distribution of the dissolved carbonate species, as a function of pH, can be calculated by substituting the mass action expressions (5.3) and (5.4) in (5.5). Carbonic acid is dominant at pH< 6.3 while CO_3^{2-} becomes dominant at pH>10.3, and in between these pH values HCO_3^- is the major dissolved carbonate species in water. Therefore, the four Equations (5.2), (5.3), (5.4) and (5.5) together define the carbonate species in groundwater. The number of variables in the equations amounts to six. Therefore, the determination of any two variables fixes the remaining variables in the aqueous carbonate system. In usual practice one of the following pairs of variables are analyzed: pH and alkalinity, TIC and alkalinity, P_{CO_2} and alkalinity, or P_{CO_2} and pH.

In most cases the carbonate speciation is determined by measuring pH and alkalinity. This is best done in the field immediately after sample retrieval, to avoid degassing of CO_2.

The alkalinity of a water sample is equal to the number of equivalents of all dissociated weak acids. Phosphoric acid and other weak acids may contribute to some extent, and in leachates from waste site various organic acids may be important. However, normally only the carbonate ions are of quantitative importance for the measured alkalinity. It is then equal to:

$$Alk = m_{HCO_3^-} + 2m_{CO_3^{2-}} \tag{5.6}$$

The CO_2 concentration in the atmosphere is low, 0.03vol%, which is equivalent to $P_{CO_2} = 3\times10^{-4}$ of $1atm=10^{-3.5}atm$. The CO_2 pressure of groundwater is easily one or two orders of magnitude higher, primarily as result of uptake of carbon dioxide during the infiltration of rainwater through the soil. In the soil, CO_2 is generated by root respiration and decay of labile organic material:

$$CH_2O+O_2 \rightarrow CO_2+H_2O \tag{5.7}$$

During winter the biological CO_2 production stops and the CO_2 gradient indicates that groundwater is degassing to the atmosphere. In summer the CO_2 production is high and the CO_2 distribution suggests both upward and downward diffusion. The CO_2 concentrations in the atmosphere also fluctuate seasonally, concentrations are lowest during the late summer (September on the northern hemisphere, and March on the southern hemisphere) mainly as result of the photosynthesis of land plants.

Soil respiration is not the only source of CO_2 entering groundwater. Also the oxidation of organic carbon present in sediments or of DOC carried with the groundwater is an important source of TIC. Furthermore, CO_2 derived from deep sources, which include degassing of magma or thermal metamorphosis of oceanic carbonate rock, may migrate upward along fracture zones and contribute to the groundwater CO_2 input.

A higher P_{CO_2} and the associated increase of dissolved carbonic acid allows more calcite to dissolve. The dissociation reaction for calcite with the reactions for aqueous carbonate is:

$$CO_{2(g)}+H_2O+CaCO_3 \leftrightarrow Ca^{2+}+2HCO_3^- \quad K=[Ca^{2+}][HCO_3^-]^2/[P_{CO_2}]=10^{-6.0} \tag{5.8}$$

For a system containing only calcite, CO_2 and H_2O, the electroneutrality equation in the solution is:

$$2m_{Ca^{2+}} + m_{H^+} = m_{HCO_3^-} + m_{OH^-} + 2m_{CO_3^{2-}} \tag{5.9}$$

Assume the pH of the soil water to be less than 8.3, then $m_{CO_3^{2-}}$ and m_{OH^-} can be neglected in Equation (5.9). Furthermore, assume that m_{H^+} is negligible in comparison to $m_{Ca^{2+}}$ and the electroneutrality is reduced to:

$$2m_{Ca^{2+}} = m_{HCO_3^-} \tag{5.10}$$

Combining with Equation (5.8), and assuming that activity equals molality yields:

$$m_{Ca^{2+}} = \sqrt[3]{10^{-6.0}[P_{CO_2}]/4} \tag{5.11}$$

A simple relationship between the Ca^{2+} concentration and the CO_2 pressure for equilibrium with calcite in pure H_2O was obtained. Once $m_{Ca^{2+}}$ is known, $[CO_3^{2-}]$ can be calculated from the solubility of calcite; the amount of HCO_3^- can be calculated with Equation (5.9). The Ca^{2+} concentration increases with the cube root of the CO_2 pressure. Equation (5.11) predicts that the Ca^{2+} concentration increases with the cube root of the CO_2 pressure as shown in Figure 5.2. The relation has an important consequence. Mixing of two water which are both at equilibrium with calcite but which have a different CO_2 pressure, produces a water composition on an almost straight line that connects A and B. The straight line lies below the equilibrium curve, indicating that the Ca^{2+} concentration is smaller than is required for equilibrium. In other words, a mixture of two water which are both saturated for calcite, becomes subsaturated for calcite. This effect may cause calcite dissolution in carbonate aquifers.

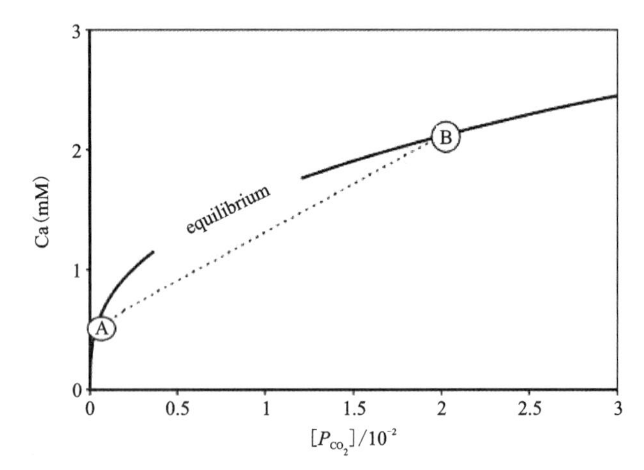

Figure 5.2 The solubility of calcite in H_2O as a function of P_{CO_2}. The curved line indicates equilibrium with calcite according to Equation (5.11). Mixtures of water A and B are positioned on the (nearly) straight line connecting A and B.

Aluminosilicate weathering releases cations into solution, according to incongruent dissolution reactions, where the primary minerals, such as feldspars, are altered to more stable clay minerals:

$$NaAlSi_3O_{8albite} + 4.5H_2O + H_2CO_3 \rightarrow Na^+ + HCO_3^- + 1/2Al_2Si_2O_5(OH)_{4kaolinite} + 2H_4SiO_4 \quad (5.12)$$

$$CaAlSi_3O_{8aporthite} + H_2O + 2H_2CO_3 \rightarrow Ca^+ + 2HCO_3^- + Al_2Si_2O_5(OH)_{4kaolinite} \quad (5.13)$$

Acid is essential to silicate hydrolysis reactions, as the product clays are hydrated with OH^-. Silica and inorganic carbon species are released into solution by the weathering of the alkali feldspars (albite and orthoclase) although not by the weathering of anorthite, in which an extra Al^{3+} replaces one Si^{4+} to maintain the charge balance with divalent Ca^{2+}.

5.2 Analytical techniques for the $^{13}C_{DIC}$ and $^{13}C_{DOC}$

As the commonly used international reference standard PDB (Pee Dee Belemnite) has been exhausted for several decades, there is a need for introducing new standards. Several other standards are in use today, nevertheless published $\delta^{13}C$-values are given relative to the V-PDB-standard (Table 5.1).

The gases used in $^{13}C/^{12}C$ measurements are CO_2 or CO obtained during pyrolysis. For CO_2 the following preparation methods exist:

(a) Carbonates are reacted with 100% phosphoric acid at temperatures between 20℃ and 90℃ (depending on the type of carbonate) to liberate CO_2 (see also "oxygen").

(b) Organic compounds are generally oxidized at high temperatures (850～1000℃) in a stream of oxygen or by an oxidizing agent like CuO. For the analysis of individual compounds in complex organic mixtures, a gas chromatography-combustion-isotope ratio mass-spectrometry (GC-C-IRMS) system is used, first described by Matthews and Hayes (1978). This device can measure individual carbon compounds in mixtures of sub-nanogram samples with a precision of better than ±0.5‰.

Table 5.1 $\delta^{13}C$-values of NBS-reference samples relative to V-PDB.

NBS-18	Carbonatite	−5.00
NBS-19	Marble	+1.95
NBS-20	Limestone	−1.06
NBS-21	Graphite	−28.10
NBS-22	Oil	−30.03

5.3 Fractionation processes

The two main terrestrial carbon reservoirs, organic matter and sedimentary carbonates, have distinctly different isotopic characteristics because of the operation of two different reaction mechanisms:

(1) Isotope equilibrium exchange reactions within the inorganic carbon system "atmospheric CO_2-dissolved bicarbonate-solid carbonate" lead to an enrichment of ^{13}C in carbonates.

(2) Kinetic isotope effects during photosynthesis concentrate the light isotope ^{12}C in the synthesized organic material.

5.3.1 Carbonate system

The inorganic carbon species in the open or close system can be calculated by Equations (5.2), (5.3), (5.4) and (5.5). In the carbonate system, the species and concentrations of inorganic carbon were controlled by the water pH and saturation index of carbonate minerals, such as calcite and dolomite. Under the oversaturation state of carbonate minerals, the inorganic carbon species (CO_3^{2-} and HCO_3^-) can combine with divalent cations to form solid minerals, calcite and aragonite being the most common:

$$Ca^{2+} + CO_3^{2-} \leftrightarrow CaCO_3 \tag{5.14}$$

An isotope fractionation is associated with each of these equilibria, the ^{13}C-differences between the species depend only on temperature, although the relative abundances of the species are strongly dependent on pH. Several authors have reported isotope fractionation factors for the system dissolved inorganic carbon (DIC)-gaseous CO_2 (Vogel et al., 1970; Mook et al., 1974; Zhang et al., 1995). The major problem in the experimental determination of the fractionation factor is the separation of the dissolved carbon phases (CO_{2aq}, HCO_3^-, CO_3^{2-}) because isotope equilibrium among these phases is reached within seconds. The generally accepted carbon isotope equilibrium values between calcium carbonate and dissolved bicarbonate are derived from inorganic precipitate data (Rubinson and Clayton, 1969; Emrich et al., 1970; Turner, 1982). What is often not adequately recognized is the fact that systematic C-isotope differences exist between calcite and aragonite. Rubinson and Clayton (1969) and Romanek et al. (1992) found calcite and aragonite to be 0.9‰ and 2.7‰ enriched in ^{13}C relative to bicarbonate at 25℃. Another complicating factor is that shell carbonate is frequently not in isotopic equilibrium with the ambient dissolved bicarbonate.

Carbon isotope fractionations under equilibrium conditions are important not only at low-temperature, but also at high temperatures within the system carbonate, CO_2, graphite, and CH_4. Of these, the calcite-graphite fractionation has become a useful geothermometer (e.g., Valley and O'Neil, 1981; Scheele and Hoefs, 1992; Kitchen and Valley, 1995). Figure 5.3 summarizes carbon isotope fractionations between various geologic materials and gaseous CO_2 (after Chacko et al., 2001).

5.3.2 Organic carbon system

The main isotope-discriminating steps during biological carbon fixation are ①the uptake and intracellular diffusion of CO_2 and ②the biosynthesis of cellular components. Such a two-step model was first proposed by Park and Epstein (1960):

$$CO_{2(external)} \leftrightarrow CO_{2(internal)} \leftrightarrow \text{organic molecule}$$

From this simplified scheme, it follows that the diffusional process is reversible,

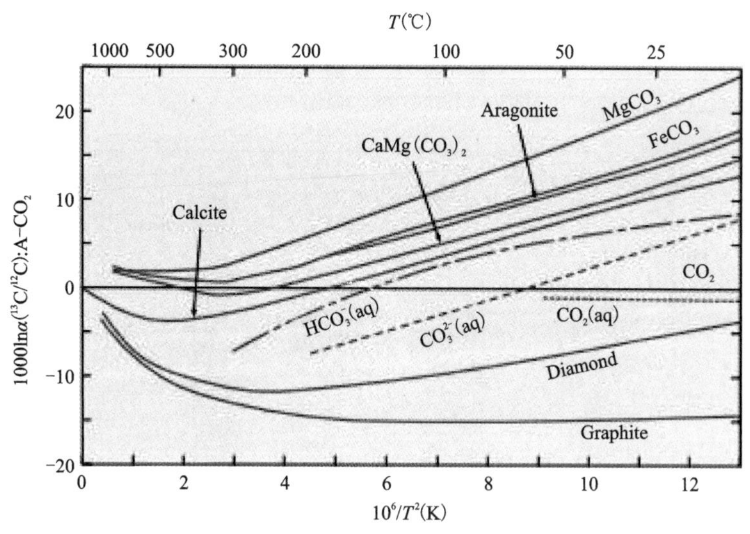

Figure 5.3 Carbon isotope fractionation between various geologic compounds and CO_2 (after Chacko et al., 2001).

whereas the enzymatic carbon fixation is irreversible. The two-step model of carbon fixation clearly suggests that isotope fractionation is dependent on the partial pressure of CO_2 of the system. With an unlimited amount of CO_2 available to a plant, the enzymatic fractionation will determine the isotopic difference between the inorganic carbon source and the final bioproduct. Under these conditions, ^{13}C fractionations may vary from $-17‰$ to $-40‰$ (O'Leary, 1981). When the concentration of CO_2 is the limiting factor, the diffusion of CO_2 into the plant is the slow step in the reaction and carbon isotope fractionation of the plant decreases.

Atmospheric CO_2 first moves through the stomata, dissolves into leaf water and enters the outer layer of photosynthetic cells, the mesophyll cell. Mesophyll CO_2 is directly converted by the enzyme ribulose biphosphate carboxylase/oxygenase ("Rubisco") to a 6 carbon molecule, that is then cleaved into 2 molecules of phosphoglycerate (PGA), each with 3 carbon atoms (plants using this photosynthetic pathway are therefore called C3 plants). Most PGA is recycled to make ribulose biphosphate, but some is used to make carbohydrates. Free exchange between external and mesophyll CO_2 makes the carbon fixation process less efficient, which causes the observed large ^{13}C-depletions of C3 plants.

C4 plants incorporate CO_2 by the carboxylation of phosphoenolpyruvate (PEP) via the enzyme PEP carboxylase to make the molecule oxaloacetate which has 4 carbon atoms (hence C4). The carboxylation product is transported from the outer layer of mesophyll cells to the inner layer of bundle sheath cells, which are able to concentrate CO_2, so that most of the CO_2 is fixed with relatively little carbon fractionation.

In conclusion, the main controls on carbon isotope fractionation in plants are the action of a particular enzyme and the "leakiness" of cells. Because mesophyll cells are permeable and bundle sheath cells are less permeable, C3 versus C4 plants have ^{13}C-depletions of about −18‰ versus −4‰ relative to atmospheric CO_2 (Figure 5.4).

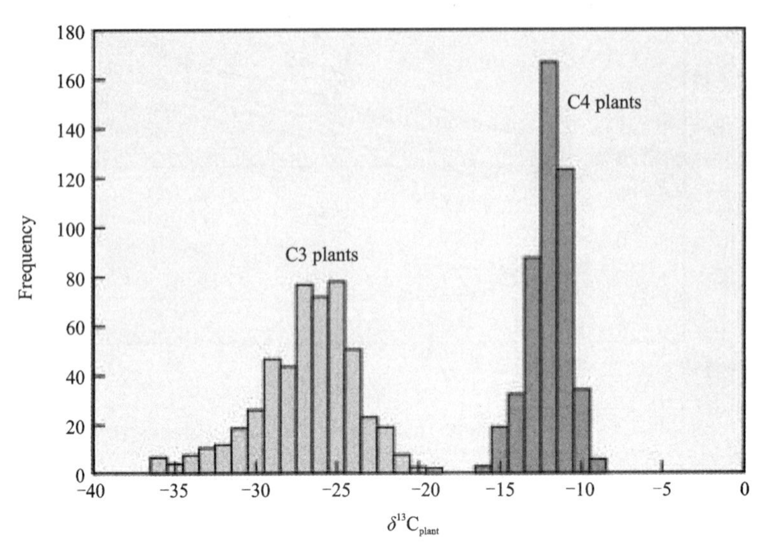

Figure 5.4 Histogram of $\delta^{13}C$ values of C3 and C4 plants (after Cerling and Harris, 1999).

The final carbon isotope composition of naturally synthesized organic matter depends on a complex set of parameters: ①The ^{13}C-content of the carbon source; ②Isotope effects associated with the assimilation of carbon; ③Isotope effects associated with metabolism and biosynthesis; ④Cellular carbon budgets.

Even more complex is C-isotope fractionation in aquatic plants. Factors that control the $\delta^{13}C$ of phytoplankton include temperature, availability of CO_2(aq), light intensity, nutrient availability, pH and physiological factors such as cell size and growth rate. In particular the relationship between the C-isotope composition of phytoplankton and the concentration of oceanic dissolved CO_2 has been subject of considerable debate because of its assumed potential as a palaeo-CO_2 barometer.

Since the pioneering work of Park and Epstein (1960) and Abelson and Hoering (1961), it is well known that ^{13}C is not uniformly distributed among the total organic matter of plant material, but varies between carbohydrates, proteins and lipids. The latter class of compounds is considerably depleted in ^{13}C relative to the other products of biosynthesis. Although the causes of these ^{13}C-differences are not entirely clear, kinetic isotope effects seem to be more plausible (DeNiro and Epstein, 1977; Monson and Hayes, 1982) than thermodynamic equilibrium effects (Galimov, 1985, 2006). The latter author argued that ^{13}C-concentrations at individual carbon positions within organic molecules are principally controlled by structural factors. Approximate calculations suggested that reduced C-H bonded positions are systematically depleted in ^{13}C, while oxidized C-O bonded positions are enriched in ^{13}C. Many of the observed relationships are qualitatively consistent with that concept. However, it is difficult to identify any general mechanism by which thermodynamic factors should be able to control chemical equilibrium within a complex organic structure.

High resolution gas source mass spectrometry opens the possibility to determine the isotope composition of a particular atom in a molecule (Eiler et al., 2014; Piasecki et al., 2016a, 2016b). Propane as the simplest organic molecule that could record site-specific carbon isotope variations shows isotope variations that can be related with the maturity of the precursor material (Piasecki et al., 2018a, 2018b).

5.3.3 Interactions Between Carbonate-Carbon and Organic Carbon

The two most important carbon reservoirs in the crust, marine carbonates and the biogenic organic matter, are characterized by very different isotopic compositions: the carbonates being isotopically heavy with a mean $\delta^{13}C$-value around 0‰ and organic matter being isotopically light with a mean $\delta^{13}C$-value around -25‰. For these two sedimentary carbon reservoirs a binary isotope mass balance must exist such that:

$$\delta^{13}C_{input} = f_{org}\delta^{13}C_{org} + (1 - f_{org})\,\delta^{13}C_{carb} \tag{5.15}$$

If δ_{input}, δ_{org}, δ_{carb} can be determined for a specific geologic time, f_{org} can be calculated, where f_{org} is the fraction of organic carbon entering the sediments. It should be noted that f_{org} is defined in terms of the global mass balance and is independent of biological productivity referring to the burial rather than the synthesis of organic material. That means that large f_{org} values might be a result of particular high productivity of organic material or of particular high levels of organic matter preservation.

The $\delta^{13}C$-value for the input carbon cannot be measured precisely but can be estimated with a high degree of certainty. As will be shown later, mantle carbon has an isotopic composition around -5‰ and estimates of the global average isotope composition for crustal carbon also fall in that range. Assigning -5‰ to $\delta^{13}C$-input, a modern value for f_{org} is calculated as 0.2 or expressed as the ratio of $C_{org}/C_{carb} = 20/80$. With each molecule of organic carbon being buried, a mole of oxygen is released to the atmosphere. Hence, knowledge of f_{org} is of great value in reconstructing the crustal redox budget.

5.3.4 Diagnosis and quantification of carbon flux pathways in soil

The measurement of $\delta^{13}C$ of individual soil carbon compounds may allow the diagnosis of particular biochemical pathways that operated until the time of sampling (Dumig et al., 2013; Lerch et al., 2011; Kodina et al., 2010). The usual approach is defining theoretically possible C flux paths that can be differentiated by their fractionation factors and testing them by $\delta^{13}C$ analysis of their substrates and products. The crucial point is that the pathways to be discriminated must display sufficiently different fractionation factors, reflected in the difference of $\delta^{13}C$ between substrate and product. Ecological studies sometimes have used tiny differences in the carbon and nitrogen isotopic values for interpretation of underlying food chains using context knowledge (reviewed in Perkins et al. ,2014;Ohkouchi et al. ,2015). In microbial ecology, on the other hand, the underlying biochemical pathways are usually much harder to resolve.

Two aspects of carbon flux in soil can be considered: ①Diagnosis and quantification of the relative contribution of different substrates to the formation of a product; ②Diagnosis and quantification of the relative contribution of different biochemical pathways to substrate conversion and product formation.

Soil organic matter usually originates from biomass that is produced by CO_2 fixation. Isotope fractionation is relatively strong, depending on the pathway of CO_2 fixation. For example, the pathway of CO_2 fixation by C4 plants has a smaller fractionation factor than that of C3 plants resulting in organic carbon that is less depleted in ^{13}C (O'Leary, 1981; Farquhar et al. , 1989) (Figures 5. 3). The isotopic signature of biomass is relatively well preserved in organic matter, since subsequent degradation by respiratory processes displays only small fractionation (Werth et al. , 2008, 2010; Deniro et al. , 1978; Mueller et al. , 2014). The same is true for microbial fermentation, which displays only a small fractionation factor (Deniro et al. , 1978; Blair et al. , 1985; Botsch et al. , 2011; Conrad et al. , 2014; Penning et al. , 2006; Zhang et al. , 2002). Therefore, different respiration or fermentation pathways cannot be distinguished using carbon isotope signals alone. However, it is possible to determine the contribution of different substrates to CO_2 production provided the $\delta^{13}C$ of the substrates is sufficiently different. For example, the contribution of the anaerobic degradation of root versus soil organic matter to CO_2 production can be quantified, if crops are changed from C3 to C4 plants or vice versa. Then, soil organic matter has still the typical ^{13}C signature while plant material has a C4 signature. Similarly, this approach can also be used for quantifying the contribution of root exudation versus degradation of soil organic matter (or straw carbon) to production of other compounds, such as microbial biomass, acetate or CH_4 (Werth et al. , 2008; Conrad et al. , 2012a,2012b; Tokida et al. , 2011; Yuan et al. , 2012, 2014). For example, it was shown that most of the CH_4 produced in flooded rice field soil is derived from root exudation (Tokida et al. , 2011; Yuan et al. , 2012). In case of aerobic degradation to CO_2 with negligible isotope fractionation the fraction of C3 (f_{C3,CO_2}) and C4 (f_{C4,CO_2}) organic carbon contributing to total CO_2 production is calculated by the following mass balance:

$$\delta^{13}C_{CO_2} = f_{C3,CO_2}\,\delta^{13}C_{C3} + (1 - f_{C3,CO_2})\,\delta^{13}C_{C4} \qquad (5.16)$$

For the anaerobic degradation to CH_4, we have to assume that the overall enrichment factor (ε_{org,CH_4}) for the conversion of organic carbon to CH_4 is the same for C3 and C4 organic carbon. This can be written as the following equation:

$$\delta^{13}C_{CH_4} = f_{C3,CO_2}\delta^{13}C_{C3} + (1 - f_{C3,CO_2})\delta^{13}C_{C4} + \varepsilon_{org,CH_4} \tag{5.17}$$

The enrichment factor ε_{org,CH_4} can be calculated from enrichment factors in the literature if the path of CH_4 production is known. The exact quantification of the relative contribution of different substrates to the degradation process requires that the degradation process has a small fractionation factor or is identical for each substrate. Therefore, application to anaerobic degradation processes is challenging, since large fractionation factors are involved in the conversion of CO_2 to acetate or CH_4.

While many respiration or fermentation pathways have only a small discrimination against ^{13}C; there are several anaerobic degradation processes, which display large fractionation factors up to $\varepsilon = -80\%$ reported for methylotrophic methanogens (Londry et al., 2008; Penger et al., 2012) that can be used for diagnosis and quantification of biochemical pathways (Figure 5.5). Examples are acetate formation by chemolithotrophic acetogenic bacteria using CO_2 as carbon substrate. This biochemical pathway exhibits a very large carbon fractionation ($\varepsilon = -68\%$ to -38%) factor allowing to determine whether acetate has been formed from CO_2 or from organic compounds (Haedrich et al., 2012; Heuer et al., 2009, 2010). Similarly, CH_4 production from CO_2 displays a much larger fractionation factor than from acetate, so that the path of CH_4 production in anoxic environments can be differentiated (Conrad et al., 2005, 2012; Mach et al., 2015; Ji et al., 2015). If soil organic matter is anaerobically degraded to CH_4 and CO_2, fermentation first produces CO_2 and acetate displaying little isotopic fractionation, so that these intermediary products have a similar $\delta^{13}C$ than the organic matter itself (Blair et al., 1985; Conrad et al., 2014; Zhang et al., 2002). Their further conversion, however, results in fractionation with CH_4 produced from CO_2 being much more depleted in ^{13}C than that from acetate. The residual CO_2 and acetate, on the other hand, become then correspondingly enriched in ^{13}C. These processes cannot only be used to diagnose the occurrence of different pathways, but can even be used to quantify their contributions, provided their fractionation

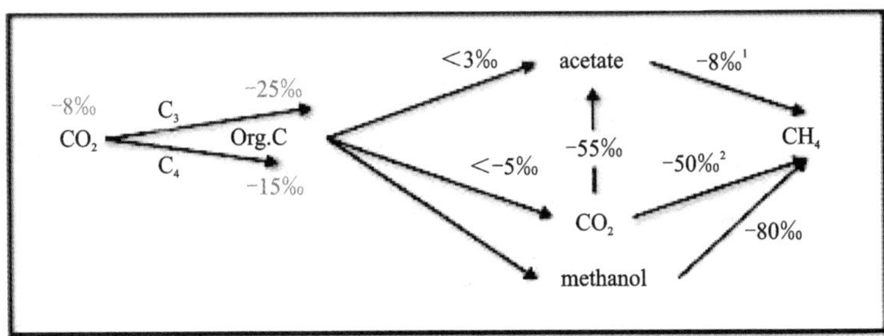

Figure 5.5 Scheme of carbon flow and stable carbon isotope enrichment factors (ε) in methanogenic environments.

factors are known and the $\delta^{13}C$ in the substrates (CO_2, acetate) and products (CH_4) can be measured (Conrad et al., 2005, 2012; Mach et al., 2015; Ji et al., 2015).

For example, for calculation of the fractions of hydrogenotrophic (f_{CO_2,CH_4}) and aceticlastic (f_{ac,CH_4}) methanogenesis of total CH_4 production it is necessary to determine the $\delta^{13}C$ in the CH_4, CO_2 and the methyl group of acetate. With these data the following mass balance equations can be used for calculation:

$$\delta^{13}C_{CH_4} = f_{CO_2,CH_4}\delta^{13}C_{C_{mc}} + (1 - f_{CO_2,CH_4})\delta^{13}C_{C_{ma}} \tag{5.18}$$

$$\delta^{13}C_{C_{mc}} = \delta^{13}C_{CO_2} + \varepsilon_{CO_2,CH_4} \tag{5.19}$$

$$\delta^{13}C_{C_{ma}} = \delta^{13}C_{ac\text{-}methyl} + \varepsilon_{ac,CH_4} \tag{5.20}$$

$$f_{CO_2,CH_4} + f_{ac,CH_4} = 1 \tag{5.21}$$

ε_{CO_2,CH_4} is the isotopic enrichment factor for the hydrogenotrophic formation of CH_4 from CO_2 (e. g. -50%), and ε_{ac,CH_4} that for the aceticlastic formation of CH_4 from the methyl group of acetate (e. g. -8%). If $\delta^{13}C_{ac\text{-}methyl}$ cannot be determined, it may be approximated from either the $\delta^{13}C$ of total acetate ($\delta^{13}C_{ac}$): $\delta^{13}C_{ac\text{-}methyl} \approx \delta^{13}C_{ac} - 8\%$ or of the $\delta^{13}C$ of organic C ($\delta^{13}C_{org}$) $\delta^{13}C_{ac} \approx \delta^{13}C_{org} - 2\%$.

Analogously, fractions of acetate produced from chemolithotrophic ($f_{CO_2,ac}$) and fermentative ($f_{org,ac}$) acetogenesis can be calculated by the following mass balance, if $\delta^{13}C$ of acetate, CO_2 and organic C have been measured.

$$\delta^{13}C_{ac} = f_{CO_2,ac}\delta^{13}C_{ac} + (1 - f_{CO_2,ac})\delta^{13}C_{ao} \tag{5.22}$$

$$\delta^{13}C_{ac} = \delta^{13}C_{CO_2} + \varepsilon_{CO_2,ac} \tag{5.23}$$

$$\delta^{13}C_{ao} = \delta^{13}C_{org} + \varepsilon_{org,ac} \tag{5.24}$$

$$f_{CO_2,ac} + f_{org,ac} = 1 \tag{5.25}$$

where $\varepsilon_{CO_2,ac}$ is the isotopic enrichment factor for the chemolithotrophic formation of acetate from CO_2 via the acetyl-COA pathway (e. g. -55%), and ε_{ac,CH_4} that for the fermentative formation of acetate from organic substrate (e. g. -2%).

5.4 Variations of Carbon isotopes in groundwater

Tracing the carbon cycle globally and in groundwater is greatly aided by the use of carbon isotopes. Fractionation of ^{13}C during organic and inorganic transformations partitions this isotope among various carbon reservoirs. These isotopes then become important tracers of carbon reactions and fluxes between reservoirs, and for past reconstructions of reservoir sizes. The range for $\delta^{13}C$ in the major carbon reservoirs is shown in Figure 5.6.

Atmospheric CO_2 today has a $\delta^{13}C$ value of about $-8.3‰$ and is gradually decreasing, mainly due to inputs of ^{13}C-depleted CO_2 from fossil fuel combustion and by enhanced soil respiration. Photosynthesis is a highly fractionating process that discriminates against ^{13}C

during transformation of CO_2 to carbohydrate, CH_2O. Most plants follow Calvin or C3 photosynthetic pathway, which involves an inefficient step of CO_2 respiration from the plant. As a result, the $\delta^{13}C$ of C3 plants is much lower than that of atmospheric CO_2, with a fractionation of close to 20‰ ($\varepsilon^{13}C_{CH_2O;CO_2} \approx -20‰$). The C3 pathway is used by almost 95% of plants, including most terrestrial trees and shrubs as well as marine plants and algae. Carbon fixed by terrestrial C3 plants has $\delta^{13}C$ values of about −27‰ (O'Leary, 1988).

C3 crops: wheat, rice, barley, oats, rye, tobacco, sugar beets, dry beans, soybeans, sunflowers, ground nuts, bluegrass and most lawn grasses, and fescue.

C4 crops: corn (maize), sugar cane, sorghum (millet), prairie and dryland grasses.

Marine organic carbon is accompanied by a greater fractionation during photosynthesis, and so has $\delta^{13}C$ values closer to −30‰. The degradation of marine organic carbon during diagenesis and hydrocarbon formation produces enriched CO_2 as a by-product, and so oil and gas have additional depletion in ^{13}C.

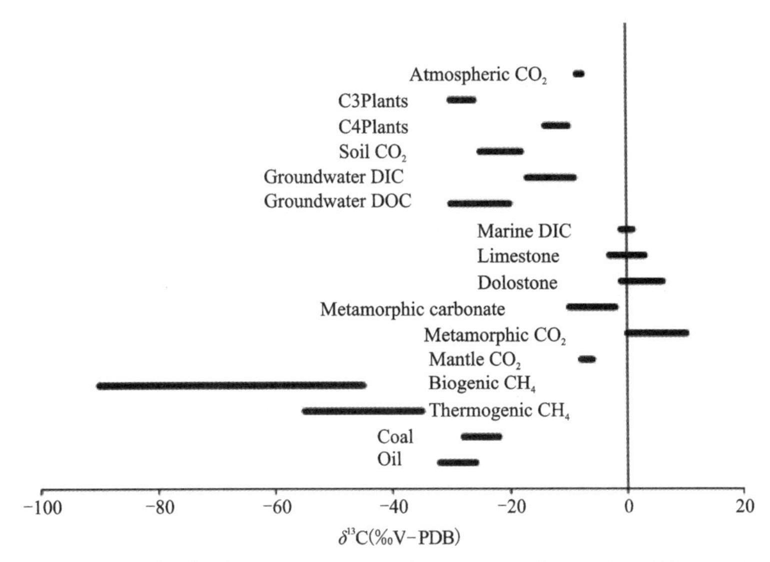

Figure 5.6 Ranges of $\delta^{13}C$ for major terrestrial reservoirs of organic and inorganic carbon.

5.4.1 $\delta^{13}C_{DIC}$ in the recharge environment

Carbon isotopes are useful for tracing the carbon cycle and evolution of DIC in water. Atmospheric CO_2 in recharge water is soon masked by the much greater reservoir of CO_2 found in soils. As the high CO_2 in soils is generated by degradation of biomass, its $\delta^{13}C$ is much more negative and closer to the ranges for C3 and C4 vegetation. Soil CO_2, however, has a slightly modified value, due to the outward diffusion from high P_{CO_2} in soils. In most C3 landscapes, the soil CO_2 has $\delta^{13}C$ in the range of −20‰ to −23‰.

During dissolution and hydration of soil CO_2, ^{13}C is preferentially partitioned into the

DIC species, with greater ^{13}C enrichment in HCO_3^- and CO_3^{2-}. The $\delta^{13}C$ of DIC is in fact a composite of the enrichment factors between CO_2 in the soil and the DIC species, weighted of course by their relative abundance. The ^{13}C content of the dissolved CO_2 [$\delta^{13}C_{CO_2(aq)}$] is about 1‰ lower than that of the soil CO_2 [$\delta^{13}C_{CO_2(g)}$], while $\delta^{13}C_{HCO_3^-}$ is about 9‰ greater than CO_2(g). As the relative abundance of these DIC species is controlled by pH, the value for $\delta^{13}C_{DIC}$ is then a function of pH, as long as the recharge water are in contact with soil CO_2 (open system conditions). The actual values for these enrichment factors, $\varepsilon^{13}C_{HCO_3;CO_2}$, for example, are given in Table 5.2. The precise value for $\delta^{13}C_{DIC}$ will depend on pH, which controls the ratio of $CO_{2(aq)}$ to HCO_3^-. Temperature also plays a role, as it affects the HCO_3^--$CO_{2(aq)}$ enrichment factor [$\varepsilon^{13}C_{HCO_3;CO_2(g)}$]. Values for the $\delta^{13}C$ of DIC during recharge at different temperatures and pH are given in Table 5.2, calculated according to the isotope mass balance equation:

$$\delta^{13}C_{DIC}=\left(\frac{m_{HCO_3^-}}{m_{DIC}}\right)\times(\delta^{13}C_{CO_2}+\varepsilon^{13}C_{H_2CO_3-CO_2})+\left(\frac{m_{HCO_3^-}}{m_{DIC}}\right)\times(\delta^{13}C_{CO_2}+\varepsilon^{13}C_{HCO_3^--CO_2}) \tag{5.26}$$

If the pH of the soil water is below about 6.4, the DIC is composed mainly of $CO_{2(aq)}$, and so $\delta^{13}C_{DIC}$ will be close to the $\delta^{13}C_{CO_2(g)}$. At greater pH values, it is mainly HCO_3^-, and so $\delta^{13}C_{DIC}$ will be closer to $\delta^{13}C_{CO_2(g)}+10‰$.

Below, this effect of pH on $\delta^{13}C_{DIC}$ during recharge becomes a useful tool to distinguish open versus closed system weathering in carbonate and silicate terrains. DIC and $\delta^{13}C_{DIC}$ in groundwater evolve to higher values as weathering proceeds. Controls on this evolution include whether open or closed system conditions prevail, and whether the parent material is silicate or carbonate. Each weathering scenario will support a different evolution for DIC and $\delta^{13}C_{DIC}$.

Table 5.2 $\delta^{13}C_{DIC}$ of recharge water in equilibrium with soil CO_2 at 10℃ for different pH values.

pH	$\frac{HCO_3^-}{H_2CO_3}$	$\delta^{13}C_{soil;CO_2}$	$\varepsilon_{H_2CO_3-CO_2}$	$\varepsilon_{HCO_3^--CO_2}$	$\delta^{13}C_{DIC}$
5.5	0.11	−20	−1.1	9.7	−20.0
6.5	1.1	−20	−1.1	9.7	−15.5
7.5	11.0	−20	−1.1	9.7	−11.2

$\varepsilon^{13}C_{H_2CO_3\text{-}CO_2(g)}=-0.000\,014\times T^2+0.004\,9\times T-1.18$

$\varepsilon^{13}C_{HCO_3^-\text{-}CO_2(g)}=0.000\,32\times T^2-0.124\times T+10.87$

$\varepsilon^{13}C_{CO_3^{2-}\text{-}CO_2(g)}=0.000\,33\times T^2-0.083\times T+8.25$

5.4.2 $\delta^{13}C_{DIC}$ and carbonate weathering

The $\delta^{13}C$ value of groundwater DIC evolves during open and closed system weathering of carbonates. This serves as a baseline for evaluating whether subsequent reactions have taken place in the subsurface. If weathering reactions take place under open system conditions during recharge in carbonate terrains, the $\delta^{13}C$ of the DIC is often controlled by exchange with the soil CO_2 and the pH. In this case, as the recharge water dissolves calcite with a $\delta^{13}C$ value of, say, 0‰, exchange between the DIC and the reservoir of soil CO_2 maintains $\delta^{13}C_{DIC}$ in equilibrium with $\delta^{13}C_{CO_2}$. Accordingly, the final $\delta^{13}C_{DIC}$ after weathering reactions can be determined from Table 5.2. As soil water moves relatively quickly toward equilibrium with calcite during open system recharge, groundwater recharged under such conditions have generally achieved calcite saturation at a circumneutral pH. The final $\delta^{13}C_{DIC}$ is then enriched by about $\varepsilon^{13}C_{DIC;CO_2}=10‰$, depending on temperature. For open system recharge through soils with $\delta^{13}C_{CO_2}$ of $-20‰$ to $-23‰$, values for $\delta^{13}C_{DIC}$ will be $-10‰$ to $-13‰$.

If carbonate weathering takes place under closed system conditions, exchange with soil CO_2 cannot occur, and so the initial $\delta^{13}C_{DIC}$ value will increase as ^{13}C-enriched DIC is gained from bedrock. According to the stoichiometry of the carbonate dissolution reaction, this results in a 50% dilution with the aquifer carbonate. Unfortunately, the result is close to that of open system dissolution with equilibrium exchange.

This will happen not only in limestone aquifers but also in silicate aquifers where calcite may be present as fracture minerals (secondary metamorphic or hydrothermal calcite), or as carbonate grains, cobbles, or cement in clastic aquifers. Most marine carbonates have $\delta^{13}C\approx 0\pm1‰$ V-PDB. Carbonate minerals in igneous and metamorphic rocks have $\delta^{13}C$ values of 0 to $-10‰$. While the open system weathering dissolves considerably more carbonate than closed system, the final $\delta^{13}C$ of the DIC is ambiguous.

5.5 Application

5.5.1 Multiple isotope (O, S and C) approach elucidates the enrichment of arsenic in the groundwater from the Datong Basin, northern China (Xie et al., 2013)

In the study, the groundwater from a known high-arsenic area in the Datong Basin was collected to examine and demonstrate the importance of the relevant geochemical/biogeochemical processes affecting arsenic mobilization in this aquifer system. The carbon

($\delta^{13}C$) and sulfur ($\delta^{34}S$) isotope data provide insight into the reaction pathways of microbially catalyzed Fe^{3+} and SO_4^{2-} reduction, organic matter mineralization and arsenic release.

The hydrologic and geologic background of Datong Basin was well described in the section of 4.4.1. The sampling location of this study was presented in Figure 5.7.

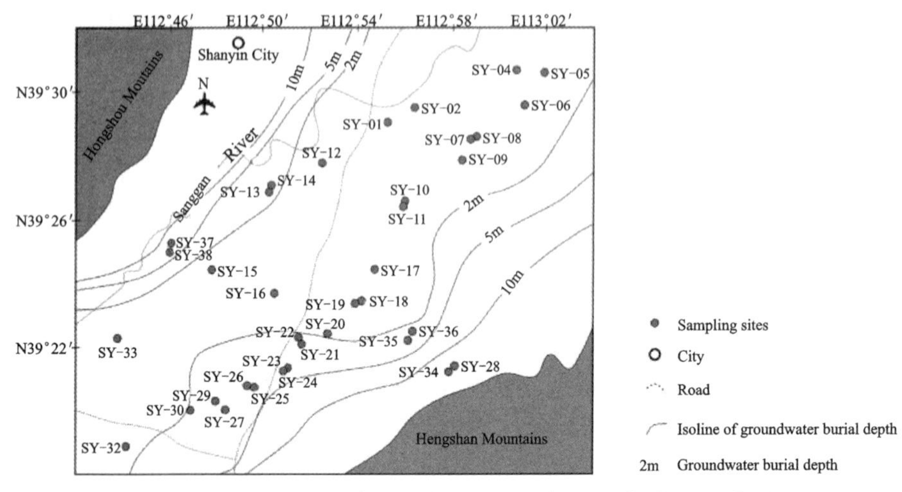

Figure 5.7 Locations of groundwater sampling at the Datong Basin.

5.5.1.1 Hydrochemistry of Datong Basin groundwater

The physico-chemical parameters of the groundwater samples are listed in Table 5.3. The high-arsenic groundwater (As $>10\mu g/L$) are typical $Na\text{-}HCO_3$ and Na-Cl types water. Low NO_3^- values, with an average of 12.1mg/L, were also detected in the groundwater. The values of ORP are generally low and range from −201.5mV to 173.9mV with an average of −24.5mV. High concentrations of NH_4^+ and HS^- were detected in the groundwater samples, with averages of 0.7mg/L and 14.7μg/L, respectively. Moreover, the groundwater has a high EC (average 2.1mS/cm) and neutral to alkaline pH (ranging between 7.5 and 8.6). The arsenic concentration of the groundwater samples ranges from 0.9μg/L to 1050μg/L with an average of 189μg/L, which far exceeds the WHO-recommended value of 10μg/L for drinking water. High Fe and Mn concentrations were detected in the water samples, with averages of 111.4μg/L and 76.7g/L, respectively. The groundwater ORP values generally indicate the anoxic nature of the aquifers (Table 5.3) and suggest that the aquifers are likely in Fe^{3+} or even SO_4^{2-} reducing conditions.

5.5.1.2 $\delta^{13}C$ values in groundwater DIC

The carbon isotopic compositions of DIC range from −6.9‰ to −22.0‰ in the present study (Table 5.4). The low $\delta^{13}C_{DIC}$ values (lower than −12‰) in the groundwater are therefore

Table 5.3 Physico-chemical and isotopic parameters of the groundwater from the Datong Basin.

Sample ID	X	Y	Depth (m)	T	pH	ORP (mV)	HCO_3^- (mg/L)	Cl^- (mg/L)	NO_3^- (mg/L)	SO_4^{2-} (mg/L)	K (mg/L)	Na (mg/L)	Ca (mg/L)	Mg (mg/L)	NH_4^+ (mg/L)	Fe(Ⅱ)/Fe_{total}	HS^- (μg/L)	As (μg/L)	Mn (μg/L)	Fe (μg/L)
SY-01	4 373 735.2	19 665 159.0	45	12.4	8.6	N.D	767	311	0.01	0.01	4.19	373	4.65	22.4	0.72	0.88	11	271.3	8.03	79.1
SY-02	4 374 766.1	19 666 896.8	50	11.4	8.56	−111.9	964	885	0.01	0.01	0.44	734	5.00	51.8	1.22	N.D	14	118.4	3.32	102
SY-04	4 376 975.2	19 672 925.6	85	12.9	8.43	22.3	901	155	0.01	0.01	1.60	410	5.35	17.5	1.89	0.21	3	168.4	4.35	64.4
SY-05	4 376 779.8	19 674 806.6	67	12.8	8.48	−21.5	761	66.7	0.01	0.01	0.01	287	5.27	17.2	N.D	N.D	2	204.9	7.84	47.5
SY-06	4 374 803.7	19 673 799.0	30	11.1	7.95	173.9	1106	302	46.6	395	3.85	649	18.3	60.4	0.14	N.D	5	6.20	16.3	61.1
SY-07	4 373 086.3	19 670 393.6	20	12.2	8.38	136.2	993	239	0.01	342	0.01	579	14.0	31.9	0.28	0.73	B	25.02	61.3	70.1
SY-08	4 373 027.3	19 670 375.7	68	12.1	8.39	−22.6	440	23.6	3.39	0.00	4.33	122	14.1	26.2	0.92	N.D	19	320.10	10.2	77.9
SY-09	4 371 599.6	19 669 720.6	19	11	8.52	−99.6	432	34.6	0.01	9.31	1.32	141	10.8	21.6	0.13	0.73	6	621.10	96.3	35.1
SY-10	4 369 032.1	19 666 295.2	58	11.6	8.56	−16.2	375	16.4	2.79	0.01	2.21	87.5	13.6	29.0	1.09	0.89	5	304.50	17.2	57.7
SY-11	4 369 002.8	19 666 370.0	30	11.3	8.27	−144.9	601	347	0.01	62.0	4.80	346	23.3	64.0	0.51	0.68	32	1052	221	49.3
SY-12	4 371 312.3	19 661 172.9	40	13.1	8.32	−25.1	429	224	0.01	49.0	0.01	274	13.1	33.4	0.56	0.88	8	216.6	42.03	59.3
SY-13	4 369 822.5	19 658 201.4	34	12.2	8.22	10.2	265	678	0.01	470	3.44	381	67.4	130	0.92	0.38	1	58.70	202	139
SY-14	4 369 806.4	19 658 168.2	50	12.6	8.23	−94.4	257	250	0.01	262	631	223	41.0	74.3	0.79	0.70	BL	64.60	90.5	447
SY-17	4 365 107.6	19 664 442.5	38	11.8	7.54	73.2	1019	855	226	846	12.0	780	128	309	0.94	0.56	BL	3.83	76.0	296
SY-18	4 363 340.2	19 663 747.2	22	10.9	7.76	22.6	731	2603	0.01	1861	14.5	1845	104	458	2.57	0.71	BL	43.13	385	533
SY-19	4 363 128.9	19 663 529.0	100	12.5	8.34	−201.5	320	12.4	0.01	0.01	1.73	76.4	20.9	20.6	1.56	0.73	26	157.6	28.28	65.1
SY-20	4 361 160.7	19 661 630.3	28	11.8	8.28	−106.7	375	17.7	0.01	0.01	1.94	75.3	21.3	30.7	0.74	0.85	38	369.0	115	50.0
SY-21	4 361 200.3	19 660 264.3	39	125	8.17	−136.6	375	44.2	2.30	0.01	8.9	96.5	179	31.8	0.43	0.65	21	537.8	96.8	58.7
SY-22	4 361 043.9	19 660 461.5	23	11.1	7.73	82	460	192	0.00	35.6	0.03	200	48.0	49.1	0.21	0.38	BL	3.94	196	64.8
SY-23	4 359 145.2	19 659 209.4	30	10.9	8.39	−144.7	294	13.2	3.42	0.01	4.80	64.2	15.9	21.3	1	0.97	64	358.6	41.2	54.3
SY-24	4 359 097.4	19 659 133.7	50	11.3	8.22	−118.3	304	11.8	0.01	6.71	2.48	65.3	20.4	19.1	0.63	0.71	60	544.8	82.5	26.6
SY-25	4 358 111.5	19 657 079.0	25	10.9	8.2	−120.2	312	18.4	1.06	2.63	3.64	69.5	15.2	24.3	0.24	0.71	BL	251.8	44.1	26.2

N.D: Not determined.

attributed to a biogenic source. Notably, the lowest $\delta^{13}C_{DIC}$ value is $-22.0‰$, which is consistent with the biogenic carbon signatures found in semi-arid climates. The high $\delta^{13}C_{DIC}$ values (higher than $-12‰$) in the groundwater are related to the mixing between biogenic carbon and limestone-originated DIC. Thus, it can be concluded that biogenic carbon is the most important contributor to the DIC in the groundwater at Datong.

Table 5.4 The $\delta^{13}C_{DIC}$, $\delta^{34}SO_4$ and $\delta^{18}OSO_4$ isotope composition of the groundwater from the Datong Basin.

Sample ID	$\delta^{13}C_{DIC}$ (PDB, ‰)	$\delta^{34}SO_4$ (V-CDT, ‰)	$\delta^{18}OSO_4$ (V-SMOW, ‰)
SY-01	−12.67	N.D	N.D
SY-02	−17.08	N.D	N.D
SY-04	−16.99	N.D	N.D
SY-05	−7.5	N.D	N.D
SY-06	−11.46	14.1	5.3
SY-07	−12.07	13.6	8.0
SY-08	−7.51	N.D	N.D
SY-09	−14.25	N.D	N.D
SY-10	−6.93	N.D	N.D
SY-11	−22.04	36.8	13.0
SY-12	−16.78	35.6	11.5
SY-13	−13.82	16.4	9.8
SY-14	−14.70	14.7	10.9
SY-17	−10.52	19.0	3.2
SY-18	−17.82	19.8	10.7
SY-19	−9.66	N.D	N.D
SY-20	−8.73	N.D	N.D
SY-21	−10.73	N.D	N.D
SY-22	N.D	20.6	7.8
SY-23	−13.15	N.D	N.D
SY-24	−16.95	N.D	N.D
SY-25	−16.91	N.D	N.D

As mentioned above, the microbial oxidation of organic matter is commonly coupled to the reduction of Fe^{3+} and SO_4^{2-} in aquifers or unsaturated zone. Similar to the sulfur isotopic composition, bacteria preferentially utilize ^{12}C, causing a depleted in the $\delta^{13}C$ values of DIC during the microbial mineralization of organic matters. As a result, the microbially mediated oxidation of organic matter (coupled to Fe^{3+} and SO_4^{2-} reduction) will deplete the ^{13}C (lowering in the $\delta^{13}C_{DIC}$) of the DIC produced, while the residual SO_4^{2-} will enrich the $\delta^{34}S$ value. In the $\delta^{13}C_{DIC}$ vs. $\delta^{34}S_{SO_4}$ plot (Figure 5.8), a negative correlation between $\delta^{13}C_{DIC}$ and $\delta^{34}S_{SO_4}$ is apparent, indicating that microbially mediated organic matter oxidation and SO_4^{2-} reduction has occurred or is occurring.

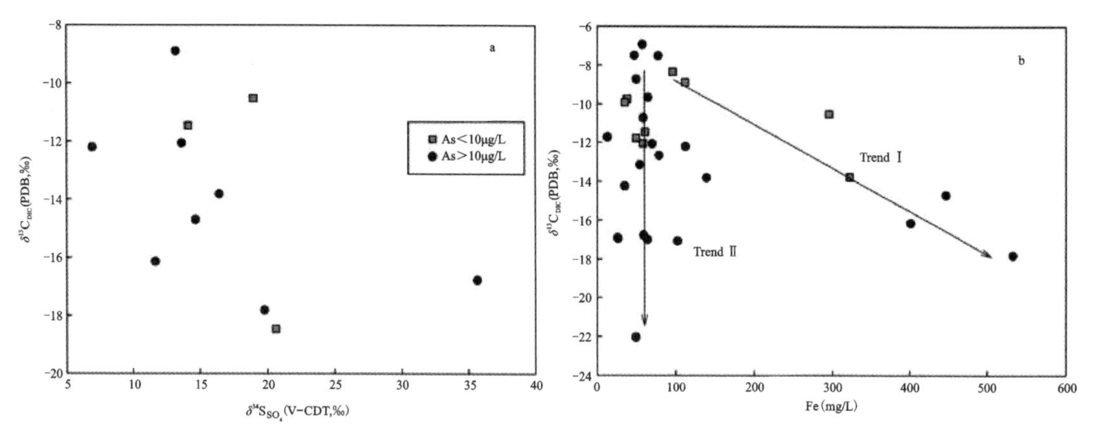

Figure 5.8 The relationships between the $\delta^{13}C_{DIC}$ values and Fe concentrations (a) and values (b) in the groundwater from the Datong Basin.

The hydrochemical data suggests that the microbially mediated Fe^{3+} and SO_4^{2-} reduction coupled to organic matter oxidation could be the key factor controlling the enrichment of arsenic in groundwater. Organic matter oxidation and SO_4^{2-} reduction can be discerned by the $\delta^{13}C_{DIC}$ and $\delta^{34}S_{SO_4}$ values of the groundwater. In the following, Fe^{3+} reduction will be further examined base on the $\delta^{13}C_{DIC}$ values. As discussed above, two origins of DIC can be discerned: biogenic DIC and DIC from carbonate dissolution. From the $\delta^{13}C_{DIC}$ vs. Fe plot (Figure 5.7b), two trends can be clearly observed. ①The $\delta^{13}C_{DIC}$ values decrease with an increase in the Fe concentration (Trend Ⅰ). Fe^{3+} and SO_4^{2-} are the primary electron acceptors for the microbial metabolism of organic matter (Appelo and Postma, 1993). Microbes preferentially utilize ^{12}C, decreasing the $\delta^{13}C$ value of DIC during the microbial oxidation of organic matters. Hence, a negative correlation can be expected between dissolved Fe concentrations and $\delta^{13}C_{DIC}$ values. Thus, Trend Ⅰ in Figure 5.8b suggests that the reduction of amorphous Fe^{3+} hydroxides has been coupled to the microbial oxidation of organic matter. ②The $\delta^{13}C_{DIC}$ values decrease significantly without a change in Fe concentration (Trend Ⅱ). Irrigation using karst water recharged surface and groundwater water is commonly practiced in the study area. Therefore, the significant variation of $\delta^{13}C_{DIC}$ values could be related to the mixing of DIC from carbonate dissolution (irrigation water) with the DIC produced via the microbial oxidation of surface-driven organic matter within the aquifers and unsaturated zone, which could cause the considerable variation observed in the $\delta^{13}C_{DIC}$ values. The upper surface is the zone of the greatest biogeochemical activity. Dissolved natural organic matter can enhance biogeochemical reactions, as observed in West Bengal aquifers (Nath et al., 2008). Inputs of organic matter from the surface can enhance microbial respiration. Large scale of corn is planted at the Datong. The $\delta^{13}C$ of organic matter originated from corn is about $-10‰$ according to Clark and Fritz (1997). Surface driven organic matters can be carried into the unsaturated zone and aquifers with the irrigation returns. The microbial mineralization of surface-driven organic matters can produce the depleted (lower than $-10‰$) and wide range of $\delta^{13}C_{DIC}$ values as observed in Figure 5.7. As discussed above, microbial oxidation of organic matter accompanying with the intensive Fe^{3+} minerals and sulfate reduction can produce iron sulfide precipitation and the low Fe concentrations. The samples fall on the trend Ⅱ with low Fe concentrations and wide range of $\delta^{13}C_{DIC}$ values can be attributed to microbial catalyzed organic matter oxidation and Fe^{3+} minerals and sulfate reduction. Notably, the samples fall on this trend have relative high arsenic concentrations indicating that microbial mediated oxidation of organic matter and reduction of Fe^{3+} and SO_4^{2-} due to surface organic matter addition may be the controlling processes for arsenic mobilization and enrichment in the Datong groundwater.

The $\delta^{13}C_{DIC}$ and Cl concentrations are plotted in Figure 5.9 to further demonstrate the source of organic matter for microbial respiration. The vertical recharge by irrigation returns can result in the large variation in Cl concentrations (due to the dissolution of halite) and the addition of young organic matter from the surface. The microbial oxidation of organic matter can produce the depleted and large variation of $\delta^{13}C_{DIC}$ values. Therefore, the vertical recharge from irrigation water with high concentrations of organic matter will result in a negative correlation between Cl concentrations and $\delta^{13}C_{DIC}$ values. From Figure 5.9, it is evident that almost all of the high arsenic water samples fall on the vertical leaching lines apart from the sample SY-11. This again confirms that vertical recharge by irrigation returns promotes the arsenic mobilization and enrichment in the Datong groundwater.

5.5.1.3 Predominant biogeochemical processes controlling arsenic mobilization

Because of the close geochemical/biogeochemical behavior of C, Fe, S and As, the carbon and sulfur isotope signatures can reveal the mechanism of arsenic mobilization within aquifers. The predominant microbial and geochemical processes controlling the mobilization of arsenic in the Datong Basin will be discussed in detail. The $\delta^{34}S_{SO_4}$ and $\delta^{18}O_{SO_4}$ values imply that dissolution of terrestrial evaporites is a significant SO_4^{2-} source to groundwater. In addition, the large contribution of sulfur from atmospheric sources and organic matter decomposition can be expected in this area. Although sulfide minerals could be a source of arsenic (Lowers et al., 2007; Verplanck et al., 2008), $\delta^{34}S_{SO_4}$ values does not reflect

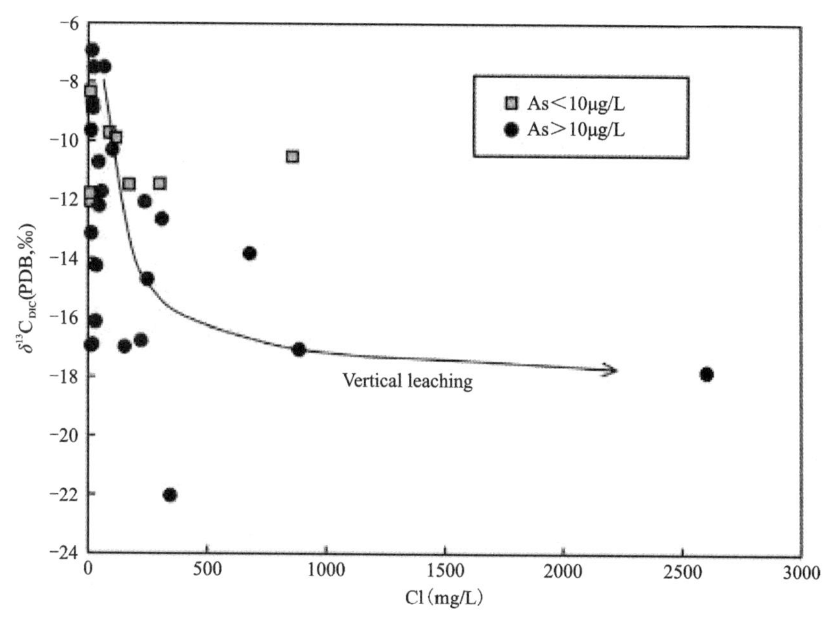

Figure 5. 9 The relationship between the values and the Cl concentration in the groundwater from the Datong Basin.

substantial oxidation of sulfide in this aquifer system. Therefore, the immediate mobilization of arsenic to groundwater seems to be caused by desorption from Fe^{3+} oxide/ hydroxides minerals and/or reductive dissolution of arsenic bearing Fe^{3+} oxide/ hydroxides minerals under strongly reducing conditions. From Figure 5. 8 and Figure 5. 9, it is apparent that most of the high-arsenic groundwater plotted on the vertical recharge and mixing line without a significant increase in the dissolved Fe concentration. As discussed above, this could be related to the microbial mediated reduction of Fe^{3+} and SO_4^{2-} coupled to the oxidation of organic matter within aquifers or unsaturated zone. According to the sequence of redox reactions (Borch et al., 2010), reduction of SO_4^{2-} to HS^- is followed by the reduction of crystalline Fe^{3+} oxide/hydroxides such as a-FeOOH(s) and the reduction of amorphous Fe^{3+} hydroxides like $Fe(OH)_3$(am)(s) occurs before the SO_4^{2-} reduction. Most samples fall on trend Ⅱ in Figure 5. 8 and fall on or closed to the vertical recharge and mixing line in Figure 5. 9 indicates that reduction of crystalline Fe^{3+} oxide/hydroxides coupled SO_4^{2-} reduction can be used to explain high-arsenic concentration in most groundwater in the study area. However, some high-arsenic groundwater samples plotted on the trend Ⅰ (Figure 5. 8) indicating amorphous Fe^{3+} oxide/hydroxides reduction without significant SO_4^{2-} reduction resulted in the enrichment of arsenic in these groundwater. Notably, most high-arsenic groundwater plotted on or closed to vertical leaching line(Figure 5. 9). Inputs of young organic matter from the surface that are associated with irrigation recharge can enhance microbially mediated Fe^{3+} and SO_4^{2-} reduction. Therefore, irrigation caused Fe^{3+} reduction can account for the high-arsenic concentrations in water samples plotted on vertical recharge line.

In general, the mobilization and distribution of arsenic in groundwater from the Datong Basin can be explained by the following models. ①The reduction of arsenic-bearing crystalline Fe^{3+} oxide/hydroxides and SO_4^{2-} coupled to the oxidation of organic matter in aquifer sediments or unsaturated zone. In spite of produced sulfide precipitation during this process, no significant amount of arsenic removal from groundwater is observed. This model accounts for the enrichment of arsenic in most of the groundwater samples from the Datong Basin. ② The reduction of amorphous Fe^{3+} oxyhydroxide without the significant sulfate reduction This can explain the high levels of arsenic found in some of the groundwater samples.

5. 5. 2 Understanding the inorganic carbon transport and carbon dioxide evasion in groundwater with multiple sulfate sources during different seasons using isotope records (Wen et al. ,2020)

Shangba village, a typical riparian zone that is seriously affected by AMD in south China, was chosen to understand the seasonal importance of different transport pathways of inorganic carbon and their responses to different sulfate sources in shallow groundwater. The objectives of this work are to determine the shifts in $\delta^{34}S_{SO_4}$, $\delta^{18}O_{SO_4}$ and $\delta^{13}C_{DIC}$ for exploring the DIC transport and transformation pathways and their responses to different sulfate sources over spatial and temporal scales and to create a conceptual model reflecting the behavior and fate of carbon.

5. 5. 2. 1 Study area

The study area covers an area of 4. 5km^2 and is located on the terraces of the Hengshi River, Wengyuan County, South China (Figure 5. 10). The study area has a subtropical humid monsoonal climate with an annual mean temperature of 19. 3～20. 6℃. The annual precipitation ranges from 1350～1750mm, most of which falls in the rainy season from April to October, and the difference of the water table depth between the wet season and dry season is approximately 1m. The paddy fields and sugar cane in the growing season from March to October are the main crops in the study area. Geologically, the study area is covered by Quaternary alluvium with a thickness of 1. 5～2. 0m overlying the bedrock that consists of granitic rocks. Topographically, the study area can be considered an independent hydrogeological unit, which is surrounded by hills on the western side and the Hengshi River on the northern and eastern sides. Hydrologically, a hydropower station has been constructed across the Hengshi River in the stretch between DB03 and DB04, which raises the river level in the upper reaches of the Hengshi River, and maintains the river level at a specific height throughout the year, so that the hydropower station results in a river level difference of approximately 4m between the upper and lower reaches of the hydropower station throughout the year. Thus, an active hydraulic connection between the Hengshi River and groundwater due to construction of the hydropower station has been confirmed through multiple-methods, including $\delta^{18}O_{H_2O}$, δD_{H_2O}, $\delta^{34}S$, and hydraulic potential, which makes the Hengshi River a losing river near DB01 and a gaining river in the lower reaches of DB04 throughout the year (Wen et al., 2018). Moreover, the potentiometric map shown in Figure 5.9 clearly depicts groundwater flow in the southeasterly direction in the aquifer. Based on Darcy's law and permeability coefficients, the groundwater residence times range from 0.5 to 21.7 years (Wen et al., 2019). In the water environment, the Hengshi River flows through the Dabaoshan mineral deposit that belongs to Jurassic granite-related Cu-Pb-Zn deposits and is the largest polymetallic sulfide meso-hypothermal deposit in South China. The minerals in the ore are mainly represented by pyrite, pyrrhotite, and chalcopyrite.

5.5.2.2 Hydrochemical evolution in groundwater

The pH and EC values in groundwater had wide ranges from 4.8~7.7μS/cm and 54.8~711μS/cm, respectively, and were similar to those in the Hengshi River, indicating the hydrochemical complexity of the groundwater environment under the recharge of river water. The major cation and anion compositions in groundwater were dominated by Ca^{2+} (0.870~102mg/L), HCO_3^- (N.D~202mg/L) and SO_4^{2-} (0.380~381mg/L). The concentrations of K^+ (0.190~12.7mg/L), Na^+ (0.720~25.3mg/L), Mg^{2+} (0.170~50.5mg/L), Cl^- (1.32~38.7mg/L) and NO_3^- (1.69~81.9mg/L) were relatively low. The order of ion abundances was $Ca^{2+} > Mg^{2+} > Na^+ > K^+$ for the cations and was $SO_4^{2-} > HCO_3^- > NO_3^- > Cl^-$ for the anions. The Hengshi River had higher SO_4^{2-} concentrations (mean of 566mg/L), lower pH values (mean of 5.8), and lower HCO_3^- concentrations (mean of 13.6mg/L).

The DIC contents increased from 7.36mgC/L to 24.2mgC/L and from 2.75 to 36.5mgC/L in the dry and wet seasons, respectively. Moreover, the CO_2 partial pressure (P_{CO_2}) of groundwater calculated by Henry's law in the northern zone was apparently higher than that in other zones, and a P_{CO_2} of 2611Pa in the dry season was higher than the 1540Pa observed in the wet season, indicating that potential changes from carbon emission to carbon sink might occur along the groundwater flow path.

The stoichiometric analysis has been useful in providing an insight into the geochemical processes in groundwater. As shown in Figure 5.11a, the relationship between $Ca^{2+} + Mg^{2+}$ and HCO_3^- of groundwater showed large variations in the study area. The ratios of $Ca^{2+} + Mg^{2+}$ and HCO_3^- gradually shifted from very high values to 2 : 1 to 1 : 1 along the flow

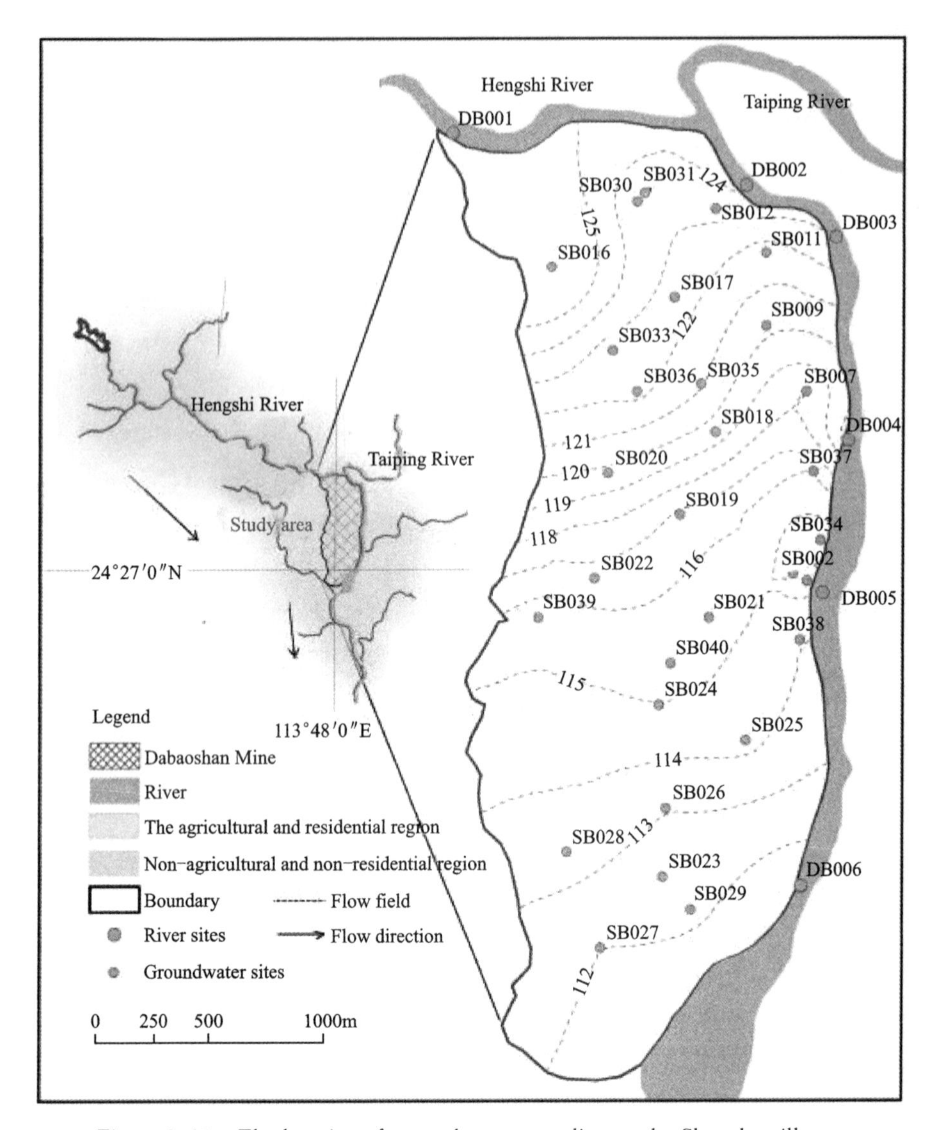

Figure 5.10 The location of groundwater sampling at the Shangba village.

direction. The relationship between $Ca^{2+}+Mg^{2+}$ and SO_4^{2-} compensated the ionic imbalance between $Ca^{2+}+Mg^{2+}$ and HCO_3^- and caused the ratios of $Ca^{2+}+Mg^{2+}$ and $HCO_3^-+SO_4^{2-}$ to be close to 1 : 1 (Figure 5.11b), suggesting that inorganic carbon transitions responded to a change in the major weathering agent from H_2SO_4 to CO_2 (soil respiration) as shown in Eqs. (5.27) and (5.28):

$$CaAl_2Si_2O_8+H_2SO_4+H_2O \rightarrow Al_2Si_2O_5(OH)_4+Ca^{2+}+SO_4^{2-} \quad (5.27)$$

$$CaAl_2Si_2O_8+2CO_2+3H_2O \rightarrow Al_2Si_2O_5(OH)_4+Ca^{2+}+2HCO_3^- \quad (5.28)$$

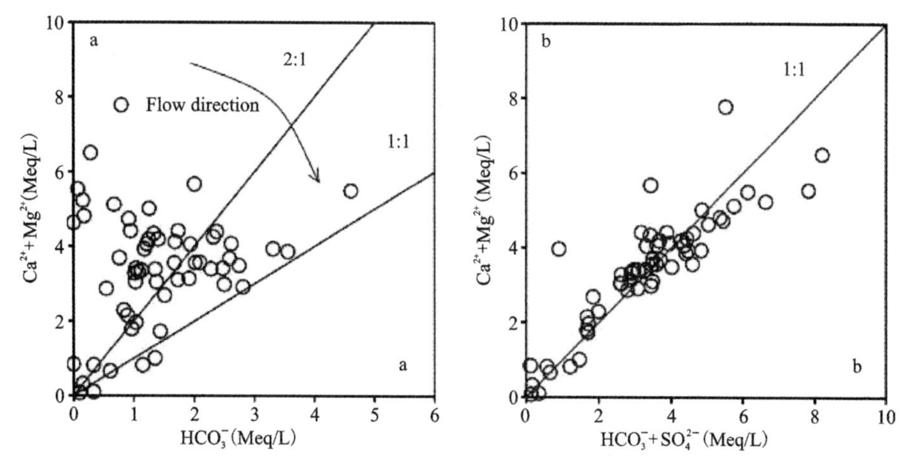

Figure 5.11 Relationship between $Ca^{2+}+Mg^{2+}$ and $HCO_3^-+SO_4^{2-}$ illustrating geochemical processes in groundwater. The 2 : 1 and 1 : 1 blue lines in the Figure 5.11a represent the dissolution of silicate minerals by H_2SO_4 and CO_2 weathering agents, respectively.

5.5.2.3 The transport of inorganic carbon and CO_2 evasion in groundwater

The $\delta^{13}C$ values of dissolved inorganic carbon in groundwater ranged from −20.9‰ to −12.1‰ with a mean value of −15.5‰ during the wet season and from −12.0‰ to −1.6‰ with a mean value of −8.6‰ during the dry season. It is clear that the $\delta^{13}C_{DIC}$ values showed considerable variation, meaning that $\delta^{13}C$ in groundwater depended on the $\delta^{13}C$ signature of the dissolving C-source and the fractionation among the carbon species in the solution.

For carbon sources, $\delta^{13}C_{DIC}$ was not detected in Hengshi River due to the intense influence of AMD, hence, the dissolved inorganic carbon in groundwater was mainly a mixture of that from soil respiration and atmospheric CO_2 (Figure 5.12). However, the $\delta^{13}C_{DIC}$ values showed an opposite spatial variation in different seasons; that is, depleted and enriched $\delta^{13}C_{DIC}$ values from −1.6‰ to −12.0‰ and from −20.9‰ to −12.1‰ with increasing DIC contents were observed in the dry and wet seasons, respectively (Figure 5.12), indicating that isotopic fractionation among different carbon species is the decisive factor compared to the contributions of these two carbon sources to groundwater $\delta^{13}C_{DIC}$ in this local area.

5.5.2.3.1 Carbon isotope fractionation during the dry season

The $\delta^{13}C_{DIC}$ during acidic environments can be used to make an even stronger case for the effects of the carbon cycle on isotope fractionation. The $\delta^{13}C$ enrichment during acidification is largely due to the escape of isotopically light CO_2 that enriches $\delta^{13}C_{DIC}$ by 9‰

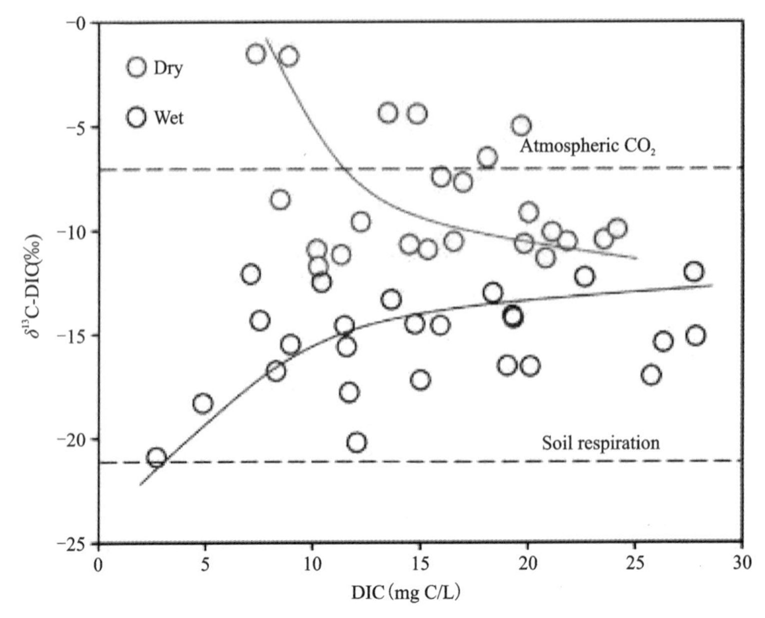

Figure 5.12 Seasonal changes in the $\delta^{13}C_{DIC}$ values of groundwater with increasing DIC contents and the relationships between the $\delta^{13}C_{DIC}$ values of groundwater and different carbon sources [The $\delta^{13}C$ of atmospheric and soil CO_2: $-7‰$ and $-21‰$ (Appelo and Postma, 2005; Hao et al., 2019; Yamanaka, 2012)].

(Rovira and Vallejo, 2008). Therefore, it is theoretically possible to calculate groundwater $\delta^{13}C_{DIC}$ based on the Rayleigh function and equilibrium isotope fractionation [Eqs. (5.29) and (5.30)] to explore whether the carbon process is consistent with the above hypothesis:

$$\delta^{13}C_{DIC} = \delta^{13}C_0 + \varepsilon^{13}C_{CO_2(g),DIC} \times \ln f \tag{5.29}$$

$$f \times \delta^{13}C_{DIC} + (1-f) \times [\delta^{13}C_{DIC} + \varepsilon^{13}C_{CO_2(g),DIC}] = \delta^{13}C_0 \tag{5.30}$$

where $\delta^{13}C_{DIC}$ is carbon isotope composition in groundwater to be predicted, $\delta^{13}C_0$ is the initial carbon isotope ratio for the whole study area as a carbon reservoir ($-12.7‰$), ε is the isotopic enrichment factor [$\varepsilon^{13}C_{CO_2(g),DIC} = -9‰$], and f is the reactant residue ratio. The results showed that the estimated $\delta^{13}C_{DIC}$ values based on Rayleigh fractionation were in accordance with the actual observations compared to the equilibrium fractionation (Figure 5.13). As the "neutralization induced theory" described, a carbon reservoir that ① is continuously decreasing due to DIC loss and is governed by first order kinetics and ②has the combined fractionation effects of diffusive loss and partial equilibration of carbon with atmospheric CO_2 during HCO_3^- dehydration (Atekwana and Fonyuy, 2009). This situation was indeed observed in the winter, when the considerably lower pH values (<4.9) and higher P_{CO_2} value (2611Pa) in groundwater were reported in the northern zone compared to those in the southern zone. Therefore, under the hydraulic connection between the river and groundwater, the transition of high P_{CO_2} values to low P_{CO_2} values along the flow path reflects the possibility of a carbon cycle path as assumed above, changing from CO_2 evasion to a carbon sink. This hypothesis indicates that a dropping water table or decreasing water content in unsaturated zones provides favorable buffer conditions for AMD neutralization and also becomes a relatively open system. Bicarbonate neutralized protons to CO_2 that is in turn spilled over the water surface, making it possible for significant enrichment of the carbon isotopes. However, this observation was not found in the southern part of the study area.

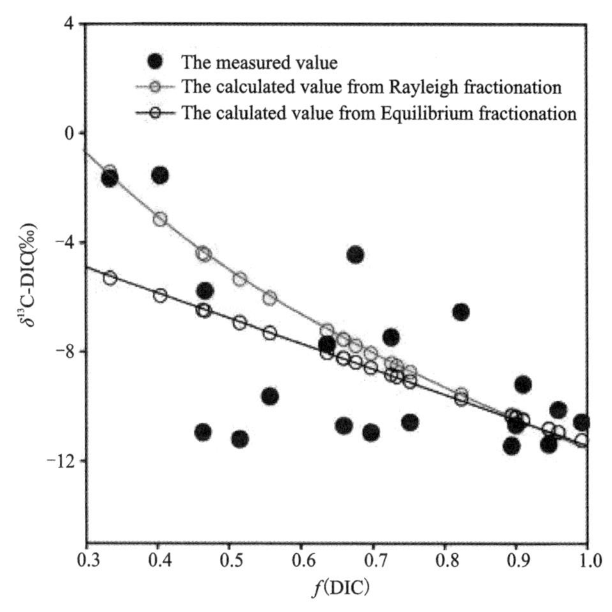

Figure 5. 13 Comparison between the measured and calculated $\delta^{13}C_{DIC}$ values of groundwater during the dry season based on the Rayleigh function and the equilibrium isotope fractionation.

5. 5. 2. 3. 2 Carbon isotope fractionation during the wet season

In the wet season, the water table and water content in unsaturated zones are both higher and the CO_2 that is consumed is resupplied by root respiration. However, the diffusion of CO_2 in the water is relatively slow so that the water-rock reaction caused by carbonic acid was hard to achieve due to a lack of CO_2 replenishment to some degree. When the water-table elevation rises, formerly unsaturated sediment is soaked so that the acidic aquifer reduces the possibility of carbon species transformation from CO_2 to carbonic acid. In other words, the CO_2 uptake by silicate weathering would be deficient due to the substitution of carbonic acid by protons from sulfuric acids. Similarly, it is theoretically possible to calculate groundwater $\delta^{13}C_{DIC}$ based on the Rayleigh function and equilibrium isotope fractionation to explore the rationality of the above hypothesis.

Considering the high water table and slow gas diffusion during the wet season, Rayleigh fractionation in both open and closed systems should be considered. The results showed that both Rayleigh and equilibrium isotope fractionation did not describe the actual situation very well (Figure 5.14). We can see that the observed $\delta^{13}C_{DIC}$ values at some sites, such as $f > 0.8$, were in line with the Rayleigh fractionation simulation in the closed system. Most $\delta^{13}C_{DIC}$ values, especially in the northern zone ($f < 0.3$), had lighter carbon isotopic compositions, which also confirmed the interruption of carbon species transformation from CO_2 to HCO_3^- (Figure 5.14). Therefore, the $\delta^{13}C_{DIC}$ values of groundwater in the northern zone recorded the $\delta^{13}C$ of soil CO_2 from the respiration of C3 and C4 plants. These two mechanisms explained the carbon behavior in the wet season: C-source mixing occurred in the AMD-affected groundwater carbon pool, whereas water-rock interaction produces carbonate that occurred in non-AMD-affected groundwater. This means that the carbon cycle path in groundwater shifted from carbon mixing to a carbon sink along the groundwater flow direction. This hypothesis indicates that the rising water table causes a similar closed system to some extent, slowing down the replenishment of bicarbonate and even CO_2 uptake, especially in the AMD-affected area so that the equilibrium and mixture of carbon sources in groundwater become dominant after AMD enters the aquifer. Therefore, in the northern part of the study area, the weathering of the aquifer caused by carbonic acid was reduced, whereas inorganic carbon species were changed with the changes in sulfuric acid sources in the southern zone. As a result, although the P_{CO_2} values of groundwater in the northern zone are high in both wet and dry seasons, the implications are different and reflect another possibility of the carbon cycle path during the wet season as assumed above, which is that the carbon cycle shifts from deficient CO_2 uptake to a carbon sink.

5.5.2.3.3 Conceptual model of DIC transport and transformation in groundwater with multiple sulfate sources during the different seasons

In the AMD-affected riparian zone, the hydraulic connection between the river and groundwater changes natural regimes and injects large amounts of exogenous compounds (e.g., dissolved oxygen, H_3O^+, and SO_4^{2-}) into the aquifer, which along with atmospheric deposition, agricultural fertilizers and domestic sewage, become the main sources of sulfuric acid, and also become the most important potential weathering agents, affecting the chemical equilibrium of inorganic carbon in aquifers and even the global carbon cycle (Figure 5.15). In this study, the changes in $\delta^{13}C_{DIC}$ values from $-1.6‰$ (dry) or $-20.9‰$ (wet) to $-12.0‰$ along the groundwater flow path were the most powerful response to the carbon cycle from a limiting carbon sink to a carbon sink. The variations in $\delta^{34}S_{SO_4}$ and $\delta^{18}O_{SO_4}$ values were accompanied by changes in $\delta^{13}C_{DIC}$ values. We can see that the $\delta^{13}C_{DIC}$ value of $-12.0‰$ influenced by atmospheric sulfates, agricultural fertilizers and domestic sewage in the southern part of the study area was significantly different from those values ($-1.6‰$ or $-20.9‰$) affected by pyrite in the northern zone. Therefore, $\delta^{13}C_{DIC}$ values of groundwater

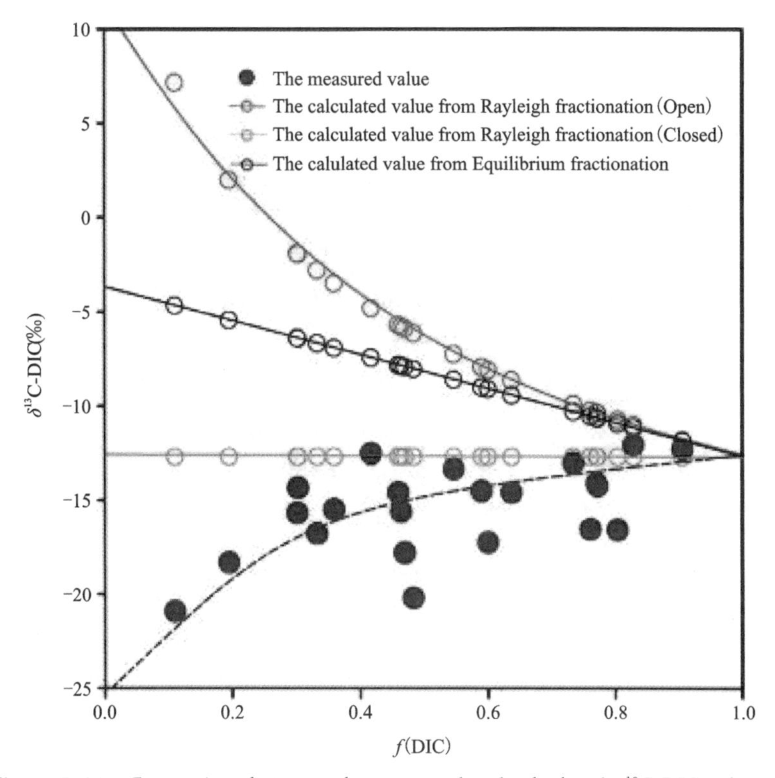

Figure 5.14 Comparison between the measured and calculated δ^{13}C-DIC values of groundwater during the wet season based on the Rayleigh function and the equilibrium isotope fractionation.

in the southern zone were mainly determined by carbonic acid equilibrium, and atmospheric and agricultural sulfates as well as domestic sewage hardly involved in the isotopic fractionation of $\delta^{13}C_{DIC}$, while obvious $\delta^{13}C_{DIC}$ enrichment or depletion in the northern zone was caused by AMD through DIC degassing or through limiting carbon transformation (Figure 5.15). Although there were longterm accumulations of precipitation and fertilization in the waterhed, they played a limited role in affecting the carbon cycle compared to the contributions of point sources from intensive human activities (e.g. AMD from pyrite oxidation), which also means that not all sulfate sources necessarily lead to CO_2 evasion or deficient CO_2 uptake. This also provides a new insight into whether different sulfate sources contribute to different weathering processes. In addition, the seasonal fluctuations of the water table affected the paths of DIC transport and transformation. In different seasons, different groundwater levels were observed and caused the different open or closed systems in the aquifer, resulting in different carbon cycle pathways. During the dry season, the lower groundwater level tended toward an open system and the carbon isotopic compositions in groundwater obeyed Rayleigh fractionation. Sulfuric acid from AMD prompted HCO_3^- to $CO_{2(g)}$, which limited the carbon sink of silicate. Thus, the δ^{13}C enrichment was largely due to the escape of isotopically light $CO_{2(g)}$. Along the flow path, carbonic acid weathering gradually returned to dominance and the weathering of silicates became a carbon sink again. However, Rayleigh fractionation and equilibrium fractionation did not describe the depleted carbon isotopic compositions during the wet season very well. The rising water table slowed the replenishment of bicarbonate and directly prevented the water-rock reaction caused by carbonic acid, especially in the AMD-affected area, which prompted $\delta^{13}C_{DIC}$ of groundwater to almost record the $\delta^{13}C$ of soil CO_2. Similarly, the carbon cycle gradually shifted towards a net consumption of CO_2 along the flow path, and $\delta^{13}C_{DIC}$ inherited the characteristics of soil CO_2 and HCO_3^- (Figure 5. 15). In short, although the PCO_2 values of groundwater in the northern zone were high in both the wet and dry seasons, the implications were different in terms of the contribution mechanisms of atmospheric CO_2, which were CO_2 evasion into the atmosphere and deficient atmospheric CO_2 uptake, respectively.

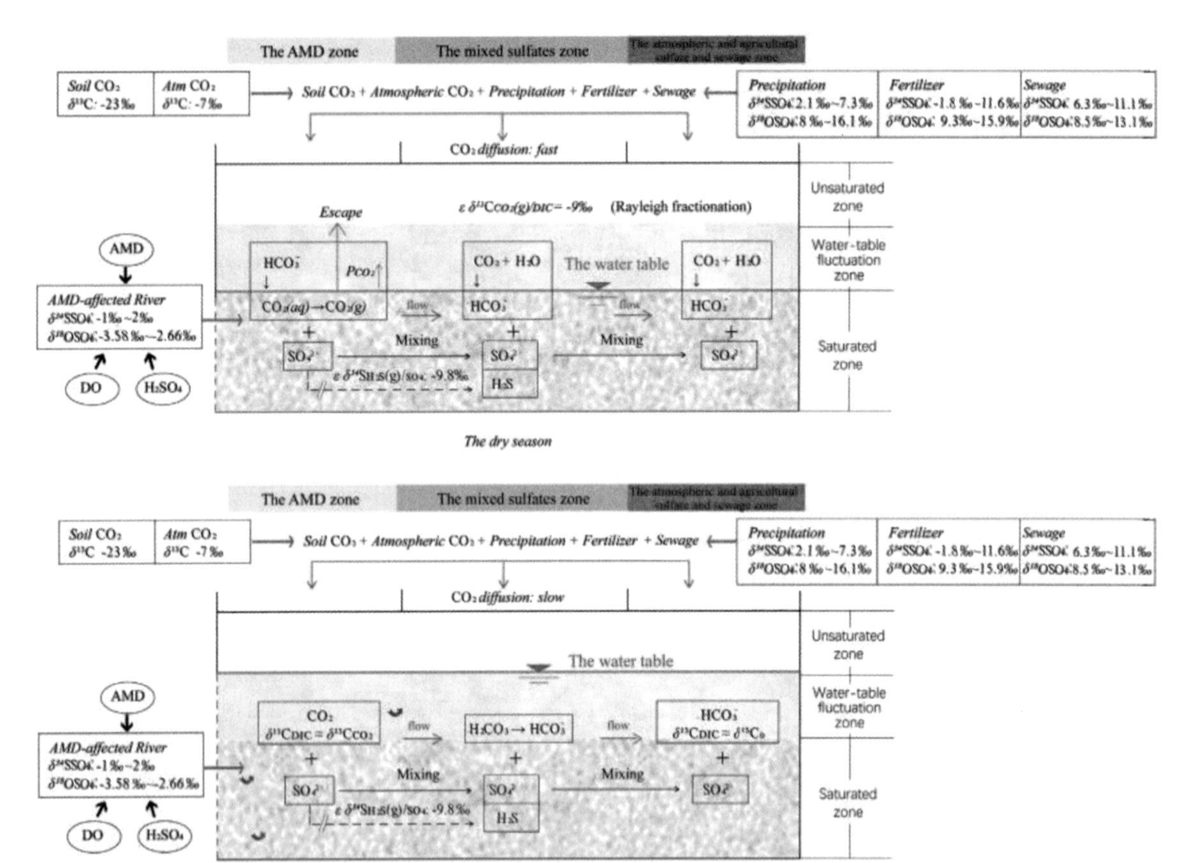

Figure 5. 15 Conceptual model of the C cycle in groundwater and its response to different sulfate sources during different seasons.

References

ABELSON P H, HOERING T C, 1961. Carbon isotope fractionation in formation of amino acids by photosynthetic organisms[J]. PNAS, 47:623.

ATEKWANA E A, FONYUY E W, 2009. Dissolved inorganic carbon concentrations and stable carbon isotope ratios in streams polluted by variable amounts of acid mine drainage[J]. J. Hydrol. ,372 (1-4):136-148.

BLAIR N, LEU A, MUNOZ E, et al. , 1985. Carbon isotopic fractionation in heterotrophic microbial-metabolism[J]. Appl. Environ. Microbiol. ,50:996-1001.

BORCH T, KRETZSCHMAR R, KAPPLER A, et al. , 2010. Biogeochemical redox processes and their impact on contaminant dynamics[J]. Environmental Science and Technology,44:15-23.

BOTSCH K C, CONRAD R, 2011. Fractionation of stable carbon isotopes during anaerobic production and degradation of propionate in defined microbial cultures[J]. Org. Geochem. ,42:289-295.

CHACKO T,COLE D R,HORITA J,2001. Equilibrium oxygen,hydrogen and carbon fractionation factors applicable to geologic systems[J]. Rev. Mineral Geochem. ,43:1-81.

CONRAD R, 2005. Quantification of methanogenic pathways using stable carbon isotopic signatures: a review and a proposal. Organic Geochemistry Stable Isotope Applications in Methane Cycle Studies[J]. Elsevier:739-752.

CONRAD R, CLAUS P, CHIDTHAISONG A, et al. , 2014. Stable carbon isotope biogeochemistry of propionate and acetate in methanogenic soils and lake sediments[J]. Org. Geochem. ,73:1-7.

CONRAD R,KLOSE M,LU Y,2012. Methanogenic pathway and archaeal communities in three different anoxic soils amended with rice straw and maize straw[J]. Frontiers Microbiol. ,3:1-12.

CONRAD R,KLOSE M,YUAN Q,et al. ,2012. Stable carbon isotope fractionation, carbon flux partitioning and priming effects in anoxic soils during methanogenic degradation of straw and soil organic matter[J]. Soil Biol. Biochem. ,49:193-199.

DENIRO M J,EPSTEIN S,1977. Mechanism of carbon isotope fractionation associated with lipid synthesis[J]. Science,197:261-263.

DENIRO M J,EPSTEIN S,1978. Influence of diet on distribution of carbon isotopes in animals[J]. Geochim. Cosmochim. Acta,42:495-506.

DUMIG A,RUMPEL C,DIGNAC M F,et al. ,2013. The role of lignin for the delta C-13 signature in C-4 grassland and C-3 forest soils[J]. Soil Biol. Biochem. ,57:1-13.

EILER J M,et al. ,2014. Frontiers of stable isotope geoscience[J]. Chem. Geol. ,372: 119-143.

EMRICH K, EHHALT D H, VOGEL J C, 1970. Carbon isotope fractionation during the precipitation of calcium carbonate[J]. Earth Planet Sci. Lett., 8: 363-371.

FARQUHAR G D, EHLERINGER J R, HUBICK K T, 1989. Carbon isotope discrimination and photosynthesis[J]. Ann. Rev. Plant Physiol. Plant Mol. Biol., 40: 503-537.

GALIMOV E M, 1985. The biological fractionation of isotopes[M]. Orlando: Academic Press Inc.

GALIMOV E M, 2006. Isotope organic geochemistry [J]. Org. Geochem., 37: 1200-1262.

HAEDRICH A, HEUER V B, HERRMANN M, et al., 2012. Origin and fate of acetate in an acidic fen[J]. FEMS Microbiol. Ecol., 81: 339-354.

HEUER V B, KRUEGER M, ELVERT M, et al., 2010. Experimental studies on the stable carbon isotope biogeochemistry of acetate in lake sediments[J]. Org. Geochem., 41: 22-30.

HEUER V B, POHLMAN J W, TORRES M E, et al., 2009. The stable carbon isotope biogeochemistry of acetate and other dissolved carbon species in deep subseafloor sediments at the northern Cascadia Margin[J]. Geochim. Cosmochim. Acta, 73: 3323-3336.

KITCHEN N E, VALLEY J W, 1995. Carbon isotope thermometry in marbles of the Adirondack Mountains, New York[J]. J. Metamorph. Geol., 13: 577-594.

KODINA L A, 2010. Carbon isotope fractionation in various forms of biogenic organic matter: I. Partitioning of carbon isotopes between the main polymers of higher plant biomass [J]. Geochem. Int., 48: 1157-1165.

JI Y, SCAVINO A F, KLOSE M, et al., 2015. Functional and structural responses of methanogenic microbial communities in Uruguayan soils to intermittent drainage[J]. Soil Biol. Biochem., 89: 238-247.

LERCH T Z, NUNAN N, DIGNAC M F, et al., 2011. Variations in microbial isotopic fractionation during soil organic matter decomposition[J]. Biogeochemistry, 106: 5-21.

LONDRY K L, DAWSON K G, GROVER H D, et al., 2008. Stable carbon isotope fractionation between substrates and products of Methanosarcina barkeri[J]. Org. Geochem., 39: 608-621.

LOWERS H A, BREIT G N, FOSTER A L, et al., 2007. Arsenic incorporation into authigenic pyrite, Bengal Basin sediment, Bangladesh[J]. Geochimica et Cosmochimica Acta, 71: 2699-2717.

MACH V, BLASER M B, CLAUS P, et al., 2015. Methane production potentials, pathways, and communities of methanogens in vertical sediment profiles of river Sitka[J]. Front Microbiol., 6: 506.

MATTHEWS D E, HAYES J M, 1978. Isotope-ratio-monitoring gas chromatography-mass spectrometry[J]. Anal. Chem., 50: 1465-1473.

MONSON K D, HAYES J M, 1982. Carbon isotopic fractionation in the biosynthesis of bacterial fatty acids, Ozonolysis of unsaturated fatty acids as a means of determining the intramolecular distribution of carbon isotopes[J]. Geochim. Cosmochim. Acta, 46: 139-149.

MOOK W G, BOMMERSON J C, STAVERMANN W H, 1974. Carbon isotope fractionation between dissolved bicarbonate and gaseous carbon dioxide[J]. Earth Planet Sci. Lett., 22: 169-174.

MUELLER C W, GUTSCH M, KOTHIERINGER K, et al., 2014. Bioavailability and isotopic composition of CO_2 released from incubated soil organic matter fractions[J]. Soil Biol. Biochem., 69: 168-178.

OHKOUCHI NONO, CHIKARAISHI Y, TANAKA H, et al., 2015. Biochemical and physiological bases for the use of carbon and nitrogen isotopes in environmental and ecological studies[J]. Prog. Earth Planetary Sci., 2: 2-17.

O'LEARY M H, 1981. Carbon isotope fractionation in plants[J]. Phytochemistry, 20: 553-567.

O'LEARY M H, 1988. Carbon isotopes in photosynthesis[J]. Bioscience, 38: 328-336.

PARK R, EPSTEIN S, 1960. Carbon isotope fractionation during photosynthesis[J]. Geochim. Cosmochim. Acta, 21: 110-126.

PENGER J, CONRAD R, BLASER M, 2012. Stable carbon isotope fractionation by methylotrophic methanogenic archaea[J]. Appl. Environ. Microbiol., 78: 7596-7602.

PENNING H, CONRAD R, 2006. Carbon isotope effects associated with mixed-acid fermentation of saccharides by Clostridium papyrosolvens[J]. Geochim. Cosmochim. Acta, 70: 2283-2297.

PERKINS M J, MCDONALD R A, VAN VEEN F J, et al., 2014. Application of nitrogen and carbon stable isotopes $\delta^{15}N$ and $\delta^{13}C$ to quantify food chain length and trophic structure[J]. PLoS One, 9(3): e93281.

PIASECKI A, SESSIONS A, LAWSON M, et al., 2016. Analysis of the site-specific carbon isotope composition of propane by gas source isotope ratio mass spectrometry[J]. Geochim. Cosmochim. Acta, 188: 58-72.

ROMANEK C S, GROSSMAN E L, MORSE J W, 1992. Carbon isotope fractionation in synthetic aragonite and calcite: effects of temperature and precipitation rate[J]. Geochim. Cosmochim. Acta, 56: 419-430.

ROVIRA P, VALLEJO V R, 2008. Changes in d13C composition of soil carbonates driven by organic matter decomposition in a Mediterranean climate: a field incubation experiment[J]. Geoderma, 144 (3-4): 517-534.

RUBINSON M, CLAYTON R N, 1969. Carbon-13 fractionation between aragonite and calcite[J]. Geochim. Cosmochim. Acta, 33: 997-1002.

SCHEELE N, HOEFS J, 1992. Carbon isotope fractionation between calcite, graphite and CO_2[J]. Contr. Mineral Petrol. ,112:35-45.

TOKIDA T,ADACHI M,CHENG W G,et al. ,2011. Methane and soil CO_2 production from currentseason photosynthates in a rice paddy exposed to elevated CO_2 concentration and soil temperature[J]. Global Change Biol. ,17:3327-3337.

TURNER J V, 1982. Kinetic fractionation of carbon-13 during calcium carbonate precipitation[J]. Geochim. Cosmochim. Acta,46:1183-1192.

VALLEY J W,O'NEIL J R,1981. $^{13}C/^{12}C$ exchange between calcite and graphite: a possible thermometer in Greville marbles[J]. Geochim. Cosmochim. Acta,45:411-419.

VERPLANCK P L, MUELLER S H, GOLDFARB R J, et al. , 2008. Geochemical controls of elevated arsenic concentrations in groundwater,Ester Dome,Fairbanks district, Alaska[J]. Chemical Geology,255:160-172.

VOGEL J C, GROOTES P M, MOOK W G, 1970. Isotopic fractionation between gaseous and dissolved carbon dioxide[J]. Z. Physik. ,230:225-238.

WERTH M, KUZYAKOV Y, 2008. Root-derived carbon in soil respiration and microbial biomass determined by C-14 and C-13[J]. Soil Biol. Biochem. ,40:625-637.

WERTH M, KUZYAKOV Y, 2010. C-13 fractionation at the root-microorganisms-soil interface: a review and outlook for partitioning studies[J]. Soil Biol. Biochem. ,42:1372-1384.

YUAN Q, PUMP J, CONRAD R, 2012. Partitioning of CH_4 and CO_2 production originating from rice straw,soil and root organic carbon in rice microcosms[J]. PLoS One: 7e49073-e49073.

YUAN Q, PUMP J, CONRAD R, 2014. Straw application in paddy soil enhances methane production also from other carbon sources[J]. Biogeosciences,11:237-246.

ZHANG C L,YE Q,REYSENBACH A L,et al. ,2002. Carbon isotopic fractionations associated with thermophilic bacteria Thermotoga maritima and Persephonella marina[J]. Environ. Microbiol. ,4:58-64.

ZHANG J,QUAY P D,WILBUR D O,1995. Carbon isotope fractionation during gas-water exchange and dissolution of CO_2[J]. Geochim. Cosmochim. Acta,59:107-114.

Chapter 6 Nitrogen Isotope

Nitrogen consists of two stable isotopes: ^{14}N and ^{15}N. More than 99% of the known nitrogen on or near the Earth's surface is present as atmospheric N_2 or as dissolved N_2 in the ocean. Atmospheric nitrogen, given by Rosman and Taylor (1998) has the following composition:

$^{14}N=99.63\%$

$^{15}N=0.37\%$

Only a minor amount is combined with other elements, mainly C, O, and H. Nevertheless, this small part plays a decisive role in the biological realm. Since nitrogen occurs in various oxidation states and in gaseous, dissolved, and solid forms (N_2, NO_3^-, NO_2^-, NH_3, NH_4^+, it is a highly suitable element for the search of natural variations in its isotopic composition. The range of reported $\delta^{15}N$-values covers 100‰, from about −50‰ to +50‰. However, most δ-values fall within the much narrower spread from −10‰ to +20‰, as described in more recent reviews of the exogenic nitrogen cycle (Heaton,1986), Owens (1987), Peterson and Fry (1987) and Kendall (1998).

6.1 Nitrogen species in groundwater

Atmospheric N_2 can be transformed between a variety of redox states and species in a complex cycle involving inorganic and bacterially mediated reactions under both aerobic and anaerobic conditions. The principal species and their redox state in groundwater was presented in Table 6.1. Of these nitrogen species it is N_2, NH_4^+, NO_3^-, NO_2^-, and N_2O that represent the largest reservoirs in groundwater.

Table 6.1 Nitrogen species in groundwater.

Species		Valence state	
Nitrate	NO_3^-	N^{5+}	Stable oxide of N, highly soluble as an anion
Nitrogen dioxide	NO_2	N^{4+}	Produced by lightning and combustion
Nitrite	NO_2^-	N^{3+}	Intermediate in conversions between NO_3^- and NH_4^+

Continue

Species		Valence state	
Nitrogen oxide	NO	N^{2+}	Produced by lightning and combustion, $NO_2+NO=NO_x$
Nitrous oxide	N_2O	N^{+}	Bacterially produced gas, in atmosphere at $<1\times10^{-6}$
Nitrogen	N_2	N	Elemental nitrogen gas
Hydroxylamine	NH_2OH	N^{-}	Intermediate species during oxidation of NH_4^+
Ammonia	NH_3	N^{3-}	Unionized ammonia gas
Ammonium	NH_4^+	N^{3-}	Ionized ammonia (dominates at pH below 9.23)
Organic N	$-NH_2$	N^{3-}	Reduced N in organic compounds including protein

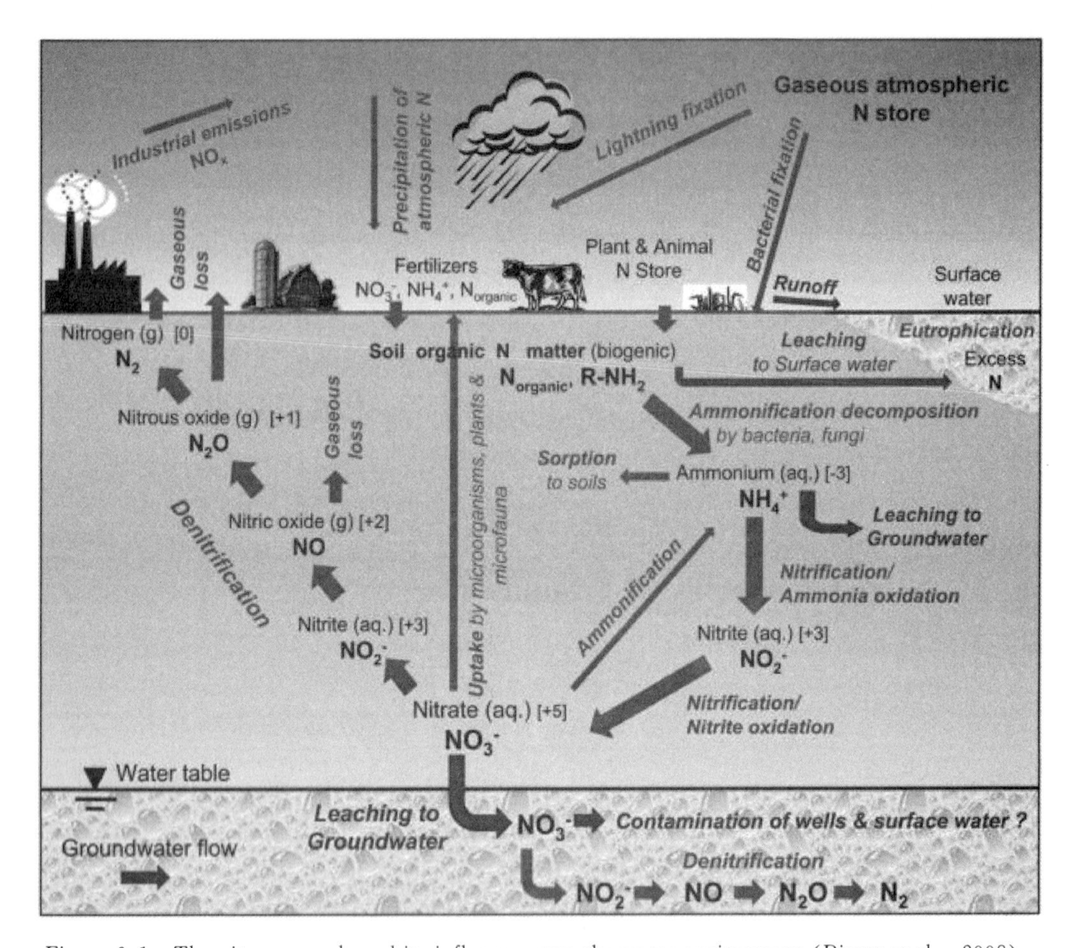

Figure 6.1 The nitrogen cycle and its influence upon the water environment (Rivett et al., 2008).

The cycle of nitrogen in the vadose zone and groundwater is shown in Figure 6.1. Denitrification is central to the nitrogen cycle with respect to the subsurface groundwater environment and involves the reduction of nitrate via a chain of microbial reduction reactions to nitrogen gas (Knowles, 1982). Nitrate can also be reduced to nitrite and nitrous oxide gas by abiotic reactions but, in the subsurface, these reactions are minor compared with biological denitrification. The organisms capable of denitrification tend to be ubiquitous in surface water, soil and groundwater (Beauchamp et al., 1989). They are found at great depths in aquifers: in clayey sands to 289m (Francis et al., 1989); in limestone to 185m (Morris et al., 1988); and in granite to 450m depth (Neilsen et al., 2006). Denitrifiers are mostly facultative anaerobic heterotrophs and hence obtain both their energy and carbon from the oxidation of organic compounds. However, some denitrifying bacteria are autotrophs, obtaining their energy from the oxidation of inorganic species. In general, the absence of oxygen and the presence of organic carbon, reduced sulphur or iron facilitates occurrence of denitrification.

The denitrification can be presented as follows:

$$5CH_2O+4NO_3^- +4H^+ \rightarrow 2N_2+5CO_2+7H_2O \quad \Delta G_r^0=-252.47\text{kJ/mol} \tag{6.1}$$

Where a low-pe electron donor exists, such as carbon or sulfide, nitrate can be used as an electron acceptor with nearly the same energy yield as O_2. As this reaction requires organic carbon, it generally will not occur in oxygenated water where aerobic bacteria outcompete denitrifiers for available carbon substrates. In anaerobic water with low nitrate concentrations (NO_3^--limited), denitrification to N_2 gas may not be complete, resulting in the production of N_2O gas.

Although denitrification has a stable endpoint at nitrogen gas, the process can be arrested at any of the intermediate stages (Figure 6.1). This is important because nitrite is significantly more toxic than nitrate (WHO, 2004). Nitrite interferes with the blood's capacity to carry oxygen. This leads to methemoglobinemia or blue baby syndrome in infants. The maximum acceptable concentration (MAC) for nitrate in drinking water is universally set at 10mg/L as N, and for NO_2^- at 1mg N/L. Nitrite is significantly more reactive than nitrate and is stable only within a limited range of redox conditions (Figure 6.2). In particular, the action of the nitrite reductase enzyme is more sensitive to oxygen concentrations than that of nitrate reductase (Hochstein et al., 1984; Korner and Zumft, 1989). The difference in energy available from the reduction reaction means that nitrate is used by denitrifiers preferentially to nitrite, even when both enzymes are present. A build-up of nitrite may then occur due to the time lag between the onset of nitrate reduction and the subsequent onset of nitrite reduction (Betlach and Tiedje, 1981; Gale et al., 1994). In natural water, nitrite rarely occurs at concentrations comparable with those of nitrate (typically 2～5 orders of magnitude lower; Environment Agency data for 2003), except temporarily under reducing conditions. Nitrite also readily reacts with dissolved organic compounds to form dissolved organic nitrogen compounds (Davidson et al., 2003), especially in low pH environments where nitrous acid (HNO_2) is the key reactant.

Nitric oxide (NO) and nitrous oxide (N_2O) are formed during denitrification but, in favorable conditions, transform rapidly to benign nitrogen gas. Both NO and N_2O contribute to acid rain, promote the formation of ground-level ozone and contribute to global warming; N_2O also destroys ozone in the upper atmosphere. N_2O is equally produced as an intermediate product in the nitrification of ammonium; this process, rather than denitrification, is the main contributor to N_2O emissions from UK Chalk groundwater (Hiscock et al. , 2003). Free NO is rarely observed because it transforms to N_2O rapidly under typical environmental conditions. It is usually observed only in small-scale laboratory studies as an intracellular intermediate (Scheible, 1993).

When oxygen levels are very low, nitrogen gas (N_2) is the end product of the denitrification process; but, where oxygen levels are more intermediate or variable, the reactions may stop with the formation of NO_x. Very high nitrate concentrations or low pH values also arrest denitrification at the N_2O stage. Indeed, N_2O is often used in wetland studies as an indicator that denitrification is taking place (Bernot et al. , 2003; Delaune and Jugsujinda, 2003). Formation of N_2 can be arrested in experimental studies by applying an excess of acetylene (C_2H_2) such that all denitrified nitrogen can be measured as N_2O (Yoshinari and Knowles, 1976). However, the presence of N_2O as an indicator of denitrification is not necessarily conclusive as it may form from partial nitrification of ammonium (Kinniburgh et al. , 1999). The denitrification process can be reactivated further along a flow line; for example, LaMontagne et al. (2002) studied an estuarine environment in which groundwater supersaturated with N_2O entered but was converted to nitrogen in anoxic benthic sediments.

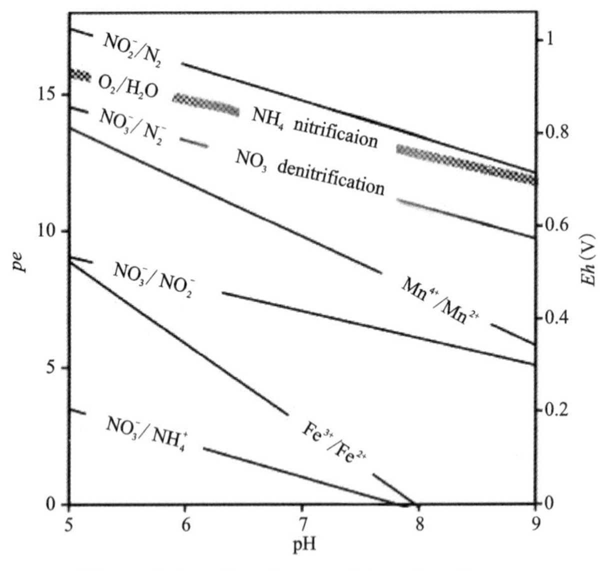

Figure 6.2 pH-redox conditions for nitrogen transformation reactions in the environment.

Ammonium is sorbed onto clay minerals in soils and aquifers with a selectivity coefficient close to that of K^+ (Figure 6.1). Accordingly, NH_4^+ migration in aquifers is significantly retarded. Erosion of NH_4-bearing soils is one of the major sources of ammonia contamination in surface water. The sorption of ammonium can be presented as follows:

$$Na-clay+NH_4^+ \rightarrow NH_4-clay+Na^+ \quad (6.2)$$

Both ionized and unionized ammonia can be oxidized to NO_3^- in water by reaction with elemental oxygen (O_2). The significant energy yield makes this favorable for bacteria. It follows a two-step reaction of oxidation to nitrite and then to nitrate. Dissolved oxygen is required, limiting this reaction to manure piles, soils, and shallow groundwater. The aerobic nitrification of ammonium can be presented as follows:

$$NH_4^+ + 2O_2 \rightarrow NO_3^- + H_2O + 2H^+ \quad \Delta G_r^0 = -266.5\text{kJ/mol} \quad (6.3)$$

Anaerobic oxidation of ammonium-anammox is found to operate in anaerobic environments where both ammonium and nitrate species are present, such as in wastewater streams, anoxic marine water, and contaminated groundwater. The anammox bacteria use NH_4^+ as an electron donor and NO_2^- generated through partial reduction of NO_3^- as an electron acceptor in a reaction that converts both species to elemental nitrogen, N_2. It is apparently the only known biologically mediated reaction for the conversion of ammonium to N_2 in natural systems. The nitrite for this reaction can be produced by partial denitrification of NO_3^-. Anaerobic oxidation of ammonium-anammox can be presented as follows:

$$NH_4^+ + NO_2^- \rightarrow N_2 + 2H_2O \quad \Delta G_r^0 = -362.8\text{kJ/mol} \quad (6.4)$$

The transformations of nitrogen species are strongly redox sensitive, with nitrate stable under oxidizing conditions, and ammonium stable under reducing conditions (Figure 6.2). This restricts the settings under which nitrification, denitrification, and anammox can take place.

6.2 Analytical techniques for the ^{15}N-NO_3^- and ^{18}O-NO_3^-

Precise, accurate, but also inexpensive and fast analysis of NO_3^- for both $\delta^{15}N$ and $\delta^{18}O$ is needed for improved NO_3^- source identification, quantification and uncertainty assessment.

In recent years, the so called "ion-exchange" or "$AgNO_3$ method" for both $\delta^{15}N$-NO_3^- and $\delta^{18}O$-NO_3^- analysis has been developed by Chang et al. (1999) and Silva et al. (2000). This method is used to concentrate and purify NO_3^- in water samples for simultaneous ^{15}N and ^{18}O determination. Briefly, NO_3^- is purified and concentrated by passing samples through cation and then anion exchange resin columns. NO_3^- is eluted using hydrochloric acid, neutralized with silver oxide and then filtered to remove the AgCl. For accurate

$\delta^{18}O$-NO_3^- analysis, all non-nitrate oxygen-bearing anions (e. g. , SO_4^{2-} , CO_3^{2-} , PO_4^{3-}) are removed from the sample by adding $BaCl_2$ to the $AgNO_3$ solution and then the precipitate is filtered out. The filtered solution is then passed through a cation exchange resin to remove excess Ba^{2+} ions and re-neutralized with Ag_2O. The resulting solution can be freeze-dried (Silva et al. , 2000) or oven-dried (Fukada et al. , 2003) to produce $AgNO_3$ salts. $\delta^{15}N$ analysis of the prepared $AgNO_3$ can be conducted by conversion to N_2 gas for IRMS analysis via mixing $AgNO_3$ with a CuO, Cu wire and CaO catalyst and combusting in a sealed tube at 850℃ (Kendall and Grim, 1990). $\delta^{18}O$ analysis can be conducted using the combustion method which generates CO_2 by adding finely ground spectrographic graphite or by using pyrolysis systems that can generate CO by addition of graphite. Alternatively, $\delta^{15}N$ and $\delta^{18}O$ can be simultaneously analyzed via TC/EA-IRMS (thermal conversion/elemental analyzer-isotope ratio mass spectrometer). Prepared $AgNO_3$ samples are converted (N_2 and CO) at 1400℃ in a molybdenum-lined, aluminum oxide combustion tube which is filled with glassy carbon and topped with a glassy carbon crucible. Upon pyrolysis of the $AgNO_3$, the reaction gases are separated via a 1m E3030 GC column (Elemental Microanalysis) and analyzed via IRMS. The "ion-exchange method" has the following advantages: ①The concentration of NO_3^- from water samples onto anion exchange resin columns can be accomplished in the field and is convenient for transporting and storing water samples for NO_3^- isotopic analysis; ②There is minimal isotopic fractionation for NO_3^- stored on anion exchange columns (Silva et al. , 2000);③The technique achieves a high level of sensitivity. The disadvantages of the "ion-exchange method" can be summarized as follows:①The sample preparation procedure is relatively labor-intensive (3～5 days for sample preparation) and cost intensive (up to 60 Euro per sample just for consumables only);②High concentrations of anions (e. g. , Cl, SO_4^{2-} , DOC, etc.) in water samples can interfere with the adsorption of NO_3^- onto anion exchange resins; ③Target sample size of 100～200mmol of NO_3^- for optimal analysis requires large sample volumes for low NO_3^- concentration samples.

Another sample preparation technique for both $\delta^{15}N$-NO_3^- and $\delta^{18}O$-NO_3^- determination is the so called "bacterial denitrification method" (Sigman et al. , 2001; Casciotti et al. , 2002; Rock and Ellert, 2007). This method allows for the simultaneous determination of $\delta^{15}N$ and $\delta^{18}O$ of N_2O produced from the conversion of NO_3^- by denitrifying bacteria which naturally lack N_2O-reductase activity (the enzyme that catalyses the reduction of N_2O to N_2). In brief, bacterial cultures are grown for 6～10 days in amended tryptic soy broth (TSB), divided into centrifuge tubes of 40mL aliquots and centrifuged. After centrifugation, the supernatant is decanted, reserved and 4mL of the TSB is pipetted back into the tubes to obtain a 10-fold concentration of bacteria. These tubes are then vortexed to ensure homogenized cultures and then transferred as 2×2mL aliquots into 20mL headspace vials. The vials are crimp-sealed with Teflon-backed silicone septa. To ensure anaerobic conditions, a reduced blank effect and removal of N_2O produced prior to sample injection, the headspace vials are purged with N_2 gas for 3h. Samples of dissolved NO_3^- (100mmol) are then injected into the headspace vials and are incubated overnight to allow for complete conversion of NO_3^- to N_2O. The next day, 0. 1mL of 10M NaOH is injected into the headspace vials to stop bacterial activity and to scrub any CO_2 gas in the vial which can interfere with the N_2O measurement. $\delta^{15}N$ and $\delta^{18}O$ analysis of the produced N_2O can be conducted by extracting N_2O offline via freezing it out in a vacuum line (cryo-concentration) and analyzing by IRMS (Sigman et al. , 2001), or extracting N_2O online via an autosampler injection system and using Nafion as an additional water trap, and then analyzing by IRMS (Casciotti et al. , 2002). The "bacterial denitrification method" has the following advantages: ①The preparation procedure is less labor-intensive (2～3 days for sample preparation) and less expensive (up to 5 Europer sample for consumables);②The technique requires a smaller sample size (three orders of magnitude smaller than the "ion-exchange method") and allows for the analysis of low NO_3^- concentration samples; ③The method achieves a high level of sensitivity. The disadvantages of the "bacterial denitrification method" are summarized as follows:①Bacterial growth takes a longer time of about 10～12 days (from Petri plate to media bottles); ②Bacterial cultivation is potentially affected by toxicity of the sample (e. g. antibiotics, heavy metal, pesticides, etc.), which is difficult to predict;③The presence of NO_2 in water samples can bias the isotopic composition of product N_2O;④Mathematical approaches are needed to quantify and correct for both fractionation of oxygen isotopes during oxygen atom loss in the reaction process and exchange of oxygen atoms in nitrite and nitric oxide with water that are inherent to the denitrifier method for $\delta^{18}O$ analysis.

Besides the now well-established bacterial conversion of NO_3^- to N_2O and subsequent measurement of $\delta^{15}N$ and $\delta^{18}O$, a new method is gaining ground; conversion of NO_3^- to N_2O by a two-step chemical reduction procedure. McIlvin and Altabet (2005) published the "Cadmium reduction method" or "Azide method" which first converts NO_3^- to N_2O using cadmium reduction followed by a subsequent reduction to N_2O using sodium azide. The resulting N_2O is then analyzed for $\delta^{15}N$ and $\delta^{18}O$ in the same way described for the denitrifier method. The first conversion step of this technique (reduction of NO_3^- to NO_2^- by cadmium) has been previously applied to the spectrometric analysis of nitrate in water (Margeson et al. , 1980). First, NO_3^- in water samples (50mL) is reduced to NO_2^- in Teflon sealed batches through the addition of 1g spongy cadmium (buffered with 1M imidazole solution to pH 9) which is shaken overnight (horizontal) at 120rpm. Freshwater samples require a sodium chloride concentration of 0. 5M. After separation of the spongy cadmium by centrifugation, the water samples are introduced into new vials. Schilman and Teplyakov (2007) have proposed an alternate technique for this first reduction step which involves passing the sample through a cooperized granular-cadmium filled-column at a rate of 5mL/min. Complete conversion of sample NO_3^- to N_2O, avoiding any isotopic fractionation, can be achieved within a few minutes. The samples are then filled into 60mL vials, sealed with Teflonlined septa and purged with N_2 or He to remove any N_2O. For the second reduction step, 2mL 1 : 1 sodium azide (2M)/acetic acid (20%) solution is injected into the sample to reduce NO_2^- to N_2O and N_2. In order to remove any NH_3, which is extremely toxic and volatile, 1mL of 6M NaOH is added. Positive aspects of the method include: ①The preparation is less laboratory intensive (results can be available within one day) and consumables are inexpensive (cadmium can also be reused); ②Low concentrations (0.5mmol/L) and small sample volumes can be analyzed; ③No interferences with high concentrations and toxic substances; ④The possibility for automation with high sample throughput. Negative aspects of the method include: ①Environmental and health protection measures due to the toxicity of the cadmium, sodium azide, and NH_3; ②Results require corrections for oxygen exchange and fractionation; ③The presence of NO_2^- in water samples can bias the isotopic composition of product N_2O and N_2.

6.3 Fractionation processes

6.3.1 ^{15}N-NO_3

Previous studies showed that denitrification and nitrification alter the original $\delta^{15}N$-NO_3^- isotopic composition of NO_3^- in groundwater under agricultural areas. Isotope effects of the considered N processes are presented in terms of their enrichment factors which show isotope enrichment of a reaction product relative to that of the substrate and are determined by means of the Rayleigh equation (Mariotti et al., 1981):

$$\varepsilon=\frac{10^3\ln\dfrac{10^{-3}\delta(NO_3^-)_{measure}+1}{10^{-3}\delta(NO_3^-)_{initial}+1}}{\ln[C(NO_3^-)_{measure}/C(NO_3^-)_{initial}]} \tag{6.5}$$

where ε is the isotopic enrichment factors for N or O, δ is the $\delta^{15}N$ and $\delta^{18}O$ values, respectively and C-NO_3^- concentration.

As mentioned above, denitrification has attracted most considerable research effort as it plays a significant role in the attenuation of NO_3^- pollution in the subsurface (Rivett et al., 2008). Experimental results suggest that it is a strongly fractionating process responsible for preferential conversion of the lighter isotope ^{14}N to N_2O and N_2. Consequently, the corresponding enrichment of theresidual (unreacted) NO_3^- with the heavy isotope ^{15}N is observed

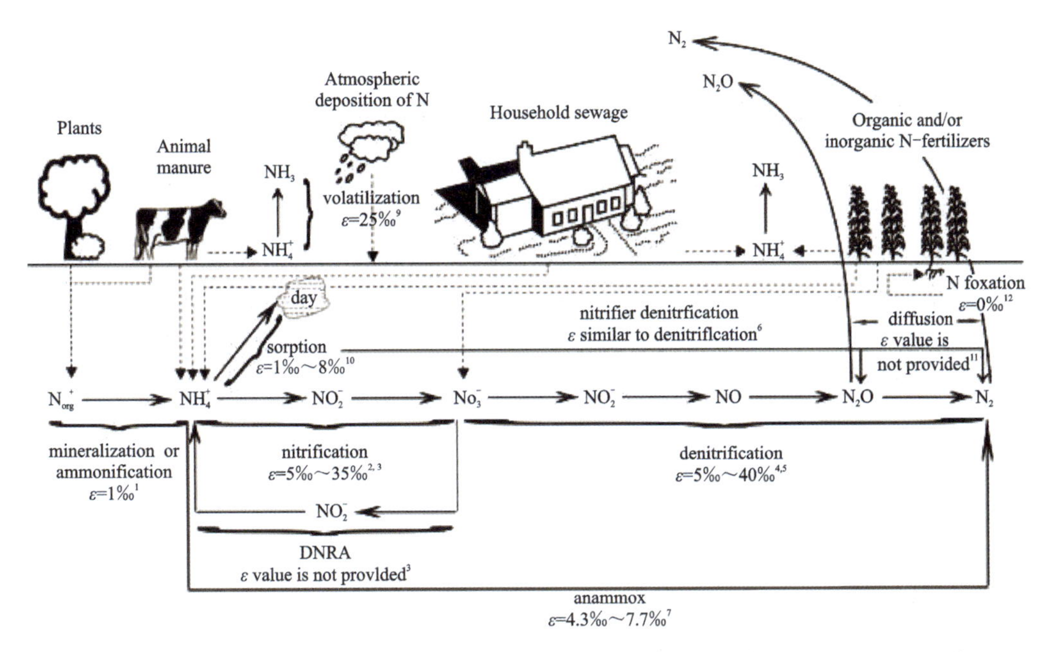

Figure 6. 3 N sources and transformation processes that affect N species in the subsurface. The enrichment values ($^{15}N-NO_3^-$, $^{15}N-NH_4^+$) of such processes are also provided. [⤏ shows the transformation of the initial N compound; ⟶ shows sources of different N species. References: 1. Sharp, 2007; 2. Kendall and Aravena, 2000; 3. Mariotti et al. , 1981; 4、7. Clark, 2015; 5. Kendall, 1998; 6. Well et al. , 2012; 8. Michener and Lajtha, 2007; 9. Bedard-Haughn et al. , 2003; 10. Hübner, 1981; 11. Minamikawa et al. , 2011; 12. Brandes and Devol, 2002]

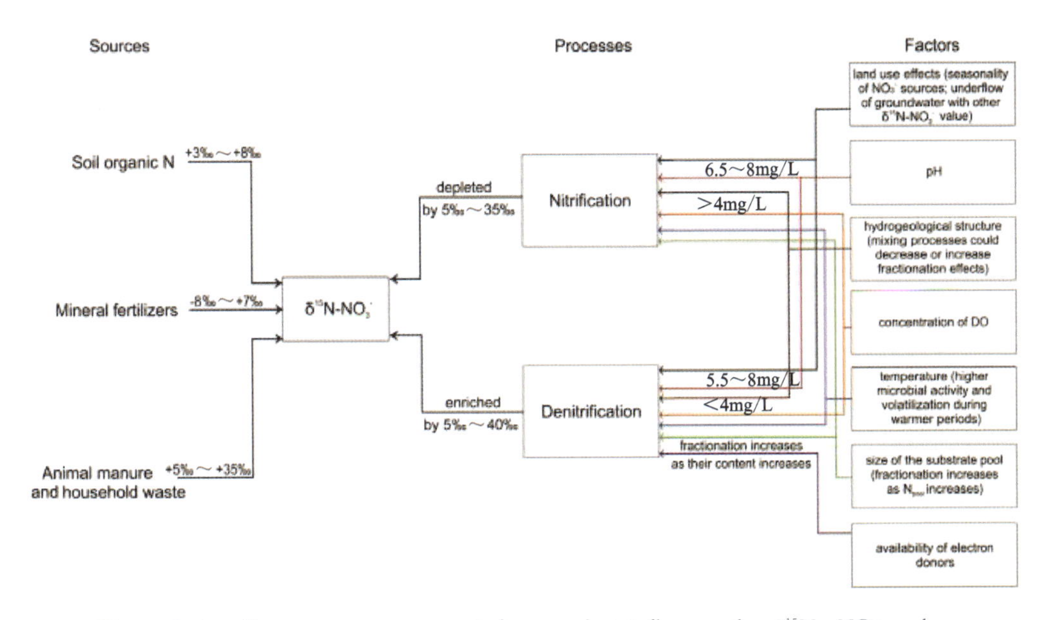

Figure 6. 4 Sources, processes and factors that influence the $\delta^{15}N-NO_3^-$ values: summary (the following arrows connect processes with factors that have decisive effect on their dynamics and, consequently, on resulting fractionation effects: → availability of electron donors; → size of the substrate pool; → temperature; → concentration of DO; → hydrogeological structure; → pH; → land use).

(Knöller et al., 2011; Fukada et al., 2003). During this process the $\delta^{15}N$ value of the initially produced NO_3^- might be enriched in comparison to N_2 or N_2O by approximately 20‰~30‰ (Clark, 2015), or 5‰~40‰ (Kendall, 1998). For example, denitrification of NO_3^- fertilizer that originally had a distinctive $\delta^{15}N$ value of +1‰ can yield residual NO_3^- with a $\delta^{15}N$ value of +15‰ which is within the range of composition expected for a NO_3^- from a manure or septic-tank source (Kendall, 1998) (Figure 6.4).

In contrast, nitrification reaction results in the preferential incorporation of the lighter isotopes into NO_3^- and often leads to decrease in the $\delta^{15}N$-NO_3^- (Barnes and Raymond, 2010). In average the difference between initial $\delta^{15}N$-NH_4^+ and produced $\delta^{15}N$-NO_3^- can reach 12‰~29‰ (Kendall and Aravena, 2000), or 5‰~35‰ (Mariotti et al., 1981) (Figure 6.4). However, evidence has been also obtained that both $\delta^{15}N$-NH_4^+ and $\delta^{15}N$-NO_3^- will increase as the NH_4^+ reservoir is converted to NO_3^-, with $\delta^{15}N$-NO_3^- evolving towards the initial $\delta^{15}N$-NH_4^+ value (Clark, 2015). In general, it appears that the final $\delta^{15}N$ of NO_3^- derived via nitrification from manure-N would be more positive than that from fertilizer-N (Choi et al., 2003).

The magnitude of fractionation related to nitrification, denitrification and anammox processes is influenced by ambient conditions of hydrogeological systems where they occur, e.g. substrate concentration, availability of electron donors, concentration of dissolved oxygen, temperature, pH, residence time, etc. (Böttcher et al., 1990) (Figure 6.4).

In particular, it has been demonstrated that the size of the substrate pool (the amount of the chemical species which reacts with a reagent to generate a specific product) determines the extent of fractionation by minimizing it in N-limited systems and maximizing in systems with constant and high supply of N compounds (Li et al., 2007). For example, nitrification processes will be more intensive under the presence of a large amount of NH_4^+ (e.g. due to application of artificial fertilizers), which would likely cause considerable fractionation (Kendall, 1998). However, as the NH_4^+ pool is consumed, the overall nitrification fractionation gradually decreases. It has also been revealed that excessive concentrations of NO_3^- might induce a termination of denitrification with the formation of N_2O (Rivett et al., 2008). The threshold concentrations for the occurrence of this effect appear to be case-specific, since in some cases it has been reported that even low concentrations affected the ratio between produced N_2O and N_2. That is why it is essential to consider the initial concentration of the substrate in order to achieve more accurate conclusions concerning the production/consumption of NO_3^- and related changes in its isotopic composition.

Availability of electron donors is mostly discussed in the context of fractionation effects caused by denitrification. In general, it is suggested that denitrification may not play an important role in increasing $\delta^{15}N$ of NO_3^- under the conditions of low contents of electron donors (Choi et al., 2003). Electrons needed for denitrification can originate from the microbial oxidation of organic C or reduced S which might be present in water as the S^{2-} state in H_2S, S^- in FeS_2, S^0 in elemental sulfur, S^{2+} in thiosulfate ($S_2O_3^{2-}$) or S^{4+} in sulfite (SO_3^{2-}), (to the S^{6+} state as sulfate) (Rivett et al., 2008). To consider the potential impact of limited availability of electron donors on isotopic composition of NO_3^- it has been proposed to monitor their concentrations throughout the periods of observation of the ^{15}N isotopic signatures. For example, the presence of DOC in water has been used as an indicator of an available carbon source for denitrification. Moreover, concentrations of sulfate ion have also been measured to test for consistency with denitrifying environment (Kellman and Hillaire-Marcel, 2003). It should be mentioned that the amount of DOC has been shown to decrease in conjunction with an increase in sulfate concentration. This effect is related to the reduced solubility of DOC under conditions of increased ionic strength and acidity of water (Evans et al., 2006; Clark et al., 2005).

Concentration of dissolved oxygen (DO) in hydrogeological systems can also have a crucial impact on observed NO_3^- isotopic signatures. It may determine the type of N biochemical transformations occurring, which can alternatively lead either to decrease or increase of $\delta^{15}N$ of NO_3^-. As a common rule, the low content of oxygen is associated with denitrification reactions which lead to the increase of $\delta^{15}N\text{-}NO_3^-$. On the contrary, higher content of oxygen usually accompanies nitrification reactions which result in low $\delta^{15}N\text{-}NO_3^-$ values. From previous studies, it has become obvious that the occurrence of denitrification and nitrification processes could not be associated with clearly defined values (or narrowly constrained intervals) of DO concentrations. In particular, there is the range of DO concentration where both nitrification and denitrification can occur. For instance, denitrification cannot occur if the content of DO is above 0.2mg/L according to Feast et al., 1998, above 2mg/L according to Rivett et al. (2008) or above 4mg/L according to Baily et al. (2011). At the same time, it has been reported that the rate of nitrification reactions is maximized for a range of DO concentrations between 0.3mg/L and 4mg/L (Stenstrom and Poduska, 1980). However, the experimental evidence is not conclusive, as in some cases it has been determined that a dissolved oxygen concentration in excess of 4.0mg/L was required to achieve the highest nitrification rates (Stenstrom and Poduska, 1980). That is why, in order to be able to distinguish these two processes it is important to consider thoroughly the data about pH, availability of electron donors etc.

As the water temperature controls microbial activity and, consequently, DO content in groundwater, any seasonal changes could affect the $\delta^{15}N$ of NO_3^-, resulting in higher values of isotopic enrichment in the summer periods in aquifers where denitrification occurs, or lower values in groundwater influenced by nitrification activity. However, evidence about the impact of water temperature is not yet conclusive, as some reports suggested that $\delta^{15}N$-NO_3^- values might not exhibit seasonal trends (Danielescu and MacQuarrie, 2013). So it is essential to study microbial communities and distribution of potential denitrifying genera, as this will allow to get better insight into the nature of NO_3^- production/consumption processes and, in particular, into the impact of temperature on their dynamics (Hernández-del Amo et al., 2018).

The pH range is another important factor that affects the intensity of microbiological reactions and influences the magnitude of fractionation effect. It has been reported that pH ranging between 6.5 and 8 is the optimal range for nitrification, and reaction rates are likely to be significantly decreased below pH 6.0 and above pH 8.5 (Buss et al., 2004). Denitrification processes typically occur under a pH range be between 5.5 and 8, but the optimal pH is site-specific because of the effects of adaptation on the microbial ecosystems. Anammox activity is observed in a pH range from 6.5 to 9.3 with the optimum pH at 8 (Tomaszewski et al., 2017; Jin et al., 2012).

Furthermore, the hydrogeological structure of the area predetermines the processes of mixing of water derived from different sources and of different age. Therefore, it also profoundly affects the dynamics of $\delta^{15}N$ isotopic signature. Therefore, comprehensive analysis of $\delta^{15}N$-NO_3^- distribution in groundwater should be supported by in-depth consideration of hydrogeological features of the examined territories, for instance-the extent of confined and unconfined zones in the subsurface system, their connection and location of the recharge areas along the aquifer.

While studying variations of $\delta^{15}N$-NO_3^- in agricultural areas, it is particularly important to consider agricultural practices and the types of adjacent land uses, as they might significantly alter the isotopic signature of NO_3^- in groundwater samples. In agricultural areas where it is common to leave crop residues on the fields over the winter period it is necessary to consider the seasonality of NO_3^- sources. Previous studies which analyzed the influx of N from inorganic fertilizers into aquifer systems under intensive row-cropping and fertilization highlighted the significance of the intermediate N cycling processes of mineralization and nitrification of soil organic matter, such as crop residue, in the overall N cycling. Since resulting winter and spring load of NO_3^- is attributed to slow mineralization and nitrification during soil organic matter degradation, it is hard to identify precisely the source of NO_3^- in groundwater using its isotopic signature, since $\delta^{15}N$-NO_3^- values are close to those typical for fertilizers. Moreover, Sebilo et al. (2013) showed that the isotopic composition of NO_3^- in groundwater might be considerably influenced by mineralization of N fertilizers incorporated into the soil organic matter pool several decades ago. Therefore, the evidence regarding the dynamics of isotopic signatures should be supported by the expert knowledge about the local agricultural practices.

To summarize, the previous studies considered in this review have demonstrated that aquifers under agricultural areas are characterized with a wide range of $\delta^{15}N\text{-}NO_3^-$ determined by the variability of N sources and N transformation processes, intensity of which is controlled by the ambient geochemical conditions and hydrogeological settings.

In general, mineral fertilizers typically show the lowest $\delta^{15}N\text{-}NO_3^-$ values, followed by the isotopic signatures of soil-derived organic NO_3^-. The highest $\delta^{15}N\text{-}NO_3^-$ are commonly observed in animal manure or household sewage. Among the microbiological and physicochemical processes influencing isotopic composition of NO_3^- in groundwater, the highest $\delta^{15}N\text{-}NO_3^-$ values are associated with the denitrification activity. On the contrary, nitrification is responsible for the occurrence of NO_3^- with the ^{15}N isotopic signature on 5‰～35‰ lower in comparison to the ^{15}N of initial NH_4^+. While exploring the variability of ^{15}N in groundwater systems, it is important to account for possibilities of physical mixing of water of different origins and the impact of multiple environmental parameters on the intensity of transformation processes as they might lead to change in the isotopic signature of initial N pollutants.

6.3.2 $^{18}O\text{-}NO_3$

Combined use of the $\delta^{18}O$ and $\delta^{15}N$ of NO_3^- may allow better separation of atmospheric and terrestrial NO_3^- sources, including the possible separation of different anthropogenic sources. In addition, oxygen isotope ratios could be used for distinguishing N_2O originating from nitrification and denitrification (Kendall, 1998). The isotopic signature of $\delta^{18}O\text{-}NO_3^-$ in groundwater might vary in the range between −8.1‰ to +48‰, which reflects the variability of NO_3^- sources.

In particular, the isotopic signature $\delta^{18}O\text{-}NO_3^-$ could help to separate NO_3^- originated from the fertilizers application from NO_3^- inflow originating from other sources which deliver NO_3^- produced by nitrification of NH_4^+ or organic N. It is observed that synthetic NO_3^- fertilizers, which are derived from the atmospheric N_2, have $\delta^{18}O$ value close to the atmospheric value of +23.5‰ (Moore et al., 2006). Meanwhile, NO_3^- from other sources tend to have lighter $\delta^{18}O$ values because the NO_3^- derived from nitrification processes incorporates only one O atom from dissolved atmospheric O_2 and the other two atoms from water (Kendall and Aravena, 2000). In general, isotopic signature of $\delta^{18}O\text{-}NO_3^-$ originated from nitrification can be calculated using the following equation:

$$\delta^{18}O_{NO_3^-} = \frac{1}{3} \times \delta^{18}O_{O_2} + \frac{2}{3} \times \delta^{18}O_{H_2O} \tag{6.6}$$

Nitrification has been associated with the $\delta^{18}O\text{-}NO_3^-$ values in a range between −2‰ to +6‰ (Liu et al., 2006; Sebilo et al., 2006; Smith et al., 2006) or approximately 0‰ (Böhlke et al., 2006). However, it should be emphasized that the isotopic composition of

NO_3^- produced by nitrification depends on a range of factors which might alter those numbers: ① H_2O might be enriched in ^{18}O isotope because of evaporation (Hoefs, 2018; Sharp, 2007); ②O isotope fractionation during respiration can increase the $\delta^{18}O$ value of soil O_2 in comparison to that of atmospheric O_2 (Mayer et al., 2001); ③The ratio of O incorporation from H_2O and O_2 is not exactly 2 : 1 (e. g. more O_2 may be derived from atmospheric O_2 when NH_4^+ is limiting) (Knöller et al., 2011; Kool et al., 2011); ④Low pH conditions might support the occurrence of another microbial process that consume atmospheric O_2 more intensively than nitrification consequently resulting in suppression of nitrification (Liu et al., 2006); ⑤Oxygen isotope exchange of intermediates (especially NO_2) with ambient water might occur (Casciotti et al., 2010; Kool et al., 2011).

Oxygen isotopes can also be used to trace denitrification in groundwater, as ^{18}O and ^{15}N become concurrently enriched in the remaining NO_3^- during bacterial denitrification (Petitta et al., 2009). During denitrification, the isotopic signature of the residual $\delta^{18}O$-NO_3^- tends to be enriched by nearly 10‰ or 8‰～18‰ in comparison to the produced N_2O (Clark, 2015). Therefore, N_2O that is instantaneously produced is depleted in ^{18}O. According to Casciotti et al. (2002), the value of $\delta^{18}O$ is also affected by oxygen exchange with water, with the exchange ratio varying across different microbial species (Well et al., 2005). It is also important to take into account that the isotopic expression of $\delta^{18}O$-NO_3^- in groundwater might be influenced by atmospheric precipitation. Its $\delta^{18}O$ values can vary within an interval between +30‰ and +70‰ (Choi et al., 2003).

In general, it is clear that typical $\delta^{18}O$ values of NO_3^- originated from nitrification (including $\delta^{18}O$ values of NO_3^- derived from NH_4^+ in fertilizers and precipitation, NO_3^- derived from soil N and NO_3^- derived from manure and sewage) are lower than that of NO_3^- from precipitation and NO_3^- from application of fertilizers. Denitrification is responsible for the simultaneous enrichment of the remaining NO_3^- with ^{18}O and ^{15}N isotopes which might be traced in accordance to certain constant ratios. Therefore, application of O isotopes analysis along with N isotopes measurement can help to understand better the nature of $\delta^{15}N$ variability in groundwater.

6.3.3 ^{15}N-NH_4^+

During the transport of contaminants within the hydrogeological system the initial $\delta^{15}N$ values of NH_4^+ pollution sources can undergo considerable changes due to mineralization, sorption, volatilization, nitrification, anammox and dissimilatory NO_3^- reduction to NH_4^+ (DNRA). So far, significant research efforts have been devoted to estimation of fractionation effects of different processes which underlie the observed $\delta^{15}N$-NH_4^+ variability (Jin et al., 2012; Michener and Lajtha, 2007; Böhlke et al., 2006, Buss et al., 2004).

The conducted analysis showed that mineralization or ammonification usually causes only small fractionation (nearly ±1‰) between soil organic matter and soil NH_4^+ (Sharp, 2007). According to Michener and Lajtha (2007), the term mineralization might be used to describe the overall process of production of NO_3^- from organic matter, which usually involves several reaction steps. Under such definition, observed fractionation ranged from −35‰ to 0‰, depending on which step was considered as the limiting one (Michener and Lajtha, 2007). However, the results of such observations should be used cautiously, since such large and variable range might be attributed not to the mineralization step itself, but rather to nitrification of NH_4^+ to NO_3^- (Figure 6.5).

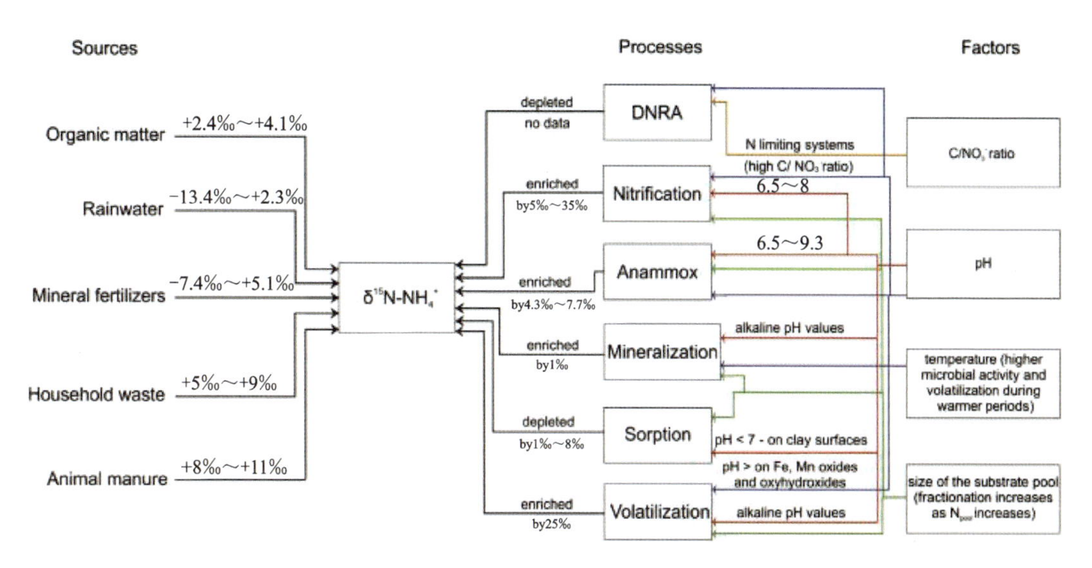

Figure 6.5 Sources, processes and factors that influence the $\delta^{15}N\text{-}NH_4^+$ values: summary (the following arrows connect processes with factors that have the decisive effect on their dynamics and, consequently, on resulting fractionation effects: →C/NO_3-ratio; →pH; →temperature; →size of the substrate pool).

Small isotopic fractionations have been reported for NH_4^+ sorption/desorption processes on charged surfaces of clays and other minerals. According to laboratory studies, NH_4^+ sorbed from solutions by clays commonly is enriched in ^{15}N relative to the NH_4^+ that remains in solution (Böhlke et al., 2006). Also, Hübner (1981) showed that ion exchange fractionations are commonly in the range of 1‰ to 8‰ and stated that the actual fractionation is dependent on concentration and the fractionation factor for the exchange with the clay material. According to Kendall (1998) the fractionation factor will probably vary with depth in the soil because of changes in clay composition and water chemistry (Kendall, 1998). These factors might retard or intensify sorption processes leading, respectively, to enrichment or depletion of $^{15}N\text{-}NH_4^+$ in groundwater.

Volatilization is a highly fractionating process in which the produced NH_3 gas has a lower $\delta^{15}N$ value than the residual NH_4^+. It involves several steps that cause fractionation, including: ①Equilibrium fractionation between NH_4^+ and NH_3 in solution, and between aqueous and gaseous NH_3; ②Kinetic fractionation caused by the diffusive loss of ^{15}N depleted NH_3. In general, the overall dynamics of the process leads to the enrichment of the remaining NH_4^+ in ^{15}N on the order of 25‰ in comparison to the volatilized NH_3. However, it is noticed that the actual fractionation could depend on the pH and temperature (Bedard-Haughn et al., 2003).

Nitrification of NH_4^+ is a two-step process which yields ^{15}N-depleted products and commonly results in a substantial increase of $\delta^{15}N$-NH_4^+ value. The oxidation of NH_4^+ to NO_3^- enriches the remaining NH_4^+ by approximately 30‰ in comparison to produced NO_3^-. In general, the total fractionation associated with nitrification depends on which step is rate determining. Because the oxidation of NO_2^- to NO_3^- is rapid in natural systems, this step is usually not considered as the rate-determining one, and most of the observed N fractionation is caused by the slower oxidation of NH_4^+ to NO_2^- (Michener and Lajtha, 2007). The extent of fractionation during nitrification is also evidently dependent on the fraction of the substrate pool that is consumed during the process.

Anammox or anaerobic oxidation of NH_4^+ to N_2 leads to a slight enrichment of the residual NH_4^+ by 4‰ ~ 8‰ (Clark, 2015). The low fractionation effect of anammox process, usually observed during field studies, could probably be caused by the presence of greater reservoir of NH_4^+ sorbed on the aquifer that buffers the enrichment of $\delta^{15}N$ in the dissolved NH_4^+ in the explored cases (Clark, 2015). So far, the anammox process was detected mostly within the long pollution plumes (i. e. from several hundred meters to 1km in length) originating from point pollution sources (septic tanks, industrial or residential effluents). For example, Böhlke et al. (2006) explored anammox activity in the contaminated groundwater plume created by land disposal of treated wastewater which appeared at the location of Cape Cod (Massachusetts, USA). Similarly, Robertson et al. (2012) explored the possibilities for occurrence of anammox conditions in a septic system plume originating from the washroom facility located on the north shore of Lake Erie (between USA and Canada).

Since it has been discovered that, under anaerobic conditions, NO_3^- may also be reduced to NH_4^+ by a process known as DNRA, it is necessary to consider its potential impact on $\delta^{15}N$-NH_4^+ as well. In general, this process occurs under the same conditions as denitrification, but is less commonly observed in practice. While, to the best of our knowledge, the reports devoted exclusively to the investigation of the N isotope fractionation occurring during DNRA are yet not available, broader studies conducted so far have demonstrated that NH_4^+ produced by DNRA has much lower $\delta^{15}N$ that the substrate

NO_3^-, which suggests an ongoing kinetic fractionation (Michener and Lajtha, 2007).

The extent of fractionation effect caused by NH_4^+ transformation processes depends on multiple environmental factors (Figure 6. 5) which, therefore, can substantially influence the observed dynamics of $\delta^{15}N$ values of NH_4^+ in the subsurface. Among these factors, pH, temperature and size of the substrate pool are the ones most discussed in the available research literature.

The pH parameter defines the intensity of not only microbiological reactions, but also affects the rate of volatilization: it is proved that this process is intensified under the alkaline soil pH. For this reason, the observed high rates of NH_3 volatilization are associated with the high carbonate content of soils (Bedard-Haughn et al., 2003). For example, in the unconfined High Plains aquifer (USA) NH_3 volatilization was promoted by the calcareous soils of the area (McMahon and Böhlke, 2006). At the same time, the pH values which support the development of DNRA are unclear. Some studies indicated that high rates of DNRA are associated with alkaline conditions, while the other ones revealed the negative correlation between DNRA occurrence and pH parameter. As for N mineralization process, it tends to become more intensive with an increase of pH values towards more alkaline range. At $pH<7$, NH_4^+ is predominantly sorbed on clay surfaces, and at higher pH values it starts to be sorbed by metal oxides and oxyhydroxides (e. g., $FeOOH$, MnO_2) (Buss et al., 2004).

The temperature variability can also have an impact on the changes in dynamics of $\delta^{15}N$-NH_4^+ values. It should be particularly noticed that higher temperatures are also associated with the increasing rate of ongoing NH_3 volatilization, since they stimulate growth and activity of bacteria. Consequently, it can be expected that the isotopic composition of N species exhibits pronounced seasonal patterns (Bedard-Haughn et al., 2003). The optimal temperature range for mineralization is 25～40℃, for nitrification −15～35℃ and for anammox −30～40℃ (Li et al., 2014; Guntiñas et al., 2012; Jin et al., 2012).

In addition, the extent of observed fractionation effects is assumed to be dependent on the size of the substrate pool (reservoir). Usually, in Nlimited systems, fractionation associated with nitrification is comparatively small. For instance, NH_4^+ concentration in groundwater of the Sichuan Basin in China were low [and even occasionally below the detection limit (0. 05mg/L)], suggesting minimal isotopic fractionation during nitrification in groundwater (Li et al., 2007).

Finally, it should also be noticed that the relative concentrations of NO_3^- to organic C (C/NO_3^- ratio) control whether NO_3^- is reduced by denitrification or DNRA. In general, DNRA, which leads to the production of isotopically depleted NH_4^+, is favored when NO_3^- is limiting, while denitrification is favored when C (electron donor) is limiting.

The presented evidence suggests that the variability in the $\delta^{15}N-NH_4^+$ in groundwater heavily depends both on the type of pollution sources as well as on the dynamics of microbiological and physicochemical processes (Figure 6.5). In general, $\delta^{15}N-NH_4^+$ values in groundwater are lower and less variable in comparison to $\delta^{15}N-NO_3^-$, which is probably explained by the high sorption potential of NH_4^+ and it intensive involvement into oxidation processes. Among the pollution sources, animal wastes and household sewage contribute to the highest enrichment of NH_4^+ in groundwater with ^{15}N isotope. As for the processes resulting in isotope fractionation and respective changes in isotopic signatures of groundwater samples, it is revealed that volatilization and nitrification significantly contribute to higher accumulation of ^{15}N in the residual NH_4^+. However, the extent of fractionation effects due to these processes may depend on the environmental conditions. On the contrary, mineralization and sorption usually show small isotopic effects. Finally, there is still not much evidence available about the quantitative alterations in the isotopic composition of NH_4^+ during DNRA (Michener and Lajtha, 2007).

6.3.4 $^{15}N-N_2O$

The experimental evidence suggests that changes in N_2O isotopic signatures are caused by both physical and microbial processes. It is generally assumed that the enrichment factors of microbial processes tend to be large than those related to physical processes (Goldberg et al., 2008). Among the bacterial transformations, denitrification, nitrification and nitrifier denitrification are the processes that seem to be the most discussed in the research literature in the context of the isotopic composition of $\delta^{15}N-N_2O$ (Jurado et al., 2017; Well et al., 2012). As for the impact of physical processes, it appears that diffusion frequently might be responsible for the alterations of detected $\delta^{15}N-N_2O$ values.

In the denitrification pathway, N_2O is produced as well as consumed during the subsequent reduction of NO_3^- to N_2 ($NO_3^- \rightarrow NO_2^- \rightarrow NO \rightarrow N_2O \rightarrow N_2$) (Figure 6.6). The $\delta^{15}N$ values of N_2O derived from denitrification depends upon the isotope fractionation during its production and consumption. N_2O originated from the reduction of NO_3^- is typically depleted in ^{15}N in comparison to the initial substrate (NO_3^-). The reduction of N_2O to N_2 results in the enrichment of the residual N_2O. It is reported that the isotope fractionation factors for N during both processes are of comparable order of magnitude. If N_2O is accumulated as the intermediate product of steady-state denitrification, it is observed that, its $\delta^{15}N$ value should become close to the value of the initial substrate NO_3^-. Correspondingly, significant N isotope discrimination between N_2O and NO_3^- in groundwater might suggest that a large portion of N_2O may originate from nitrification. Nitrification, which is also a multistep reaction ($NH_3/NH_4^+ \rightarrow H_2NOH \rightarrow NO_2^- \rightarrow NO_3^-$), yields N_2O which is isotopically light in comparison to its precursors. N_2O derived during this process could be produced as a byproduct from the complete or partial direct oxidation of H_2NOH to NO or N_2O (Schmidt et al., 2004) (Figure 6.6).

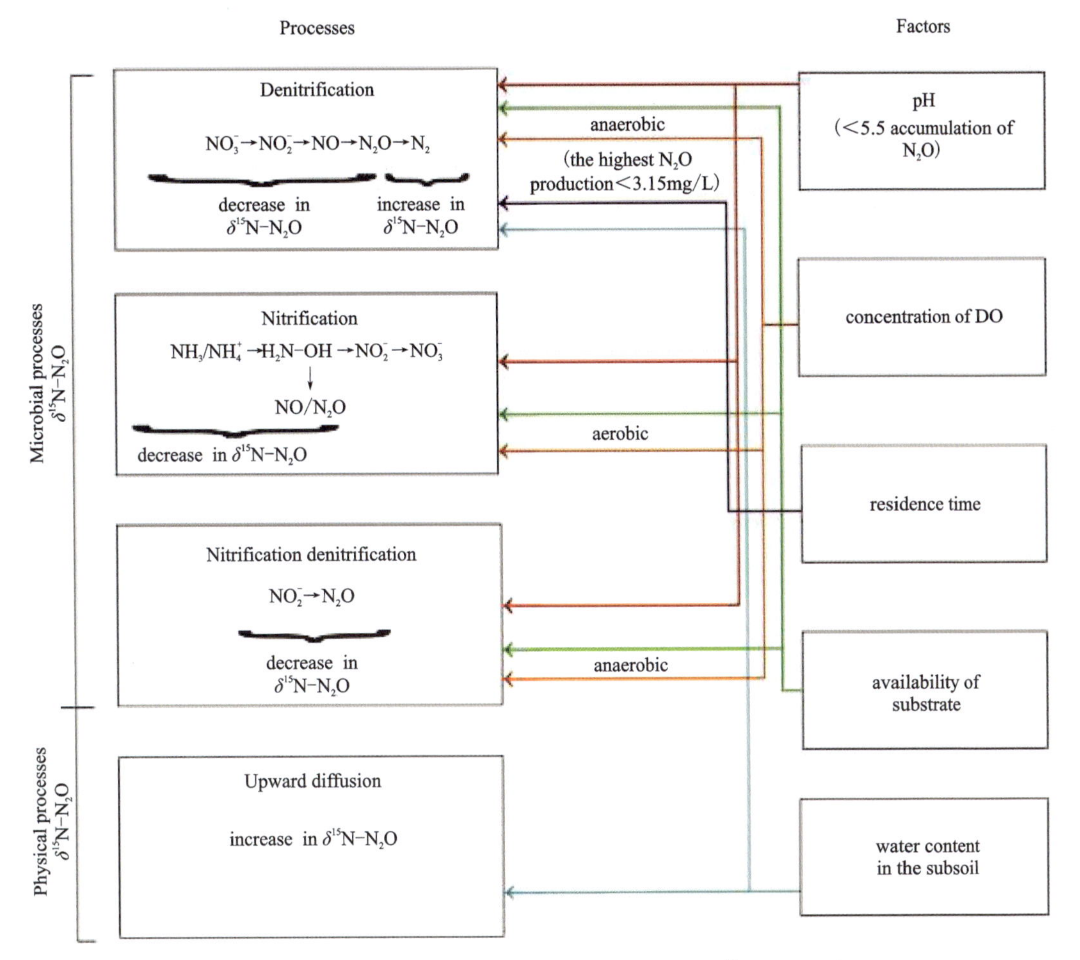

Figure 6.6 Sources, processes and factors that influence the $\delta^{15}N-N_2O$ values: summary (the following arrows connect processes with factors that have decisive effect on their dynamics and, consequently, on resulting fractionation effects: → water content in the subsoil; → availability of substrate; → residence time; → concentration of DO; → pH).

In addition, at low DO level, N_2O production is likely to proceed via nitrifier denitrification, i.e. NO_2^- reduction to N_2O, which yields isotopic signatures similar to bacterial denitrification (Well et al., 2012). Consequently, these two processes cannot be distinguished using solely the data regarding ^{15}N isotope natural abundance, and additional evidence is necessary (Well et al., 2012).

The isotopic composition of N_2O detected in the groundwater samples can also be significantly influenced by its upward diffusion and volatilization from shallow groundwater to the atmosphere (Minamikawa et al., 2011). Available experimental data indicate that in the subsoil environment characterized with high diffusivity exchange with atmospheric N_2O may diminish the effects of isotopic fractionations expected from the previously described microbial processes (Goldberg et al., 2008). The rate of occurring diffusion depends mainly on the water content in the subsoil. The higher water content suggests that the time required for N_2O to diffuse from the soil profile to the surface is also increased, since diffusion of N_2O in water is approximately 4 orders of magnitude lower than in air (Clough et al., 2005). In addition, it should be highlighted that the macropores and cracks can also enhance the upward N_2O diffusion (Minamikawa et al., 2011).

To summarize, the research accomplished so far has demonstrated that both nitrification and denitrification processes are responsible for the depletion of ^{15}N value of N_2O in comparison to its substrates (Toyoda et al., 2017; Schmidt et al., 2004). However, further reduction of N_2O to N_2 during denitrification leads to the enrichment of the remaining N_2O with ^{15}N (Clark, 2015; Knöller et al., 2011). In comparison to biochemical processes occurring in aquifers, diffusion usually results in less pronounced isotopic effects. However, the distribution of the $\delta^{15}N$-N_2O values in groundwater cannot be comprehensively analyzed and clearly interpreted without referring to the heterogeneity of environmental factors of the studied hydrogeological systems.

Among the factors controlling the dynamics of N_2O production/consumption processes and resulting variations in $\delta^{15}N$-N_2O values, the residence time, DO concentration, availability of substrate and pH are typically considered as the most decisive in the literature.

As the concentration of NO_3^- within a denitrifying layer diminishes with increasing residence time of groundwater, it appears, that with longer residence time, NO_3^- reduction to N_2 is more likely to be complete (provided the is no additional supply of NO_3^- and a sufficient amount of electron donors), which means that the isotopic compositions of $\delta^{15}N$-NO_3^- and $\delta^{15}N$-N_2O become closer. At the same time, the instantaneously produced N_2O is typically depleted with respect to the NO_3^- signature (Well et al., 2005).

The DO concentration significantly impacts the isotopic signatures of N_2O in groundwater, because it determines the type of dominant microbial processes in the aquifer and it also affects the completeness of their reaction steps. In particular, under anaerobic conditions, microbial nitrification is unlikely to occur, at least the groundwater table (Goldberg et al., 2008), and denitrification usually prevails under such conditions. In particular, it is reported that denitrification might yield the highest N_2O amounts at intermediate O_2 concentrations (below 3.15mg/L to 4mg/L) as most denitrifiers are facultative anaerobes (Deurer et al., 2008). That is why it is frequently reported that the NO_3^- consumption, which is associated with the formation of excess N_2 and intermediate accumulation of N_2O, increases with the depth (Well et al., 2012).

In sequential reaction processes, such as denitrification, the supply of the members of the denitrification pathways, i. e., NO_3^-, NO_3^-, NO, N_2O, N_2, depends on the rate of previous reaction steps, except for NO_3^- which can be introduced to the system from the external sources. The availability of substrate, therefore, seems to have considerable impact on the magnitude of isotopic fractionation occurring during N_2O production/consumption processes. In particular, if NO_3^- supply is high in relation to reduction capacity of the subsurface system, substantial isotope fractionation effect occurs, whereas the effect is low or negligible in the opposite case. Overall, the same fractionation control principle appears to be relevant for the other N species subject to reduction during further stages of denitrification, namely NO_3^-, NO, and N_2O. However, for these species the situation is even more complicated, not only because their respective pool sizes depend on the rates of the previous reactions, but also because some microbes might lack enzymes for some of the reduction steps, which implies that transport within denitrifying species will be a necessary precondition for further reduction in such cases (Well et al., 2005). As a result, the isotopic signature of N_2O as an intermediate is influenced both by the kinetics of its production during NO reduction and consumption during N_2O reduction to N_2 affected by the availability of reaction substrates on the corresponding transformation steps.

It has been found that pH values below 5.5 seem to promote accumulation of N_2O, most probably because N_2O reductase is mostly inhibited by acid conditions that enable the build-up of N_2O in the subsurface environment, and the denitrification process does not proceed to the final step. Overall, since N_2O is an intermediate product of microbial reactions, its isotopic composition is determined by the rates of previous reactions as well as biological and physicochemical conditions of the aquifer (Figure 6.6). It could be summarized that production processes of N_2O (e. g., nitrification, denitrification, etc.) lead to its depletion in the $\delta^{15}N$ value, whereas consumption processes, such as reduction of N_2O to N_2, enrich it with ^{15}N. Residence time, DO concentration, substrate availability and pH are important parameters that affect the intensity of N_2O isotope fractionation processes. The large variability of $\delta^{15}N$ value of N_2O in the groundwater implies that N_2O production and consumption processes in the hydrogeological system occur simultaneously. However, the isotopic fractionation effects of these processes might be diminished by the effects of upward diffusion.

6.4 Variations of Nitrogen isotopes in groundwater

The isotopic signatures of N species (NO_3^-, N_2O, NH_4^+) in groundwater under agricultural lands exhibit different ranges depending on variability of N sources, transformation processes and migration pathways (Well et al., 2012; Liu et al., 2006). In the cases when observed isotopic signatures of NO_3^-, N_2O, NH_4^+ in groundwater are simultaneously influenced by multiple sources and occurrence of several N-cycle processes, interpretation of $\delta^{15}N$ values demands thorough attention.

6.4.1 Variability of $\delta^{15}N$-NO_3^- in groundwater and its sources

The isotopic signature of $\delta^{15}N$-NO_3 in groundwater under agricultural areas shows a considerably wide range from −8.3‰ to +65.5‰ (Figure 6.7), depending on the heterogeneity of N sources, geochemical conditions and groundwater flow patterns as well as on the peculiarities of agricultural practices in the explored regions.

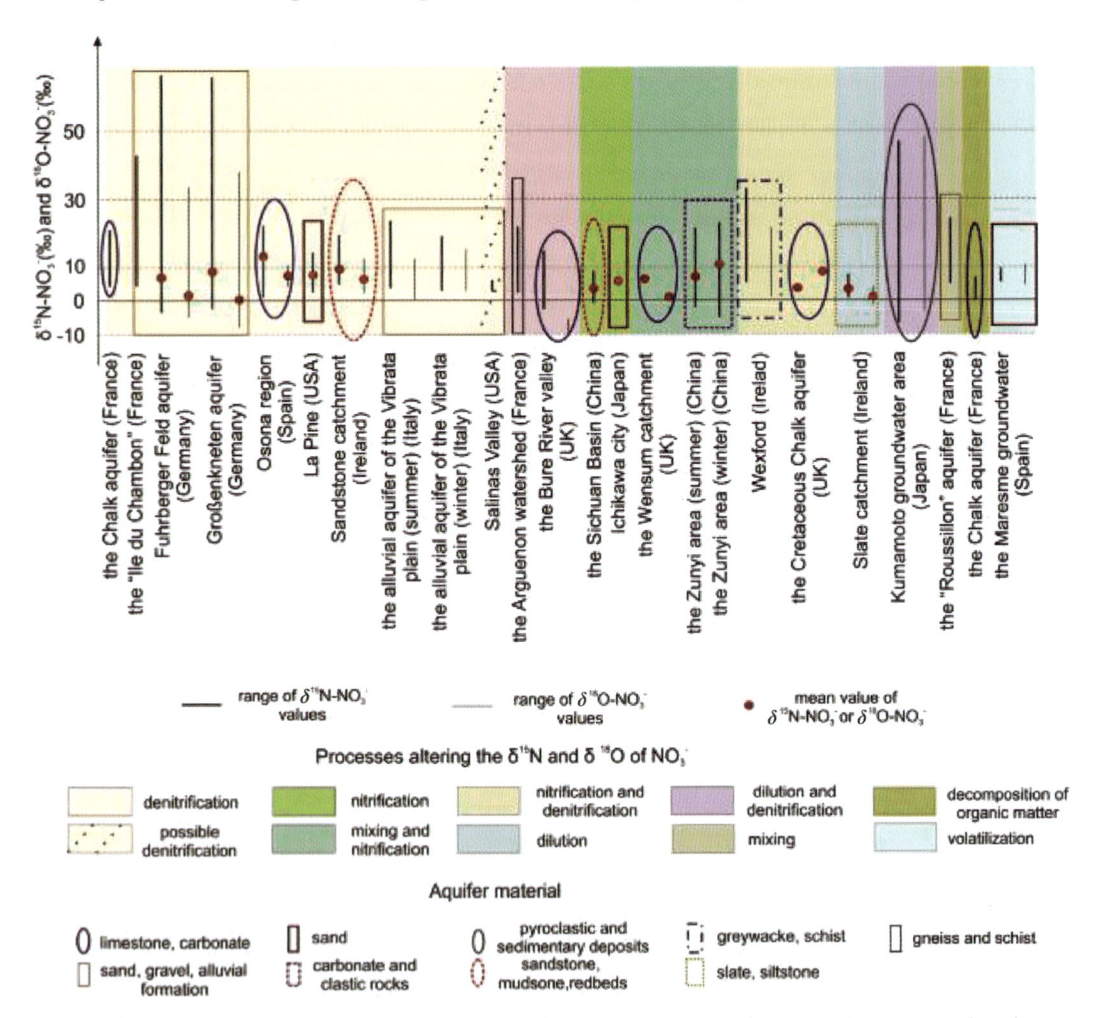

Figure 6.7 $\delta^{15}N$-NO_3^- isotopic signatures in groundwater: a summary of case studies in agricultural areas.

The observed inflow of N into groundwater in agricultural areas can be attributed to multiple sources such as organic and inorganic fertilizers, manure, soil organic N, sewage

(e. g. septic wastewater), and atmospheric precipitations. N originating from each source is characterized with distinct intervals of $^{15}N\text{-}NO_3^-$ enrichment values (Figure 6. 4), which can be used to determine the origin of observed NO_3^- and estimate the relative contribution of NO_3^- sources to its content in the groundwater.

In particular, it has been observed that the organic and inorganic fertilizers are characterized with different isotopic signatures, which is explained by their production processes. For example, synthetic fertilizers, such as urea or NH_4^+ and NO_3^- fertilizers, are usually produced by fixation of atmospheric N_2 which has $\delta^{15}N$ of $0 \pm 3‰$ (Kendall, 1998). This process only slightly fractionates the isotope composition resulting in low $\delta^{15}N$ range of inorganic fertilizers, from $-4‰$ to $+4‰$ (Sharp, 2007), $-8‰$ to $+7‰$ (Kendall, 1998) or $-6‰$ to $+6‰$ (Xue et al. , 2009). However, in groundwater, this typical isotopic composition of inorganic fertilizers frequently changes because of N isotope fractionation during various physicochemical or biochemical reactions (e. g. , NH_3 volatilization, nitrification or denitrification).

In line with these suggestions, further studies demonstrated that the $\delta^{15}N\text{-}NO_3$ in groundwater of cropping areas with mineral fertilizer application may be in the range of $+4.5‰ \sim 8.5‰$ (Choi et al. , 2007) or $-7‰ \sim 5‰$ (Danielescu and MacQuarrie, 2013). At the same time, organic fertilizers, such as plant compost or liquid and solid animal waste, generally are characterized with higher initial $\delta^{15}N$ values and a broader range of isotopic composition ($6‰ \sim 30‰$) than inorganic fertilizers. This is explained by the processes occurring in animal wastes such as excretion of isotopically light N in urine and accumulation of heavy ^{15}N isotope in the residual waste as well as volatilization of ^{15}N depleted ammonia with subsequent oxidation of the residual waste (Sharp, 2007).

In comparison to both organic and inorganic fertilizers, NO_3^- produced by nitrification of manure-N has higher $\delta^{15}N\text{-}NO_3^-$, since during its storage, treatment and application, the volatilization of NH_3 causes significant enrichment of ^{15}N in the residual NH_4^+, while most of this NH_4^+ is subsequently oxidized to ^{15}N-enriched NO_3 (Widory et al. , 2004). Consequently, $\delta^{15}N$ values of NO_3^- originating from manure usually range between $5‰ \sim 25‰$ (Xue et al. , 2009), $10‰ \sim 22‰$ (Bateman et al. , 2005), $5‰ \sim 35‰$ (Widory et al. , 2005).

Soil organic-derived NO_3^- is a product of bacterial decomposition of organic matter originated from degradation of plants and animal wastes. The $\delta^{15}N\text{-}NO_3^-$ of soil NO_3^- may be between $+3‰$ and $+8‰$ (Kendall and Aravena, 2000). It is also particularly important to consider, in groundwater polluted by fertilizers, the possible mixing of N originating from the addition of fertilizers and N mineralized from soil organic matter which might not be taken up by crops if their demands are already satisfied (Li et al. , 2007). On the contrary, the isotopic signature of NO_3^- originated from animal or sewage waste is commonly less influenced by interaction with soil N because the distribution of waste is often localized at point sources with high concentrations. In some cases, the observation of the distribution of point and non-point sources of pollution can help to identify the origin of NO_3^- more precisely.

Another significant source of NO_3^- in groundwater under agricultural lands is household sewage whose $\delta^{15}N$-NO_3^- range vary between +4‰ and +19‰ (Xue et al., 2009). In many cases, experimental studies have revealed similar ranges of $\delta^{15}N$ for both animal manure and sewage. Consequently, it is often difficult to determine exactly the origin of NO_3^- in areas characterized with simultaneous occurrence of groundwater pollution from livestock manure and household wastes.

The amount of N contained in atmospheric precipitation is influenced by several factors: volatilization of NH_3, nitrification and denitrification occurring in the soils and the impact of various anthropogenic sources. In general, the $\delta^{15}N$-NO_3^- composition of rain is higher than that of the co-existing $\delta^{15}N$-NH_4^+ (Bedard-Haughn et al., 2003). The $\delta^{15}N$-NO_3^- isotopic signature of rain might vary between −10‰ and +9‰ based on various case studies (Sharp, 2007), −11.8‰ and +11.4‰ reported for eastern Canada (Savard et al., 2010), −10.2‰ and −4.4‰ reported for central China (Li et al., 2007).

This overview demonstrates that the sources of NO_3^- pollution are characterized with relatively different $\delta^{15}N$-NO_3^- isotope ranges: rain water from −12‰ to +11‰, inorganic fertilizers from −8‰ to +7‰, organic fertilizers from +6‰ to +30‰, soil organic matter from +3‰ to +8‰, manure from +5‰ to +35‰, and household sewage from +3‰ to +25‰. The lowest values of $\delta^{15}N$-NO_3^- are typical for inorganic fertilizers followed by NO_3^- derived from soil organic matter, while the highest values are usually related to the impact of manure or household wastes, both of which may overlap. However, the isotope composition of NO_3^- from different sources might be subject to considerable alterations due to fractionation processes occurring under certain biochemical or physicochemical reactions during the migration to or within the aquifer.

6.4.2 Variability of $\delta^{15}N$-NH_4^+ in groundwater and its sources

In comparison to the amount of information regarding $\delta^{15}N$-NO_3^- in groundwater under the agricultural areas, the data about distribution of $\delta^{15}N$-NH_4^+ are less abundant. In general, conducted studies revealed that the $\delta^{15}N$ values of NH_4^+ in aquifers cover the range from −8.5‰ to +23.8‰, being significantly lower than the corresponding $\delta^{15}N$ values of NO_3^- (Li et al., 2007; Hinkle et al., 2007; Liu et al., 2006).

Overall, fertilizers, manure and sewage effluent are the principal anthropogenic sources of the NH_4^+ in groundwater under agricultural areas. Rainwater and organic matter may also substantially contribute to NH_4^+ concentration in groundwater (Hinkle et al., 2007). The comparison of $\delta^{15}N$-NH_4^+ values of different pollution sources with the isotopic signatures of groundwater samples is widely used for identification of the origin of detected NH_4^+.

NH_4^+ fertilizers usually have $\delta^{15}N$ values of 0‰ or lower (Kendall, 1998). Available data provide the following ranges: from −1.5‰ to −0.7‰ (Wassenaar, 1995); from −7.4‰ to +3.6‰ (median value −0.6‰) (Vitòria et al., 2004); from +2.7‰ to +5.1‰ (mean value +4.2‰±0.8‰) (Li et al., 2007); −3.9‰±0.3‰ (Choi et al., 2007), −0.91‰±1.88‰ (Kendall, 1998). In general, the isotopic signature of $\delta^{15}N$-NH_4^+ is reported to be 2.5‰ lower than the isotopic signatures of $\delta^{15}N$-NO_3^- of synthetic fertilizers.

Application of manure in agricultural fields or animal waste effluents from farms might increase the isotopic signature of $\delta^{15}N$-NH_4^+ in the groundwater located under such areas in comparison to the aquifers effected by the fertilizer use, as animal waste is characterized by higher level of $\delta^{15}N$ enrichment of NH_4^+ (Figure 6.4). It appears that the higher $\delta^{15}N$ values observed in animal wastes are related to the increase in $\delta^{15}N$ by 3‰~4‰ at each successive trophic level (step in a nutritive series, or food chain, of an ecosystem). The most important factor contributing to this increase is the excretion of isotopically light urine: animal waste gets further enriched in ^{15}N by the subsequent volatilization of isotopically light NH_3 (Sharp, 2007). The initial $\delta^{15}N$-NH_4^+ values of manure may vary between +8‰ and +10‰ for pig waste (Vitòria et al., 2004) and around +7.4‰±3.8‰ for cow waste (Maeda et al., 2016).

NH_4^+ is also one of the major components in groundwater contamination plumes originating from septic tank effluents or wastewater release from treatment plants. In untreated sewage, the isotopic signature of $\delta^{15}N$-NH_4^+ is typically between +5‰ and +9‰ (Cole et al., 2006). The sewage effluent in Guiyang (China) showed the mean value of $\delta^{15}N$-NH_4^+ at +5.3‰ (Liu et al., 2006), and Robertson et al. (2012) detected thc $\delta^{15}N$-NH_4^+ value of +4.4‰±4.6‰ in the septic system of the Long Point campground located on the shore of Lake Erie (USA and Canada). Usually, the contamination plumes exhibit clear stratification between the differently enriched NH_4^+ species. The top of the plume is typically characterized with more enriched $\delta^{15}N$-NH_4^+ values, caused by ongoing nitrification, in comparison to the core of the plume, where NO_3^- and NH_4^+ coexist and anammox reaction enriches both compounds, and below plume where only NO_3^- attenuated by denitrification remains (Clark, 2015).

NH_4^+ is also the most abundant N compound in rainwater which commonly exhibits negative $\delta^{15}N$ values. In particular, experimental data provided by Li et al. (2007) in the Sichuan river basin (China) showed that $\delta^{15}N$-NH_4^+ in atmospheric precipitation vary from

−13.4‰ to +2.3‰ (mean value −6.6‰ ± 4.0‰). Isotope analyses conducted on rainwater samples from Zunyi in China, also demonstrated negative (approximately −12‰) $\delta^{15}N\text{-}NH_4^+$ values (Li et al., 2010). The inflow of NH_4^+ originating from decomposition of organic matter in sediments and soils may also influence the isotopic signature of $\delta^{15}N\text{-}NH_4^+$ in groundwater. In general, $\delta^{15}N\text{-}NH_4^+$ in soil or sediments usually differs from the isotopic composition of total organic N in such samples only by ±1‰ (Kendall, 1998). This is explained by the small magnitude of fractionation effect occurring during mineralization of organic matter.

To sum up, the most negative values of $\delta^{15}N\text{-}NH_4^+$ could be observed in rainwater, while the highest positive isotopic signatures are typical for animal manure and sewage. At the same time, organic matter exhibits slightly higher $\delta^{15}N\text{-}NH_4^+$ isotopic composition in comparison to synthetic fertilizers. However, the available experimental evidence also suggests that in practice the isotopic signals of various NH_4^+ sources might overlap due to the peculiarities of environmental settings in certain areas.

6.4.3 Variability of $\delta^{15}N\text{-}N_2O$ in groundwater

The information about the isotopic composition of $\delta^{15}N\text{-}N_2O$ in aquifers affected by agricultural activity is also scarce, as in the case of data regarding the natural abundance of $^{15}N\text{-}NH_4^+$. In general, it has been reported that the values of $\delta^{15}N\text{-}N_2O$ could vary from −55.4‰ to +89.4‰. So the isotopic signatures of N_2O in groundwater samples demonstrate the largest variability among different isotopic compositions of N compounds considered in this review. It appears that such wide range of observed $\delta^{15}N\text{-}N_2O$ values is related to the fact that the production of N_2O involves many reactions steps which presume diverse fractionation effects depending on chemical processes kinetics and heterogeneous conditions of the subsurface environment along the vertical and lateral groundwater flow paths. Evidently, it also reflects the impact of the diversity of isotopic signatures of the initial substrates (e.g., NO_3^-, NH_4^+) and their involvement into microbial processes. In particular, according to previous studies, $\delta^{15}N$ values of N_2O emitted from fertilized soils are predominantly negative, which is explained by ^{15}N depletion during N_2O production by nitrification and denitrification. At the same time, positive $\delta^{15}N\text{-}N_2O$ values are likely to be attributed to ongoing N_2O reduction during denitrification (Well et al., 2005). Further discussion of the factors influencing variability of $\delta^{15}N\text{-}N_2O$ in groundwater will be devoted predominantly to shifting dynamics of various hydrobiogeochemical processes that affect the isotopic composition of N_2O.

6.5 Application

6.5.1 Nitrate source apportionment in groundwater using Bayesian isotope mixing model based on nitrogen isotope fractionation (Yu et al., 2020)

Here, we take the Dagu River groundwater reservoir located in the Qingdao, Shandong Province, China as an example to present the application of nitrogen isotopes on the identification of groundwater nitrogen sources.

6.5.1.1 Study area

The Dagu River is the main water source for surrounding agriculture and industry uses in Qingdao (Figure 6.8). The study area contains a shallow aquifer and strong permeability because of the thinner overlying sticky sand layer. Although infiltration is beneficial to the recharge of groundwater, it also increases the risk of leaching soluble pollutants. Groundwater recharge in the region mainly occurs via precipitation infiltration, surface water infiltration, agricultural irrigation, and lateral runoff recharge. The land use of the study area includes agricultural land, residential areas, forests, water, roads, and grasslands with area proportions of 70.95%, 17.13%, 5.69%, 5.23%, 0.81%, and 0.19%, respectively (Figure 6.8). Therefore, the study area was divided into the vegetable cultivation area (VCA) and grain cultivation area (GCA).

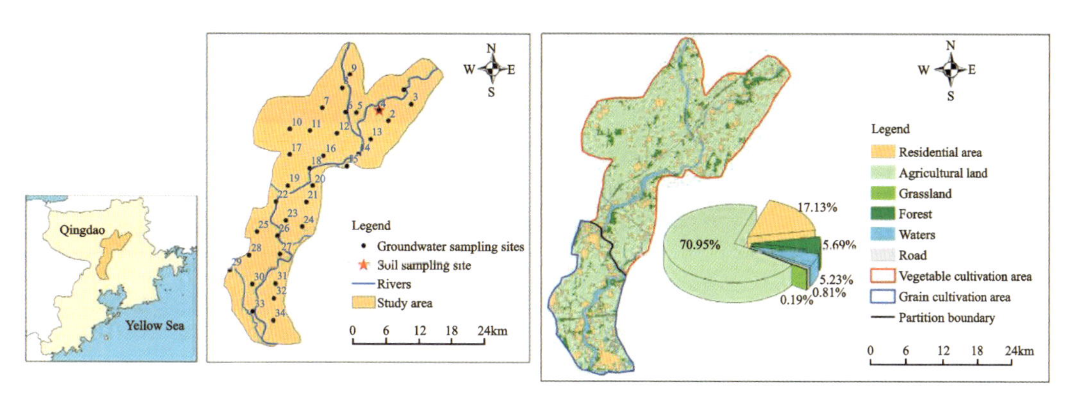

Figure 6.8 Location of the Dagu River groundwater reservoir and sampling sites of groundwater and soil sample, and the pattern and proportion of land use in the study area.

A total of 34 groundwater samples from representative wells of the study area were collected in March 2019, including 24 sites 1～24 in the VCA and 10 sites 25～34 in the GCA (Figure 6.8). The potential sources in the study area, containing atmospheric deposition (AD), soil nitrogen (SN), chemical fertilizers (CF), and manure and sewage (M&S), were sampled and analyzed. Six precipitation samples (both rain and snow) were collected. Eight soil samples (0 ~ 20cm) were collected from fallow farmlands, near the groundwater sampling sites 4, 7, 14, 17, 20, 23, 28, and 33, respectively. Eight CF samples, including ammonium, urea, and nitrogenous compound fertilizers, were collected from local farmers and markets. Fourteen M&S samples were collected from local farmers and residential areas, including cow manure, pig manure, sheep manure, chicken manure, goose manure, and domestic sewage.

6.5.1.2 Isotope characteristics in groundwater and nitrate sources

The NO_3^--N concentrations ranged from 0.37mg/L to 172.57mg/L with a mean value of 61.56mg/L. In particular, only the NO_3^--N concentrations in sampling sites 7, 20, 33, and 34 were lower than 5mg/L. About 88% of the groundwater samples exceeded 《the groundwater quality standard of China》 (GB/T 14848—2017) for drinking water ($\leqslant$ 20mg/L). This indicated that most groundwater in the study area had been seriously contaminated with NO_3^-. As shown in Figure 6.9, there was significant spatial variation in the NO_3^--N concentrations under different cropping system. The NO_3^--N concentrations in the VCA were distinctly higher than those in the GCA. In the VCA, the NO_3^--N concentrations ranged from 2.62mg/L to 172.57mg/L with a mean value of 76.03mg/L. In the GCA, the NO_3^--N concentrations ranged from 0.37mg/L to 38.57mg/L, except for that in sampling site 30 (64.73mg/L). The average NO_3^--N concentration in the VCA was 2.8 times higher than that in the GCA (26.82mg/L).

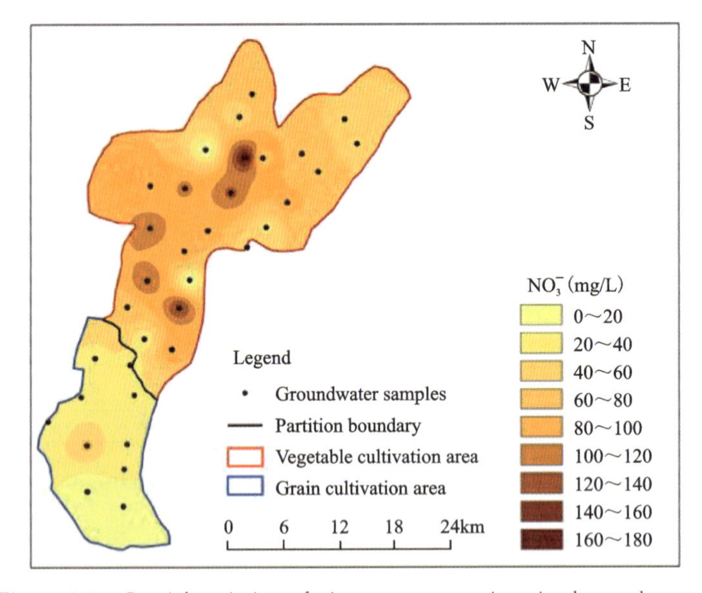

Figure 6.9 Spatial variation of nitrate concentrations in the study area.

The $\delta^{15}N$ and $\delta^{18}O$ values of potential sources in the study area were shown in Table 6.2. The isotopic distributions of AD, SN, and CF in the samples were all within the typical range summarized in the literature. However, the $\delta^{15}N$ values of M&S were lower than the literature values, which was because the manure samples were fresh livestock feces collected from local farms.

Table 6.2 The $\delta^{15}N$ and $\delta^{18}O$ values of various nitrate sources.

Source	$\delta^{15}N$ (‰)				$\delta^{18}O$ (‰)			
	Min	Max	Mean±SD	Literature	Min	Max	Mean±SD	Literature
AD	−5.93	−2.76	−4.56±1.19	−12~11	56.29	70.96	64.75±5.99	25~75
SN	3.45	5.01	4.48±0.70	0~8	2.71	3.12	2.85±0.16	−10~15
CF	−2.51	3.36	−0.29±1.96	−8~7	2.71	3.12	2.85±0.16	−10~15
M&S	0.06	6.32	2.52±1.92	3~35	2.71	3.12	2.85±0.16	−10~15

Table 6.3 Chemical parameters of groundwater samples in the Dagu River groundwater reservoir.

Sample	Land use	NO_3^--N (mg/L)	$\delta^{15}N$-NO_3^- (‰)	$\delta^{18}O$-NO_3^- (‰)	Sample	Land use	NO_3^--N (mg/L)	$\delta^{15}N$-NO_3^- (‰)	$\delta^{18}O$-NO_3^- (‰)
1	VCA	49.82	−1.90	−0.14	18	VCA	91.22	−4.81	−1.56
2	VCA	63.43	−3.22	0.09	19	VCA	121.69	−16.62	−23.14
3	VCA	41.40	2.83	0.91	20	VCA	2.62	4.80	6.16
4	VCA	80.50	−6.59	−5.57	21	VCA	142.14	−6.25	−7.62
5	VCA	71.71	−1.63	−0.75	22	VCA	71.23	−6.66	−15.65
6	VCA	172.57	−8.25	−12.16	23	VCA	25.38	−0.40	−12.80
7	VCA	2.93	−6.20	−7.28	24	VCA	96.28	−2.57	−0.13
8	VCA	46.69	−0.65	4.66	25	GCA	33.51	1.52	4.22
9	VCA	62.89	−1.50	−1.33	26	GCA	30.50	−0.78	5.32
10	VCA	80.74	−9.88	−6.98	27	GCA	31.82	1.14	−5.09
11	VCA	104.59	−14.96	−21.00	28	GCA	38.57	−0.14	0.27
12	VCA	125.78	−13.41	−16.95	29	GCA	21.77	0.96	12.34
13	VCA	87.01	−11.44	−12.79	30	GCA	64.73	−6.21	−5.02
14	VCA	34.35	−1.26	1.15	31	GCA	22.38	−1.09	1.87
15	VCA	46.87	−6.57	−4.70	32	GCA	22.25	6.00	3.39
16	VCA	87.01	−10.18	−9.44	33	GCA	2.27	4.45	7.90
17	VCA	115.91	−15.14	−21.20	34	GCA	0.37	0.49	7.56

The $\delta^{15}N-NO_3^-$ values of groundwater samples ranged from −16.62‰ to 6.00‰ with a mean value of −4.00‰, and the $\delta^{18}O-NO_3^-$ values ranged from −23.14‰ to 12.34‰ with a mean value of −3.98‰ (Table 6.3). The cross-plot of $\delta^{15}N-NO_3^-$ and $\delta^{18}O-NO_3^-$ (Figure 6.10) showed that most of the groundwater samples were within the isotopic ranges of CF, SN, and M&S. In the VCA, some of the $\delta^{15}N-NO_3^-$ and $\delta^{18}O-NO_3^-$ values were out the isotopic ranges of the listed potential sources. This may come from the significant isotopic fractionation effect in the NO_3^- transformation processes. Moreover, the lower the temperature, the greater the fractionation (Mariotti et al., 1981; Yun et al., 2011). During the sampling period, garlic and greenhouse vegetables (e.g. cucumber and leek) entered the topdressing period, and spring vegetables, such as potatoes and carrots, were planted with base fertilizer. Owing to the lower temperature in March (average temperature of 7℃) and the slower nitrification rate, ^{14}N and ^{16}O were given priority for utilization in the transformation of N fertilizer into NO_3^-. Therefore, the NO_3^- leached into the groundwater had lower $\delta^{15}N$ and $\delta^{18}O$ values. The spatial variation of isotopes in groundwater is shown in Figure 6.11. The results indicated that the $\delta^{15}N-NO_3^-$ and $\delta^{18}O-NO_3^-$ values in the VCA were significantly lower than those in the GCA. This was the opposite of the spatial distribution of the NO_3^--N concentrations. The reason for this phenomenon was that more chemical fertilizers with lower isotopic values had been used for vegetable cultivation during the sampling period.

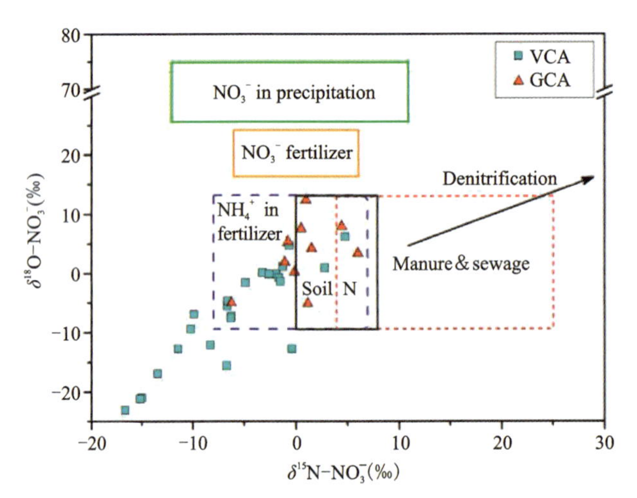

Figure 6.10 Cross-plot of $\delta^{15}N-NO_3^-$ and $\delta^{18}O-NO_3^-$ values in groundwater. The $\delta^{15}N$ and $\delta^{18}O$ compositions of various potential sources were summarized from Xue et al. (2009) and Nikolenko et al. (2018).

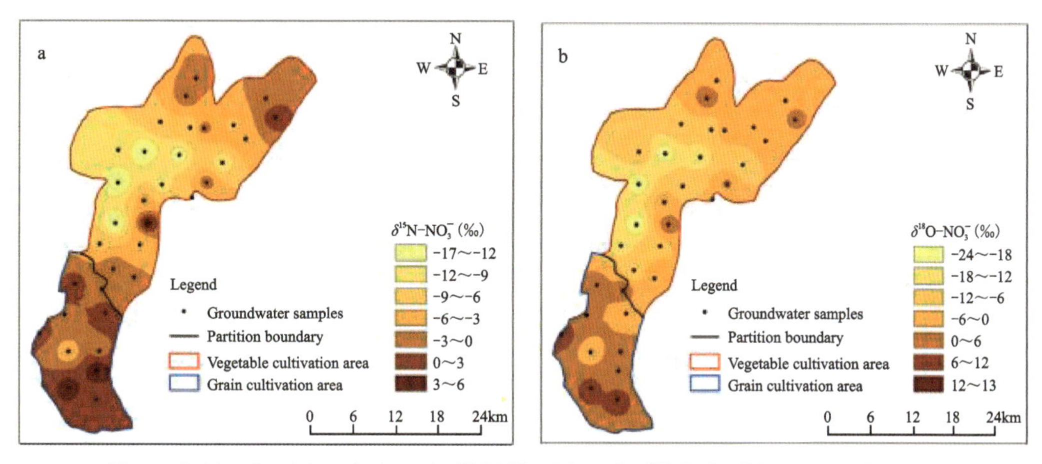

Figure 6.11 Spatial variation of $\delta^{15}N\text{-}NO_3^-$ (a) and $\delta^{18}O\text{-}NO_3^-$ (b) in groundwater.

6.5.1.3 Source apportionment using Bayesian isotope mixing model (SIAR)

6.5.1.3.1 Bayesian isotope mixing model (SIAR)

In order to estimate the proportional contributions of NO_3^- sources, Xue et al. (2016) successfully applied $\delta^{15}N$, $\delta^{18}O$, and Bayesian isotope mixing model (SIAR) to NO_3^- source apportionment for the first time. The SIAR model could estimate the probability distribution of the proportional contribution based on Bayesian, which overcomes the limitations of the linear isotopic mixing model.

The SIAR model is considered a reliable tool for determining NO_3^- source apportionment (Xue et al., 2016). Set as N measurements of J isotopes with K sources, the SIAR model can be expressed as follows (Parnell et al., 2010):

$$X_{ij} = \sum_{k=1}^{K} P_k (S_{jk} + \varepsilon_{jk}) + \nu_{ij} \tag{6.7}$$

$$S_{jk} \sim N(\mu_{jk}, \omega_{jk}^2) \tag{6.8}$$

$$\varepsilon_{jk} \sim N(\lambda_{jk}, \tau_{jk}^2) \tag{6.9}$$

$$\nu_{jk} \sim N(O, \sigma_j^2) \tag{6.10}$$

where X_{ij} is the isotopic value j of sample i ($i=1, 2, 3, \cdots, N$ and $j=1, 2, 3, \cdots, J$); P_k is the proportion of source k ($k=1, 2, 3, \cdots, K$), which is estimated by the SIAR model; S_{jk} is the source value k of isotope j, which is normally distributed with a mean of μ_{jk} and standard deviation (SD) of ω_{jk}^2; ε_{jk} is the enrichment factor of isotope j on source k, which is normally distributed with a mean of λ_{jk} and SD of τ_{jk}^2; and ν_{ij} is the residual error, which is normally distributed with a mean of zero and SD of σ_j^2.

6.5.1.3.2 Source apportionment using SIAR

The proportional contributions of the NO_3^- sources in groundwater were identified by the SIAR model. Two isotopes ($j=2$) ($\delta^{15}N\text{-}NO_3^-$ and $\delta^{18}O\text{-}NO_3^-$) and four potential sources ($k=4$) (AD, SN, CF, and M&S) were used in this study. The $\delta^{15}N\text{-}NO_3^-$ and $\delta^{18}O\text{-}NO_3^-$ of groundwater samples (Table 6.3) and sources (both Mean and SD in Table 6.2) were input the SIAR model. The $\varepsilon_{p/s}$ of various sources were $-8.7‰ \pm 2.7‰$. The SIAR model was conducted in the SIAR package of the R Programming Language.

The results of source apportionment were showed in Figure 6.12. The source contributions showed significant spatial variation under different cropping system. In the VCA, CF was the dominant NO_3^- source with a contribution of 54.32%. Next was SN, which contributed 37.75%, while M&S and AD contributed only 6.63% and 1.30%, respectively. In the GCA, the source contributions were in the order of SN (33.67%),N CF (33.27%),N M&S (30.16%),N AD (2.90%).

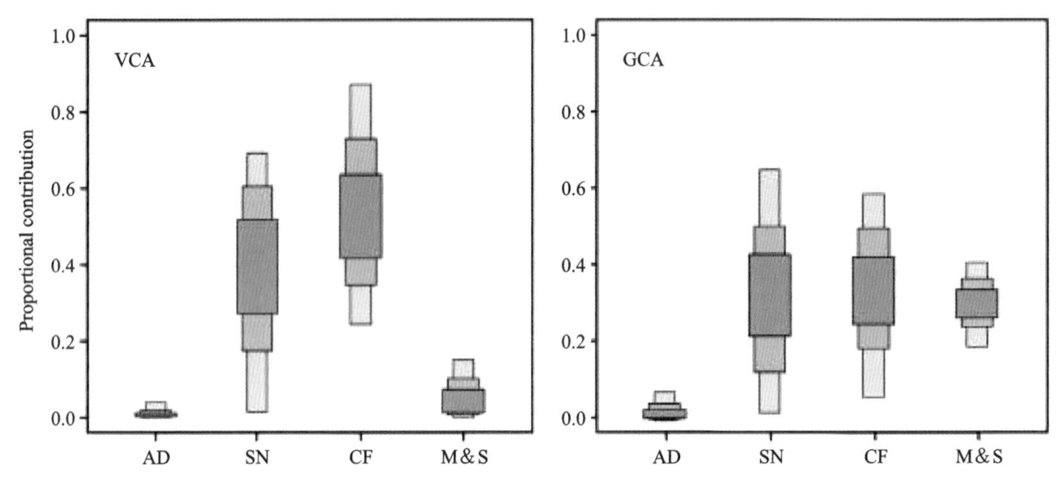

Figure 6.12 Spatial proportional contribution of NO_3^- sources estimated by SIAR in the VCA and GCA. Boxplots illustrate the 50th, 75th, and 95th percentiles from dark to light.

The results of NO_3^- source apportionment with SIAR were consistent with our field investigation. During the sampling period, vegetables were grown with large amounts of CF, thereby increasing the risk of NO_3^- contamination in groundwater. In the GCA, winter wheat was the main crop. For most winter wheat, only base fertilizer was applied during sowing, and topdressing was not applied during growing. This was the main reason why the contribution of CF in the VCA was significantly higher than that in the GCA.

According to the field survey, there were few intensive livestock and poultry farms in the study area, so the production and usage of manure were low. Therefore, the contribution of M&S was lower in the VCA. The residential areas in the GCA are more concentrated than those in the VCA (Figure 6.9). Moreover, the residential areas in the study region are mainly villages and towns with a low urbanization level and imperfect sewage treatment facilities. The denser residential areas resulted in more domestic sewage produced in the GCA. Since the $\delta^{15}N$ and $\delta^{18}O$ values of manure and domestic sewage were similar, the high contribution of M&S in the GCA may have been caused by domestic sewage. Soil N is another major source of NO_3^- pollution in groundwater. As there was no significant difference in soil types between the VCA and GCA, the contributions of SN did not show significant spatial variation (37.75% for the VCA and 33.67% for the GCA). According to the analysis of precipitation samples, the NO_3^- concentration of AD was only around 2mg/L. Moreover, sampling activities were conducted during the dry season with low precipitation. Therefore, the contributions of AD were the lowest (<3%).

References

BAILY A, ROCK L, WATSON C J, et al., 2011. Spatial and temporal variations in groundwater nitrate at an intensive dairy farm in south-east Ireland: insights from stable isotope data[J]. Agric. Ecosysts. Environ., 144:308-318.

BATEMAN A S, KELLY S D, JICKELLS T D, 2005. Nitrogen isotope relationships between crops and fertilizer: implications for using nitrogen isotope analysis as an indicator of agricultural regime[J]. J. Agric. Food Chem., 53 (14): 5760-5765.

BEAUCHAMP E G, TREVORS J T, PAUL J W, 1989. Carbon sources for bacterial denitrification[J]. Adv. Soil Sci., 10:113-142.

BEDARD-HAUGHN A A, VAN GROENIGEN J W, VAN KESSEL C, 2003. Tracing ^{15}N through the landscapes: potential uses and precautions[J]. J. Hydrol., 272: 175-190.

BERNOT M J, DODDS W K, GARDNER W S, et al., 2003. Comparing denitrification estimates for a Texas estuary by using acetylene inhibition and membrane inlet mass spectrometry[J]. Appl. Environ. Microb., 69 (10):5950-5956.

BETLACH M R, TIEDJE J M, 1981. Kinetic explanation for accumulation of nitrite, nitric oxide, and nitrous oxide during bacterial denitrification[J]. Appl. Environ. Microb., 42 (6):1074-1084.

BöHLKE J K, SMITH R L, MILLER D N, 2006. Ammonium transport and reaction in contaminated groundwater: application of isotope tracers and isotope fractionation studies [J]. Water Resour. Res., 42 (5).

BöTTCHER J, STREBEL O, VOERKELIUS S, et al., 1990. Using isotope fractionation of nitrate-nitrogen and nitrate-oxygen for evaluation of microbial denitrification in a sandy aquifer[J]. J. Hydrol., 114 (3-4):413-424.

BRANDES J A, DEVOL A H, 2002. A global marine-fixed nitrogen isotopic budget: implications for holocene nitrogen cycling[J]. Glob. Biogeochem. Cycles, 16 (4):67-1-67-4.

BUSS S R, HERBERT A W, MORGAN P, et al., 2004. A review of ammonium attenuation in soil and groundwater[J]. Q. J. Eng. Geol. Hydrogeol., 37 (4): 347-359.

CASCIOTTI K L, SIGMAN D M, GALANTER HASTINGS M, et al., 2002.

Measurement of the oxygen isotopic composition of nitrate in seawater and freshwater using the denitrifier method[J]. Anal. Chem. ,74:4905-4912.

CHANG C C Y, LANGSTON J, RIGGS M, et al. , 1999. A method for nitrate collection for ^{15}N and ^{18}O analysis from water with low nitrate concentrations[J]. Can. J. Fish. Aquat. Sci. ,56:1856-1864.

CHOI W J, LEE S M, RO H M, 2003. Evaluation of contamination sources of groundwater NO_3^- using nitrogen isotope data: a review[J]. Geosci. J. ,7 (1):81-87.

DANIELESCU S, MACQUARRIE K T, 2013. Nitrogen and oxygen isotopes in nitrate in the groundwater and surface water discharge from two rural catchments: implications for nitrogen loading to coastal water[J]. Biogeochemistry,115 (1-3):111-127.

DAVIDSON E A, CHOROVER J, DAIL D B, 2003. A mechanism of abiotic immobilization of nitrate in forest ecosystems:the ferric wheel hypothesis[J]. Glob. Change Biol. ,9 (2):228-236.

DELAUNE R D, JUGSUJINDA A, 2003. Denitrification potential in a Louisiana wetland receiving diverted Mississippi River water[J]. Chem. Ecol. ,19 (6):411-418.

EVANS C D, CHAPMAN P J, CLARK J M, et al. ,2006. Alternative explanations forrising dissolved organic carbon export from organic soils[J]. Glob. Chang. Biol. , 12 (11):2044-2053.

FRANCIS A J, SLATER J M, DODGE C J, 1989. Denitrification in deep sub-surface sediments[J]. Geomicrobiol. J. , 7 (1-2): 103-116.

FUKADA T, HISCOCK K M, DENNIS P F,et al. , 2003. A dual isotope approach to identify denitrification in ground water at a river bank infiltration site[J]. Water Res. , 37: 3070-3078.

GALE I N, MARKS R J, DARLING W G,et al. , 1994. Bacterial denitrification in aquifers: evidence from the unsaturated zone and the unconfined chalk and Sherwood sandstone aquifers[J]. National Rivers Authority R&D:215.

GOLDBERG S D, KNORR K H, GEBAUER G, 2008. N_2O concentration and isotope signature along profiles provide deeper insight into the fate of N_2O in soils[J]. Isot. Environ. Health Stud. ,44 (4): 377-391.

HEATON THE, 1986. Isotopic studies of nitrogen pollution in the hydrosphere and atmosphere: a review[J]. Chem. Geol. ,59:87-102.

HERNáNDEZ-DEL AMO E, MENCIó A, GICH F,et al. , 2018. Isotope and microbiome data provide complementary information to identify natural nitrate attenuation processes in groundwater[J]. Sci. Total Environ. ,613:579-591.

HISCOCK K M, LLOYD J W, LERNER D N, 1991. Review of natural andartificial denitrification of groundwater[J]. Water Res. ,25 (9):1099-1111.

HINKLE S R, BöHLKE J K, DUFF J H,et al. , 2007. Aquifer-scale controls on the distribution of nitrate and ammonium in ground water near La Pine, Oregon, USA[J]. J. Hydrol. ,333:486-503.

HOCHSTEIN L I, BETLACH M R, KITIKOS G, 1984. The effect of oxygen on denitrification during steady-state growth of Paracoccus halodenitrificans[J]. Arch. Microbiol. ,137 (1):74-78.

HOEFS J,2018. Stable Isotope Geochemistry[M]. Springer International Publishing.

JIN R C, YANG G F, YU J J,et al. , 2012. The inhibition of the anammox process: a review[J]. Chem. Eng. J. ,197:67-79.

JURADO A, BORGES A V, BROUYèRE S, 2017. Dynamics and emissions of N_2O in groundwater: a review[J]. Sci. Total Environ. ,584:207-218.

KELLMAN L M, HILLAIRE-MARCEL C, 2003. Evaluation of nitrogen isotopes as indicators of nitrate contamination sources in an agricultural waterhed[J]. Agric. Ecosyst. Environ. ,95 (1):87-102.

KENDALL C,1998. Tracing nitrogen sources and cycling in catchments[J]//Kendall C, McDonnell J J. Isotope tracers in catchment hydrology[M]. Amsterdam:Elsevier Science: 519-576.

KENDALL C, Aravena R, 2000. Nitrate isotopes in groundwater systems[J]//Environmental Tracers in Subsurface Hydrology[M]. Springer, U S:261-297.

KENDALL C, GRIM E, 1990. Combustion tube method for measurement of nitrogen isotope ratios using calcium oxide for total removal of carbon dioxide and water[J]. Anal. Chem. , 62:526-529.

KINNIBURGH D G, GALE I N, GOODDY D C, et al. , 1999. Denitrification in the unsaturated zones of the British Chalk and Sherwood Sandstone aquifers [R]. British Geological Survey Technical Report WD/99/2.

KIRSTEN CHRISTOFFERSEN, HANNE KAAS, 2000. Toxic Cyanobacteria in Water: A Guide to their Public Health Consequences, Monitoring and Management[J]. Limnology and Oceanography,45(5):1212.

KNOWLES R, 1982. Denitrification[J]. Microbiol. Rev. ,46 (1):43-70.

KNöLLER K, VOGT C, HAUPT M, et al. , 2011. Experimental investigation of nitrogen and oxygen isotope fractionation in nitrate and nitrite during denitrification[J]. Biogeochemistry,103 (1-3):371-384.

KOOL D M, WRAGE N, OENEMA O,et al. , 2011. Oxygen exchange with water alters the oxygen isotopic signature of nitrate in soil ecosystems[J]. Soil Biol. Biochem. ,43 (6):1180-1185.

KORNER H, ZUMFT W G, 1989. Expression of denitrification enzymes in response to the dissolved oxygen level and respiratory substrate in continuous culture of Pseudomonas stutzeri[J]. Appl. Environ. Microb. ,55 (7):1670-1676.

LAMONTAGNE M G, DURAN R, VALIELA I, 2002. Nitrous oxide sources and sinks in coastal aquifers and coupled estuarine water[J]. Sci. Total Environ. ,309 (1-3):139 - 149.

LI X D, MASUDA H, KOBA K, et al. , 2007. Nitrogen isotope study on nitrate contaminated groundwater in the Sichuan Basin, China[J]. Water Air Soil Pollut. ,178:145-156.

LI X, TANG C, HAN Z, et al. , 2014. Spatial and seasonal variation of dissolved nitrous oxide in wetland groundwater[J]. Environ. Pollut. ,3 (1):21.

LIU C Q, LI S L, LANG Y C, et al. , 2006. Using $\delta^{15}N$ and $\delta^{18}O$ values to identify nitrate sources in karst ground water, Guiyang, Southwest China [J]. Environ. Sci. Technol. ,40:6928-6933.

MAEDA K, TOYODA S, YANO M, et al. , 2016. Isotopically enriched ammonium shows high nitrogen transformation in the pile top zone of dairy manure compost [J]. Biogeosciences,13 (4):1341-1349.

MARGESON J H, SUGGS J C, MIDGETT M R, 1980. Reduction of nitrate to nitrite with cadmium[J]. Anal. Chem. , 52: 1955-1957.

MARIOTTI A, GERMON J C, HUBERT P, et al. , 1981. Experimental determination of nitrogen kinetic isotope fractionation: some principles; illustration for the denitrification and nitrification processes[J]. Plant Soil,62 (3):413-430.

MAYER B, BOLLWERK S M, MANSFELDT T, et al. , 2001. The oxygen isotope composition of nitrate generated by nitrification in acid forest floors[J]. Geochim. Cosmochim. Acta,65 (16):2743-2756.

MCILVIN M R, ALTABET M A, 2005. Chemical conversion of nitrate and nitrite to nitrous oxide for nitrogen and oxygen isotopic analysis in freshwater and seawater[J]. Anal. Chem. ,77:5589-5595.

MCMAHON P B, BöHLKE J K, 2006. Regional patterns in the isotopic composition of natural and anthropogenic nitrate in groundwater, High Plains, USA[J]. Environ. Sci. Technol. ,40 (9):2965-2970.

MICHENER R, LAJTHA K, 2007. Tracing anthropogenic inputs of nitrogen to ecosystems[J]//Kendall C, Elliott E M, Wankel S D. Stable Isotopes in Ecology and Environmental Science(2nd ed)[M]. Carlton, Victoria:Blackwell Publishing Ltd. :375-449.

MINAMIKAWA K, NISHIMURA S, NAKAJIMA Y, et al. , 2011. Upward diffusion of nitrous oxide produced by denitrification near shallow groundwater table in the summer: a lysimeter experiment[J]. Soil Sci. Plant Nutr. ,57 (5):719-732.

MOORE K B, EKWURZEL B, ESSER B K, et al. , 2006. Sources of groundwater nitrate revealed using residence time and isotope methods[J]. Appl. Geochem. , 21 (6): 1016-1029.

MORRIS J T, WHITING G J, CHAPELLE F H, 1988. Potential denitrification rates in deep sediments from the Southeastern Coastal Plain[J]. Environ. Sci. Technol. ,22 (7): 832-836.

NEILSEN M E, FISK M R, ISTOK J D, et al. , 2006. Microbial nitrate respiration of lactate at in situ conditions in ground water from a granitic aquifer situated 450m underground [J]. Geobiology, 4 (1): 43-52.

OWENS N J P, 1987. Natural variations in ^{15}N in the marine environment[J]. Adv. Mar. Biol. ,24:390-451.

PARNELL A C, INGER R, BEARHOP S, et al. , 2010. Source partitioning using stable isotopes: coping with too much variation[J]. PLoS One, 5 (3):1-5.

PETERSON B J, FRY B, 1987. Stable isotopes in ecosystem studies[J]. Ann. Rev. Ecol. Syst. ,18:293-320.

PETITTA M, FRACCHIOLLA D, ARAVENA R, et al. , 2009. Application of isotopic and geochemical tools for the evaluation of nitrogen cycling in an agricultural basin, the Fucino Plain, Central Italy[J]. J. Hydrol. ,372 (1):124-135.

RIVETT M O, BUSS S R, MORGAN P, et al. , 2008. Nitrate attenuation in groundwater: a review of biogeochemical controlling processes[J]. Water Res. ,42 (16):4215-4232.

ROCK L, ELLERT B H, 2007. Nitrogen-15 and oxygen-18 natural abundance of potassium chloride extractable soil nitrate using the denitrifier method[J]. Soil Sci. Soc. Am. J. ,71: 355-361.

ROSMAN J R, TAYLOR P D, 1998. Isotopic compositions of the elements (technical report): commission on atomic weights and isotopic abundances[J]. Pure Appl. Chem. ,70: 217-235.

SEBILO M, BILLEN G, MAYER B, et al. , 2006. Assessing nitrification and denitrification in the Seine River and estuary using chemical and isotopic techniques [J]. Ecosystems, 9 (4):564-577.

SEBILO M, MAYER B, NICOLARDOT B, et al. , 2013. Long-term fate of nitrate fertilizer in agricultural soils[J]. Proc. Natl. Acad. Sci. ,110 (45):18 185-18 189.

SCHEIBLE O K, 1993. Manual: Nitrogen Control[R]. United Stated Environmental Protection Agency. Risk Reduction Engineering Laboratory Report EPA/625/R-93/010.

SCHILMAN B, TEPLYAKOV N, 2007. Detailed protocol for nitrate chemical reduction to nitrous oxide for delta ^{15}N and delta ^{18}O analysis of nitrate in fresh and marine water[R]. Annual Report Submitted to the Earth Science Research Administration, Ministry of National Infrastructures Jerusalem, December 2007 TR-GSI/15/2007.

SCHMIDT H L, WERNER R A, YOSHIDA N, et al. ,2004. Is the isotopic composition of nitrous oxide an indicator for its origin from nitrification or denitrification? A theoretical approach from referred data and microbiological and enzyme kinetic aspects [J]. Rapid Commun. Mass Spectrom. ,18 (18):2036-2040.

Sharp Z, 2007. Principles of Stable Isotope Geochemistry: Nitrogen[J]. Pearson Prentice Hal. Upper Saddle River, N J:206-219.

SIGMAN D M, CASCIOTTI K L, ANDREANI M, et al., 2001. A bacterial method for the nitrogen isotopic analysis of nitrate in seawater and freshwater[J]. Anal. Chem., 73:4145-4153.

SILVA S R, KENDALL C, WILKISON D H, et al., 2000. A new method for collection of nitrate from fresh water and the analysis of nitrogen and oxygen isotope ratios [J]. J. Hydrol., 228:22-36.

SMITH R L, BAUMGARTNER L K, MILLER D N, et al., 2006. Assessment of nitrification potential in ground water using short term, single-well injection experiments [J]. Microb. Ecol., 51 (1):22-35.

STENSTROM M K, PODUSKA R A, 1980. The effect of dissolved oxygen concentration on nitrification[J]. Water Res., 14 (6):643-649.

TOMASZEWSKI M, CEMA G, ZIEMBIńSKA-BUCZYńSKA A, 2017. Influence of temperature and pH on the anammox process: a review and meta-analysis[J]. Chemosphere, 182:203-214.

XUE Y, SONG J, ZHANG Y, et al., 2016. Nitrate pollution and preliminary source identification of surface water in a semi-arid river basin, using isotopic and hydrochemical approaches[J]. Water, 8 (8):328.

YOSHINARI T, KNOWLES R, 1976. Acetylene inhibition of nitrous oxide reduction by denitrifying bacteria[J]. Biochem. Biophys. Res. Co., 69:705-710.

YU L, ZHENG T, ZHENG X, et al., 2020. Nitrate source apportionment in groundwater using Bayesian isotope mixing model based on nitrogen isotope fractionation[J]. Sci. of the Total Environ., 718(C): 137-242.

YUN S I, RO H M, CHOI W J, et al., 2011. Interpreting the temperature-induced response of ammonia oxidizing microorganisms in soil using nitrogen isotope fractionation [J]. J. Soils Sediments, 11 (7):1253-1261.

VITòRIA L, OTERO N, SOLER A, et al., 2004. Fertilizer characterization: isotopic data (N, S, O, C, and Sr)[J]. Environ. Sci. Technol., 38 (12):3254-3262.

WASSENAAR L I, 1995. Evaluation of the origin and fate of nitrate in the Abbotsford Aquifer using the isotopes of ^{15}N and ^{18}O in NO_3^- [J]. Appl. Geochem., 10:391-405.

WELL R, ESCHENBACH W, FLESSA H, et al., 2012. Are dual isotope and isotopomer ratios of N_2O useful indicators for N_2O turnover during denitrification in nitrate-contaminated aquifers? [J] Geochim. Cosmochim. Acta, 90:265-282.

WIDORY D, PETELET-GIRAUD E, NEGREL P, et al., 2005. Tracking the sources of nitrate in groundwater using coupled nitrogen and boron isotopes: a synthesis [J]. Environ. Sci. Technol., 39 (2):539-548.

Chapter 7 Sulfur Isotope

Sulfur is present in nearly all natural environments. It may be a major component in ore deposits, where sulfur is the dominant nonmetal, and as sulfates in evaporites. It occurs as a minor component in igneous and metamorphic rocks, throughout the biosphere in organic substances, in marine water and sediments as both sulfide and sulfate. These occurrences cover the whole temperature range of geological interest. Thus, it is quite clear that sulfur is of special interest in stable isotope geochemistry.

Sulfur has four stable isotopes with the following abundances (De Laeter et al. ,2003).

$^{32}S = 95.04\%$

$^{33}S = 0.75\%$

$^{34}S = 4.20\%$

$^{36}S = 0.01\%$

For many years the reference standard commonly referred to was sulfur from troilite of the Canyon Diablo iron meteorite (CDT). As Beaudoin et al. (1994) have pointed out, the original CDT is not homogeneous and may display variations in ^{34}S up to 0.4‰. Therefore, a new reference scale, Vienna-CDT (V-CDT) has been introduced by an advisory committee of IAEA in 1993, recommending an artificially prepared Ag_2S (IAEA-S-1) with a $\delta^{34}S_{V\text{-}CDT}$ of −0.3‰ as the new international standard reference material.

For various processes in nature, it is generally accepted that the sulfur isotope fractionation induced by them follows mass-dependent fractionation, so that there is a quantitative relationship between the $\delta^{33}S$, $\delta^{34}S$ and $\delta^{36}S$ values of the sulfur isotopic composition of most substances (i. e. $\delta^{33}S = 0.515\delta^{34}S$; $\delta^{36}S = 1.9\delta^{33}S$). However, since 2000 Earth scientists have successively discovered non-mass correlated sulfur isotope fractionation in extraterrestrial materials such as Martian meteorites and the Moon, and have linked this phenomenon to a series of major Earth science questions such as the early solar system atmospheric composition and its evolution, ancient atmospheric oxidation conditions, the interaction of the Earth's circles, and the early Earth sulfur cycle, providing a new idea for the interpretation of many important hypotheses.

Thode et al. (1949) and Trofimov (1949) were the first to observe wide variations in the abundances of sulfur isotopes. Variations on the order of 180‰ have been documented with the "heaviest" sulfates having $\delta^{34}S$-values of greater than +120‰, and the "lightest" sulfides having $\delta^{34}S$-values of around −65‰. Some of the naturally occurring S-isotope variations are summarized in Figure 7.1 reviews of the isotope geochemistry of sulfur have been published by Rye and Ohmoto (1974), Ohmoto and Goldhaber (1997), Seal et al. (2000), Canfield (2001a) and Seal (2006).

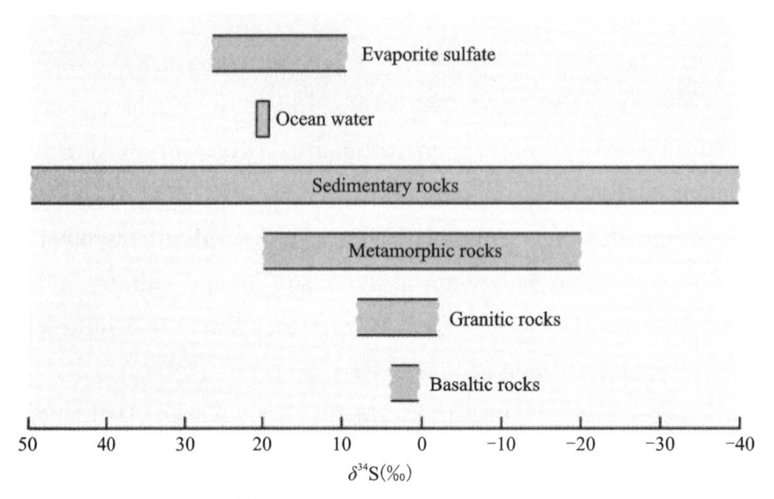

Figure 7.1 $\delta^{34}S$-values of important geological reservoirs.

7.1 Sulfur species in groundwater

Sulfur and its compounds are widely distributed in nature, such as pyrite, sulfur and gypsum. It is also used in large quantities in industrial production, such as sulfuric acid in batteries, gunpowder, bleaching agents in the paper industry, fixing agents in photography, fertilizers, etc. In the natural environment, sulfur has five valence states: S^{6+}, S^{4+}, S^0, $[S_2]^{2-}$, S^{2-}, it has a variety of valence states such as SO_4^{2-}, SO_2, $S_2O_3^{2-}$, S_x, H_2S, HS^-, S^{2-}, FeS_2 different chemical states. SO_x (variable sulfur oxide) add to the atmospheric levels of sulfur and has been identified as an indirect to low level of threat to lives. Water containing SO_x tastes bitter and in severe cases, because of its characteristic laxative property, results in dehydration. This foul taste and associated medical conditions serve as an identification of potential sulfur-contaminated groundwater and needs immediate isolation and treatment. Hydrogen sulfide is another such secondary contaminant, which has a distinct odour and corrosive property and is a threat to life forms.

Sulfur is ubiquitous in surface and ground water, and can have a wide variety of both natural and anthropogenic sources. Common natural sources include rainwater (which may derive from marine aerosols, atmospheric pollutants or volcanic gases), geological weathering of sulfide and evaporite minerals, mineralization of organic sulfur, and anthropogenic sources include surface runoff from fertilizer or sewage, acid precipitation from oil and coal burning (Chivas et al., 1991; Moncaster et al., 2000; Otero et al., 2008; Turchyn et al., 2013). The relative contribution from each of these sources varies regionally, depending on the geology, climate and land use, and can be affected by anthropogenic processes such as mining and groundwater extraction (Bottrell et al., 2008; Otero and Soler, 2002; Samborska and Halas, 2010). The abundance and speciation of sulfur in groundwater has important implications for redox and pH balance, ecology and drinking water quality.

Sulfide minerals are readily oxidised when exposed to atmospheric oxygen, and as groundwater circulates through shallow aquifers, leaching of sulfur-bearing minerals adds to the dissolved sulfur load (Figure 7.2). These sulfur-bearing minerals may include evaporites, sedimentary sulfides or magmatic sulfides, depending on the regional geology. The mineral leaching processes that add sulfur into surface water and rivers, ultimately delivering sulfur to the world's oceans, vary regionally and remain poorly constrained on a global scale (Calmels et al., 2007; Karim and Veizer, 2000; Otero et al., 2008; Turchyn et al., 2013; Yuan and Mayer, 2012).

Deep groundwater is typically relatively reduced, and as such, dissolved sulfur occurs as various sulfide ions, which can lead to deposition of secondary authigenic sulfide minerals in the hosting aquifers. These authigenic sulfide minerals should record information about ancient groundwater sulfide. Shallower groundwater, and surface water in river and lakes, are typically oxidising, and contain sulfur as dissolved sulfate ions. Here, oxidation of geological sulfide minerals and mixing with atmospheric sulfur may occur. Oxidation of sulfide minerals may continue in deep anoxic groundwater, via reduction of Fe^{3+} minerals. Evaporitic sulfate minerals may precipitate from surface water where the dissolved sulfur load is sufficiently high. The boundary between oxidising and reducing conditions is controlled primarily by the position of the water table and associated extent of the vadose zone (Sophocleous, 2002).

Sulfate extensively comes in water from both natural and anthropogenic sources. The natural sources include sulfur mineral dissolution, atmospheric deposition and sulfide oxidation from mineral. Human-induced sources are power plant, coal mines and metallurgical refinery. In many potential sources, gypsum is an important source in many aquifers having large amount of sulfate. In the last few decades, atmospheric deposition has become an important source of sulphate to soil and ultimately it goes to groundwater. Since sulfate is mobile in soil, addition into the soil will impact on shallow aquifer.

Sulfate, one of the major constituents of groundwater, can range from 1.00mg/L to 1000mg/L depending on the geographic and economic scenario of the place. The WHO level of sulphate in groundwater has been set as 250mg/L and has been permitted up to a higher limit of 400mg/L. Sulfate is not contributing to toxicity, but when ingested in higher

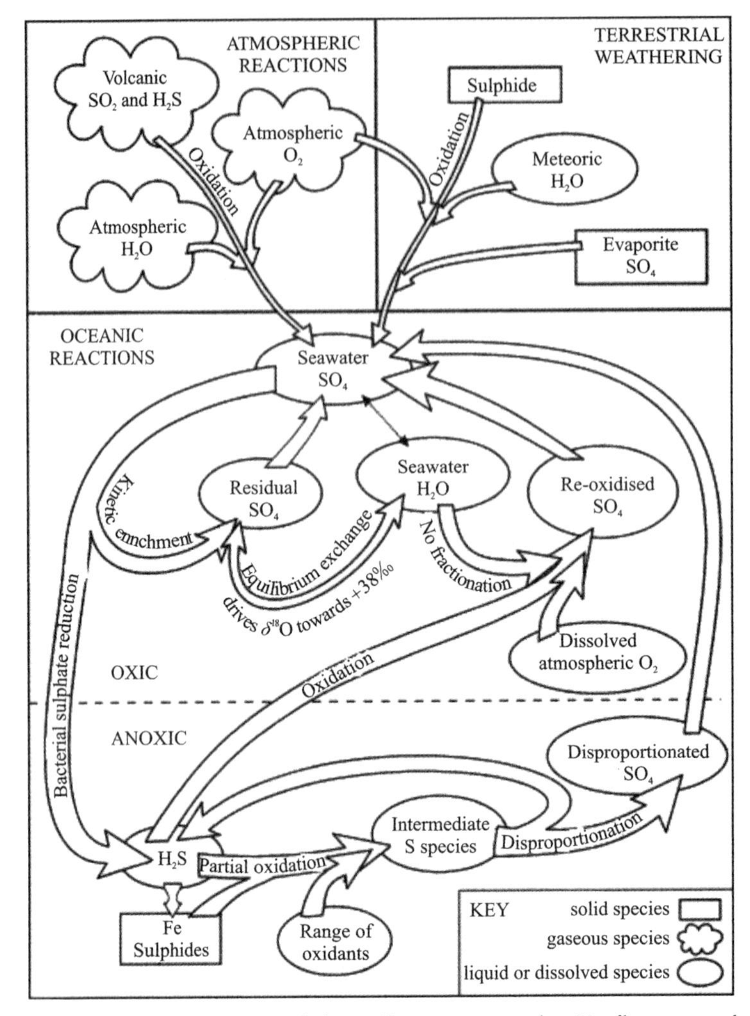

Figure 7.2 A schematic diagram of the sulfate-oxygen cycle. No fluxes are shown for the hydrothermal and volcanic fluxes since they represent constant addition. Estimate of hydrothermal sulfur isotopic composition from Walker (1986).

amount, it can cause catharsis, dehydration and diarrhoea, and sometimes the levels of methemoglobin and sulfaemoglobin are changed in humans' body system. High level of sulfate leads to change the taste of water, makes it bitter especially when the concentration exceeds to 250mg/L. Sulfate may have laxative effect that result to dehydration and high risk for infants. In addition to that, if level exceeding to 400mg/L, water is not used for baby usages (drinking, food, etc.). Sulfate can be responsible for salinization of freshwater bodies and that's why it takes as an urgent environmental problem. Sulfate can cause scaling in water pipes, which results in reduced diameter or blocked pipes. Sulfate with combination of chlorine bleach makes cleaning cloths more difficult. High sulfate levels in groundwater can lead to salinization of groundwater, with implications for drinking water safety. Thus, elevated sulfate makes problem to health as well materials.

Sulfur founds in many oxidation states and changes its states with geochemical changes when water interacts with rocks. Sulfate plays very important role in controlling redox reaction in almost all aquatic system. Sulfate reduction occurs extensively in groundwater system. Sulfate can be reduced by consumption of hydrogen ions and produced hydrogen sulfide as by-product:

$$SO_4^{2-} + 9H^+ + 8e^- \longrightarrow HS^- + 4H_2O \tag{7.1}$$

Sulfur is intimately linked with the recycling of organic matter, as sulfate reducing bacteria respire sulphate in the absence of oxygen, producing sulfide (Equation 7.1). Dissimilatory sulfate reduction is a major biogeochemical process and is globally important to the carbon and oxygen cycles. Sulfate reducers are responsible for over 50% of organic carbon remineralization in marine sediments and the sulfide produced by the process constitutes one of the major reduced burial products (Garrels and Lerman, 1981; Walker, 1986; Canfield, 2004; Johnston et al., 2009).

$$SO_4^{2-} + \text{Organic matter} \xrightarrow{\text{anaerobic bacteria}} H_2O + CO_2 + S^{2-} \tag{7.2}$$

In the presence of organic matter in the reducing environment, desulfate bacteria reduce SO_4^{2-} in the water to H_2S, thereby reducing the amount of SO_4^{2-} in the water or even making it small, an action known as desulfidation. It can be expressed as

$$SO_4^{2-} + 2C + 2H_2O \xrightarrow{\text{desulfate bacteria}} H_2S\uparrow + 2HCO_3^- \tag{7.3}$$

The result of the desulfidation is an increase in the HCO_3^- concentration of the water and an increase in pH.

In confined geological formations, desulfidation often occurs, resultingin lower SO_4^{2-} and higher free H_2S content in the groundwater within them. This feature can be used as a secondary marker in the search for oil fields.

There are many different methods for remediation of sulphate in groundwater, and these sulphate remediation methods can be classified under three categories as given below:

(1) Physical method: pump-and-treat (PAT), membrane removal of sulphate [reverse osmosis (RO), electrodialysis (ED), filtration], ion exchange.

(2) Chemical method: adsorption, gypsum formation.

(3) Biological method: in situ bioreactor.

7.2 Analytical techniques

The analytical testing techniques for stable sulfur isotopes mainly include the determination of $\delta^{34}S$ values for different sulfur-containing substances (monomeric sulfur, sulfides or sulphates etc.) and the determination of the isotopic composition of polysulfides ($\delta^{33}S$ and $\delta^{36}S$). The analytical test procedure can be basically divided into two parts:

sample preparation and isotope mass spectrometry.

7.2.1 Sulfur isotope sample preparation methods

The feed gas for sulfur isotope mass spectrometry is usually either SO_2 or SF_6. SO_2 is currently used as the feed gas for the determination of $\delta^{34}S$ values, but SF_6 has proven to be a more desirable gas for mass spectrometry when $\delta^{33}S$, $\delta^{34}S$ and $\delta^{36}S$ values are measured simultaneously.

7.2.1.1 SO_2 Method

Chemical oxidation is used to convert sulfide samples to SO_2 gas, which is the most common method. The commonly used oxidants are CuO, Cu_2O or V_2O_5, and their conversion reactions are quantitative. The reaction temperature has the greatest influence on the conversion rate. For example, the optimum temperature for the reaction of magnetic pyrite with CuO is 1050℃. The amount ratio of sulfide to oxidant and the reaction time also has an important influence. It has been suggested that the optimum conditions for the above reaction are 1/6 sulfide/oxidant and 10mins reaction time. This conversion reaction is accompanied by a side reaction to produce SO_3. SO_3 is mainly produced in the low temperature phase, therefore SO_2 is prepared by placing the reaction tube containing the sample and oxidant mixture into the reactor suddenly after the furnace has been heated to a predetermined reaction temperature to achieve a rapid temperature rise and skip the low temperature phase. Attention should also be paid to factors such as tube adsorption, kinetic fractionation effects within the tube, and oxidant purification.

7.2.1.2 SF_6 Method

The sulfate needs to be converted to sulfide form before the fluorination reaction. The $BaSO_4$ precipitate was converted to gaseous H_2S using a mixture of $HI+H_3PO_2+HCl$ as the reducing agent, after which the H_2S gas was passed into the $AgNO_3$ solution using N_2 as the carrier gas and the reaction was carried out in the dark for one week to complete the Ag_2S precipitation. After removal of the supernatant by centrifugation, the Ag_2S precipitation was washed using 20mL of 1mol/L NH_4OH followed by continuous flushing with Milli-Q water.

Sulfide is reacted with the strong oxidizer BrF_5 to form SF_6, the chemical equation is:

$$5MS+8BrF_5=4Br_2+5MF_2+5SF_6 \tag{7.4}$$

The SF_6 method can be used to extract sulfur from sulfides either as an off-line separation device or as an on-line separation using gas chromatography. The significant advantages of this method include: the large molecular weight of SF_6, the small background value of the mass spectrometry, the high sensitivity, the ability to determine trace samples as small as 30mg; the preparation process does not generate SO_3, which ensures no isotopic fractionation; the viscous effect of the mass spectrometry tube wall on SF_6 is much smaller than that of SO_2; the high accuracy of the mass spectrometry, up to ±0.07‰; fluorine has only one isotope, so there is no interference with the spectral peaks of the test. There is only one isotope of fluorine, so there is no interference with the spectral peaks of the test, and there is almost no memory effect. However, the SF_6 method requires a fluorinated vacuum line and few laboratories can achieve this condition, thus limiting the application of this method.

7.2.2 Stable Isotope Mass Spectrometry

IRMS is the most established and widespread method for the determination of sulfur isotope ratios. The method requires the sample to be fed as a gas (SO_2 or SF_6). Traditionally, off-line redox methods have been used to convert sulfides to SO_2 gas with a two-pass injection. It can now also be coupled with a sample pre-treatment unit for continuous flow injection, simplifying the complex pre-treatment process and reducing experimental error due to human intervention. The method can be used to determine sulfur isotopes in mixed gases, solids, liquids, liquid organics or inorganic gases. This makes EA-IRMS the most convenient and efficient test method, while reducing sample requirements to as low as 300μg and testing accuracy to 0.2‰.

Microanalytical techniques such as laser microprobe (Kelley and Fallick, 1990; Crowe et al., 1990; Hu et al., 2003; Ono et al., 2006) and ion microprobe (SIMS) (Eldridge et al., 1988; Kozdon et al., 2010) are a solid in situ analytical technique. The SIMS technique for testing sulfur isotopes in sulfides is well established as it accomplishes the ionisation and later transport and analysis processes under high vacuum conditions, and therefore lacks the interference of polyatomic ions from airborne oxygen and hydrogen on the sulfur signal.

The use of MC-ICP-MS techniques has been described by Bendall et al. (2006), Craddock et al. (2008) and Paris et al. (2013). MC-ICP-MS requires pre-treatment of the sample prior to injection as the injection is mainly a solution sample compared to EA-IRMS. For soluble sulfates (e.g. gypsum), it is sufficient to dissolve them by heating them directly in Milli-Q deionised water and then purify them with a cation exchange resin (Craddock et al., 2016). The determination of sulfur isotopes using MC-ICP-MS with a solution injection gives a much higher internal precision (RSE better than 0.1‰) than conventional methods (Craddock et al., 2008).

Sulfur isotope tests are calibrated for instrumental mass discrimination by the standard sample bracketing (SSB) method. In practice, the SSB method is widely used for the mass discrimination correction of stable isotopes. The assumption is that the mass discrimination of the sample and the standard sample will respond equally to time during the test, provided that the sample has the same properties as the standard. Based on this, it is possible to correct for the mass discrimination of the sample by means of the specimen before and after the sample. It is therefore essential to ensure that the sample and the specimen have the same properties during the test. Consistency of properties between sample and specimen mainly includes consistency of matrix and concentration.

7.3 Fractionation processes

Sulfur is a variable-valence element that can form negative-valence sulfides, zero-valence sulfur and up to n-hexavalent sulfates in different environments, a property that facilitates the fractionation of sulfur isotopes. The fractionation of sulfur isotopes is closely related to the biogeochemical cycling of sulfur. Two types of fractionation mechanisms are responsible for the naturally occurring sulfur isotope variations:

(1) Kinetic isotope effects during microbial processes. Micro-organisms have long been known to fractionate isotopes during their sulfur metabolism, particularly during dissimilatory sulfate reduction, which produces the largest fractionations in the sulfur cycle. Fractionation enriches the light isotope ^{32}S in the product.

(2) Various chemical exchange reactions between both sulfate and sulfides and the different sulfides themselves. Usually occurring in closed, high-temperature environments deep within the Earth, the exchange results in the enrichment of heavy isotopes in oxygenated compounds with high chemical valence, such as SO_4^{2-}.

7.3.1 Thermodynamic equilibrium Reactions

Thermodynamic equilibrium fractionation is one of the main forms of sulfur isotope partitioning effects. This fractionation mainly occurs in various hydrothermal systems in nature. In isotope exchange reactions, the highly oxidized state of sulfur is always enriched in ^{34}S, e. g.

$$^{32}SO_4^{2-} + ^{34}SO_2 \rightleftharpoons ^{34}SO_4^{2-} + ^{32}SO_2, \ \alpha = 1.015\ (250℃) \quad (7.5)$$

When the different valence states of sulfide are in isotopic equilibrium, the order of ^{34}S enrichment is

$$SO_4^{2-} > SO_3^{2-} > SO_2 > SCO > S_x > H_2S \approx HS > S^{2-} \quad (7.6)$$

There have been a number of theoretical and experimental determinations of sulfur isotope fractionations between coexisting sulfide phases as a function of temperature. Theoretical studies of fractionations among sulfides have been undertaken by Sakai (1968) and Bachinski (1969), who reported reduced partition function ratios and bond strengths of sulfide minerals and described the relationship of these parameters to isotope fractionation. In a manner similar to that for oxygen in silicates, there is a relative ordering of ^{34}S-enrichment among coexisting sulfide minerals. Mineral bond strength order based on studies of heat of generation, free energy, lattice energy and structural type of minerals (Table 7.1). Considering the three most common sulfides (pyrite, sphalerite and galena) under conditions of isotope equilibrium, pyrite is always the most ^{34}S enriched mineral and galena the most ^{34}S depleted, sphalerite displays an intermediate enrichment in ^{34}S.

Table 7.1 Equilibrium isotope fractionation factors of sulfides with respect to H_2S.

Mineral	Chemical composition	A
Pyrite	FeS_2	0.40
Sphalerite	ZnS	0.10
Pyrrhotite	FeS	0.10
Chalcopyrite	$CuFeS_2$	−0.05
Covellite	CuS	−0.40
Galena	PbS	−0.63
Chalcosite	Cu_2S	−0.75
Argentite	Ag_2S	−0.80

The temperature dependence is given by A/T^2 (after Ohmoto and Rye, 1979).

The experimental determinations of sulfur isotope fractionations between various sulfides do not exhibit good agreement. The most suitable mineral pair for temperature determination is the sphalerite-galena pair. Rye (1974) has argued that the Czamanske and Rye (1974) fractionation curve gives the best agreement with filling temperatures of fluid inclusions over the temperature range from 370℃ to 125℃. By contrast, pyrite-galena pairs do not appear to be suitable for a temperature determination, because pyrite tends to precipitate over larger time periods of ore deposition than galena, implying that these two minerals may frequently not be contemporaneous. The equilibrium isotope fractionations for other sulfide pairs are generally so small that they are not useful as geothermometers. Ohmoto and Rye (1979) critically examined the available experimental data and presented a summary of what they believe to be the best S-isotope fractionation data. They concluded that the isotopic composition of sulfur-containing compounds in hydrothermal systems is a function of the average isotopic composition of the total sulfur in the system, temperature, oxygen escape, acidity and alkali metal ion strength. Temperature affects the isotopic composition of sulfur-containing compounds in two ways. Firstly, temperature affects the rate and extent of isotopic exchange between sulfur-containing compounds; secondly, temperature changes the isotopic composition of the sulfur-containing compounds in the system accordingly. pH, oxygen fugacity and alkali metal ion strength, commonly referred to as physicochemical conditions, are factors that affect the molar fraction of each sulfur-containing compound in the system and hence its sulfur isotope composition. These

S-isotope fractionations relative to H_2S are shown in Figure 7.3.

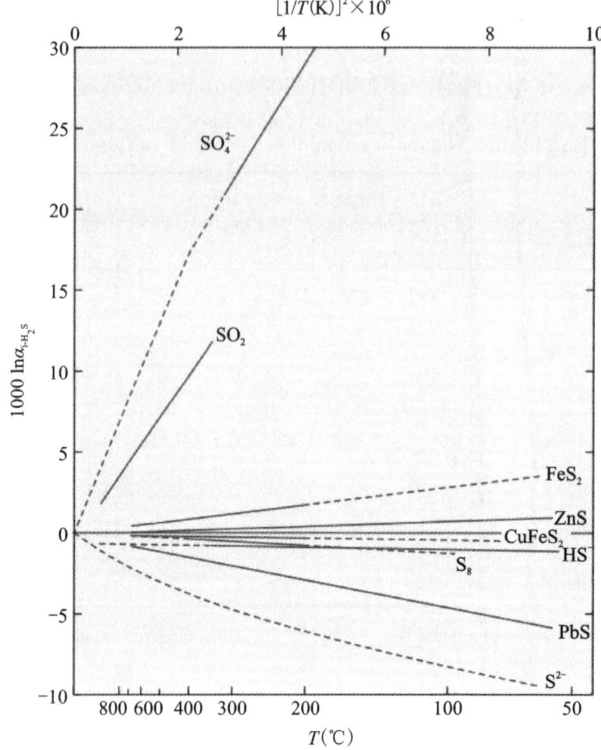

Figure 7.3 Equilibrium fractionations among sulfur compounds relative to H_2S (solid lines experimentally determined, dashed lines extrapolated or theoretically calculated) (after Ohmoto and Rye, 1979).

7.3.2 Bacterial Sulfate Reduction

Allotropic reduction of bacterial sulphate is the main cause of significant fractionation of sulfur-oxygen isotopes in sulphate at low temperatures and is usually carried out by a large group of microorganisms (more than 100 known to date; Canfield, 2001). These organisms gain energy for growth by reducing sulphate alongside oxidized reduced carbon (or H_2). Sulfate reducers are widely distributed in anoxic environments. They can tolerate temperatures from −1.5℃ to over 100℃ and salinities ranging from fresh to saline water.

The degree of isotopic fractionation of bacterial sulfate in reduction reactions is highly variable. Early laboratory studies with pure cultures of mesophilic sulfate reducing bacteria produced sulfide depleted in ^{34}S by 4‰ up to 47‰ (McCready, 1975; Bolliger et al., 2001) and for decades this maximum value was considered to be a possible limit for the microbial dissimilatory process (Canfield and Teske, 1996). More recently, sulfur isotope fractionations have been determined from incubations with sediments containing natural populations covering a wide spectrum of environments (from rapidly metabolizing microbial mats to slowly metabolizing coastal sediments). Sim et al. (2011) found that the type of organic electron donor is essential in controlling the magnitude of sulfur isotope fractionations of pure culture sulfate reducing bacteria, with complex substrates leading to sulfur isotope discrimination exceeding 47‰. Naturally occurring sulfides in sediments and euxinic water are commonly depleted in ^{34}S by up to 65‰ (Jørgensen et al., 2004), covering the range of experiments with sulfate reducing bacteria (Sim et al., 2011). Recent studies have demonstrated that natural populations are able to fractionate S-isotopes by up to 70‰ under in situ conditions (Wortmann et al., 2001; Rudnicki et al., 2001; Canfield et al., 2010).

The pathway of the bacterial sulphate reduction reaction consists of four enzyme-catalyzed steps: ①Sulfate (SO_4^{2-}) enters the bacterial cell; ②Inside the cell, sulfate is activated by adenosine triphosphate (ATP) to produce adenosine phosphosulfate (APS); ③APS is then reduced to sulfite (SO_3^{2-}); ④Sulfite is reduced to sulfide (H_2S) by the action of reductase. The result of kinetic fractionation of sulfur isotopes leads to a significant enrichment of ^{32}S in the sulfides formed by the reduction and an enrichment of ^{34}S in the remaining sulphate (Kaplan and Rittenberg, 1964). The breaking of the first S-O bond during bacterial sulfate reduction limits the rate of the overall reaction and controls the degree of isotopic fractionation throughout the process. It is generally accepted that the magnitude of sulfur isotope fractionation during bacterial sulphate reduction is related to the bacterial species reacting, the type of electron donor, the rate of reaction, the degree of opening and closing of the system, and temperature.

(1) The effect of bacterial species. Detmers et al. (2001) studied the sulfur isotope fractionation of 32 species of sulphate-reducing bacteria and found that those sulfur-reducing bacteria that completely oxidized the carbon source to CO_2 resulted in more significant sulfur isotope fractionation relative to those that ultimately oxidized the carbon source to acetate. The former showed an isotopic fractionation of 25‰, while the latter showed an isotopic fractionation of 9.5‰. They suggest that this is related to their metabolic pathways and the way in which sulphate travels through the cell membrane.

(2) The effect of electron donor. Different types of carbon sources affect the number and activity of reducing bacteria and further affect the sulfur isotope fractionation. The rate varied considerably, with isotopic enrichment factors ranging from 16.1‰ to 36.0‰; strains obtained in pure culture caused greater fractionation than strains cultured enriched from the environment under the same carbon source; and sulfur isotope fractionation was greater in media inoculated with petroleum hydrocarbon-rich substrates than in media inoculated with organic acid types. Rees (1978) noted that lactate, a bacterial nutrient, when concentration varied, it affected the rate of hydrogen sulfide production, but did not significantly alter the overall sulfur isotope fractionation effect.

(3) The effect of reaction rate on the degree of fractionation. The degree of fractionation is highest at low rates and lowest at high rates. Kaplan and Rittenberg (1964) and Habicht and Canfield (1997) suggested that fractionations depend on the specific rate ($cell^{-1}$ $time^{-1}$) and not so much on absolute rates ($volume^{-1}$ $time^{-1}$). Habicht and Canfield (1997) demonstrated that the degree of sulfur isotope partitioning has a significant negative correlation with the reaction rate. One of the intermediate products of the bacterial sulphate reduction reaction, SO_3^{2-}, undergoes the following disproportionation reactions in response to bacterial action:

$$4SO_3^{2-} + 2H_2O \rightleftharpoons H_2S + 3SO_4^{2-} \tag{7.7}$$

The rate of sulfite disproportionation affects the sulfur isotope fractionation and the size of the fractionation is inversely related to the rate of disproportionation. The smaller the rate of disproportionation reaction of the sulfite, the greater the isotopic fractionation between SO_3^{2-} and H_2S and between SO_3^{2-} and SO_4^{2-}. Habicht et al. (1997) measured an average sulfur isotopic fractionation of 28‰ between sulfite and hydrogen sulfide and 9‰ between sulfate and sulfite during disproportionation.

(4) The effect of the open or closed state of the system. An "open" system has an infinite reservoir of sulfate in which continuous removal from the source produces no detectable loss of material. Typical examples are the Black Sea and local oceanic deeps. In such cases, H_2S is extremely depleted in ^{34}S while consumption and change in ^{34}S remain negligible for the sulfate (Neretin et al., 2003). As long as environmental conditions do not change significantly, isotope fractionation will remain variable within a certain range and the isotopic fractionation factor is often a constant value. In a "closed" system, the preferential loss of the lighter isotope from the reservoir has feedback on the isotopic composition of the unreacted source material. The changes in the ^{34}S-content of residual sulfate and of the H_2S are modeled in Figure 7.4, which shows that $\delta^{34}S$-values of the residual sulfate steadily increase with sulfate consumption (a linear relationship on the log-normal plot). The curve for the derivative H_2S is parallel to the sulfate curve at a distance which depends on the magnitude of the fractionation factor. As shown in Figure 7.4, H_2S may become isotopically heavier than the original sulfate when about 2/3 of the reservoir has been consumed. The $\delta^{34}S$-curve for "total" sulfide asymptotically approaches the initial value of the original sulfate. It should be noted, however, that apparent "closed-system" behavior of covarying sulfate and sulfide $\delta^{34}S$-values might be also explained by "open-system" differential diffusion of the different sulfur isotope species (Jørgensen et al., 2004).

(5) The effect of temperature on fractionation. Temperature can regulate sulfate-reducing communities in natural populations (Kaplan and Rittenberg, 1964; Brüchert et al., 2001). Furthermore, differences in fractionation with temperature relate to differences in the specific temperature response to internal enzyme kinetics as well as cellular properties

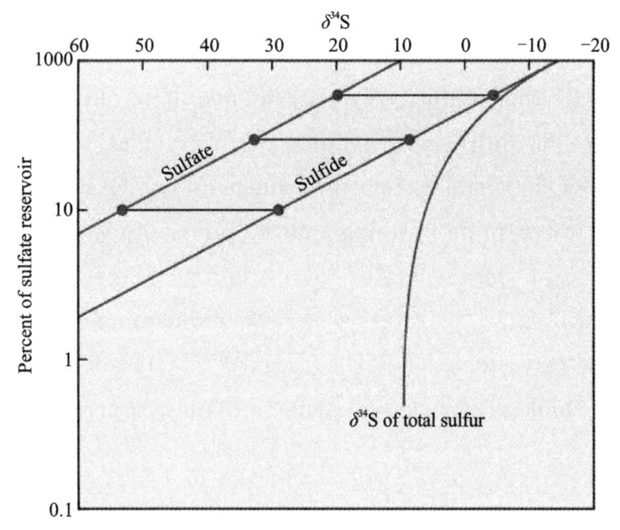

Figure 7.4 Rayleigh plot for sulfur isotope fractions during reduction of sulfate in a closed system. Assumed fractionation factor 1.025, assumed composition of initial sulfate: +10‰.

and corresponding exchange rates of sulfate in and out of the cell of mesophilic sulfate reducing bacteria. Considering different types (including thermophilic) of sulfate-reducers, Canfield et al. (2006), however, found in contrast to earlier assumptions high fractionations in the low and high temperature range, but lowest fractionations in the intermediate temperature range.

In marine sediments typically 90% of the sulfide produced during sulfate reduction is reoxidized (Canfield and Teske, 1996). The pathways of sulfide oxidation are still poorly known, and include both biological and abiological oxidation to sulfate, elemental sulfur and other intermediate compounds (Fry et al., 1988). Reoxidation of sulfide often occurs via compounds in which sulfur has intermediate oxidation states (sulfite, thiosulfate, elemental sulfur, polythionates) that do not accumulate, but are readily transformed and can be anaerobically disproportionated by bacteria. Therefore, Canfield and Thamdrup (1994) suggested that through a repeated cycle of sulfide oxidation to sulfur intermediates like elemental sulfur and subsequent disproportionation, bacteria can additionally generate ^{34}S depletions that may add on the isotopic composition of marine sulfides. Distinct sulfur isotope fractionation between generated sulfide and sulfate was also shown to occur upon bacterial disproportionation of other intermediates like thiosulfate and sulfite (Habicht et al., 1998).

Finally, it should be mentioned that sulfate is labeled with two biogeochemical isotope systems, sulfur and oxygen. Böttcher et al. (1998) and Brunner et al. (2005) argued that a single characteristic $\delta^{34}S$-$\delta^{18}O$ fractionation slope for this process does not exist, but that the isotope covariations depend on cell-specific sulfate reduction rates and associated oxygen isotope exchange rates with cellular water. Despite the extremely slow abiotic oxygen isotope exchange of sulfate with ambient water, $\delta^{18}O$ in sulfate obviously depend on the $\delta^{18}O$ of water via an exchange of sulfite with water. Böttcher et al. (1998) and Antler et al. (2013) demonstrated how the fractionation slopes depend on the net sulfate reduction rate: higher rates result in a lower slope meaning that sulfur isotopes increase faster relative to oxygen isotopes. The critical parameter for the evolution of oxygen and sulfur isotopes in sulfate is the relative difference in rates of sulfate reduction and of intracellular sulfite oxidation. Furthermore, Böttcher et al. (2001) argued that the disproportionation of sulfur intermediates in highly biologically active sediments may superimpose on the dominant sulfate reduction trend.

7.3.3 Thermochemical Reduction of Sulfate

In contrast to bacterial reduction, thermochemical sulfate reduction is an abiotic process with sulfate being reduced to sulfide under the influence of heat rather than bacteria(Krouse et al. ,1988). A high activation energy is required for the reduction to proceed and the main controlling factor is not the microorganism but the temperature. The lower temperature limit at which thermochemical sulphate reduction can occur has been a key issue of controversy. That is, whether thermochemical sulphate reduction can proceed at a slightly higher limit than microbial reduction. There is increasing evidence from natural occurrences that the reduction of aqueous sulfates by organic compounds can occur at temperatures as low as 100℃, given enough time for the reduction to proceed (Krouse et al. ,1988; Machel et al. , 1995). Sulfur isotope fractionations during thermochemical reduction generally should be smaller than during bacterial sulfate reduction, although experiments by Kiyosu and Krouse (1990) have indicated S-isotope fractionations of 10‰～20‰ in the temperature range of 200～100℃.

To summarize, bacterial sulfate reduction is characterized by large and heterogeneous ^{34}S-depletions over very small spatial scales, whereas thermogenic sulfate reduction leads to smaller and "more homogeneous" ^{34}S-depletions.

7.4 Variations of sulfur isotopes in groundwater

Different possible sources of sulfur in groundwater have characteristic isotope signatures (Figure 7.5). While some of these signatures may cover wide ranges that partially overlap, and a single sample may reflect an integrated and complex history, $\delta^{34}S$ has been used to distinguish between marine, metamorphic, anthropogenic and bacteriogenic sources of sulfur in terrestrial environments (Bottrell et al. , 2008; Calmels et al. , 2007;

Karim and Veizer, 2000; Otero et al., 2008; Pawellek et al., 2002; Samborska et al., 2013; Turchyn et al., 2013; Tuttle et al., 2009; Yuan and Mayer, 2012). Sulfur in alluvial aquifer systems is imported and exported in the form of groundwater and surface water flow, sediment and biomass transport, atmospheric deposition and gaseous diffusion. During these processes, atmospheric deposition, dissolution and precipitation of sulphate minerals, mineralization of organic sulfur and uptake of SO_4^{2-} by organisms and oxidation of sulfides do not undergo significant sulfur isotopic fractionation, so the sulfur isotopic composition of SO_4^{2-} in groundwater reflects the result of mixing of different sources of sulfur. However, bacterial sulphate reduction significantly alters the sulfur isotopic composition of SO_4^{2-} in groundwater, resulting in a significant increase in its $\delta^{34}S$ value, a phenomenon that often occurs in deep groundwater in reducing environments.

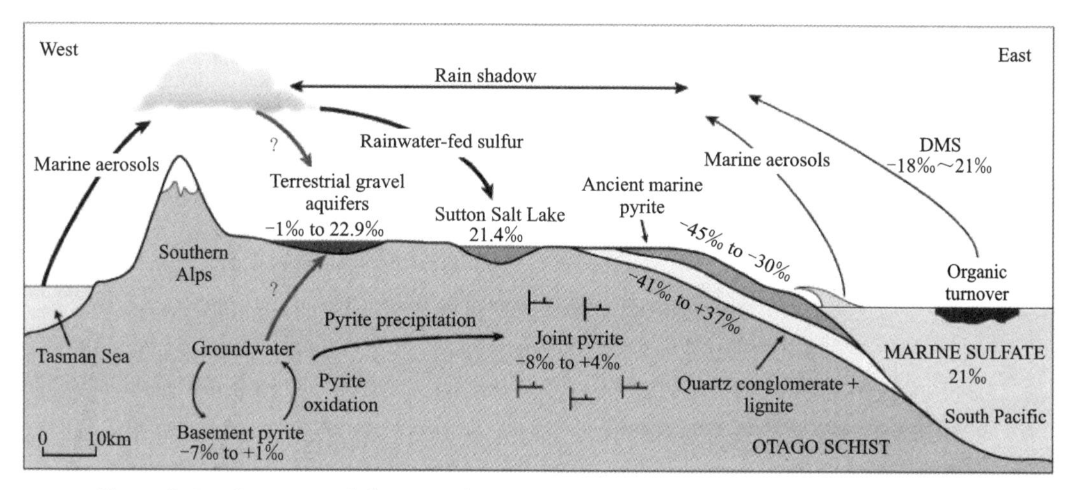

Figure 7.5 Summary of fluxes and isotope composition of processes in the terrestrial sulfur cycle across southern New Zealand. Not to scale (Rosalie, 2016).

Geological sulfide sources include magmatic and sedimentary sulfides, both of which may be homogenized or fractionated during metamorphism, and evaporite minerals (Figure 7.6). Magmatic sulfide $\delta^{34}S$ can range from $-10‰$ to $+10‰$, and metavolcanics should fall within this range. Metasediments may show a wider $\delta^{34}S$ range, depending on the $\delta^{34}S$ of any original bacteriogenic sulfide, but metamorphism tends to homogenize $\delta^{34}S$ (Canfield, 2004; Garrels and Lerman, 1981). Oxidative weathering of metamorphic pyrite may occur directly in the presence of oxygen, or via reduction of oxidized Fe^{3+} minerals (e.g. hematite). Fractionation during weathering is minimal ($<1‰$; Balci et al., 2007; Heidel and Tichomirowa, 2011), and so the oxidized products should reflect the $\delta^{34}S$ of the leached minerals. Additionally, weathering of evaporite minerals may be a significant geological sulfur source in some regions (Stallard and Edmond, 1983). Marine sulfate $\delta^{34}S$ has ranged between $\sim 15‰$ and $\sim 25‰$ over the Mesozoic and Cenozoic (Paytan et al., 1998).

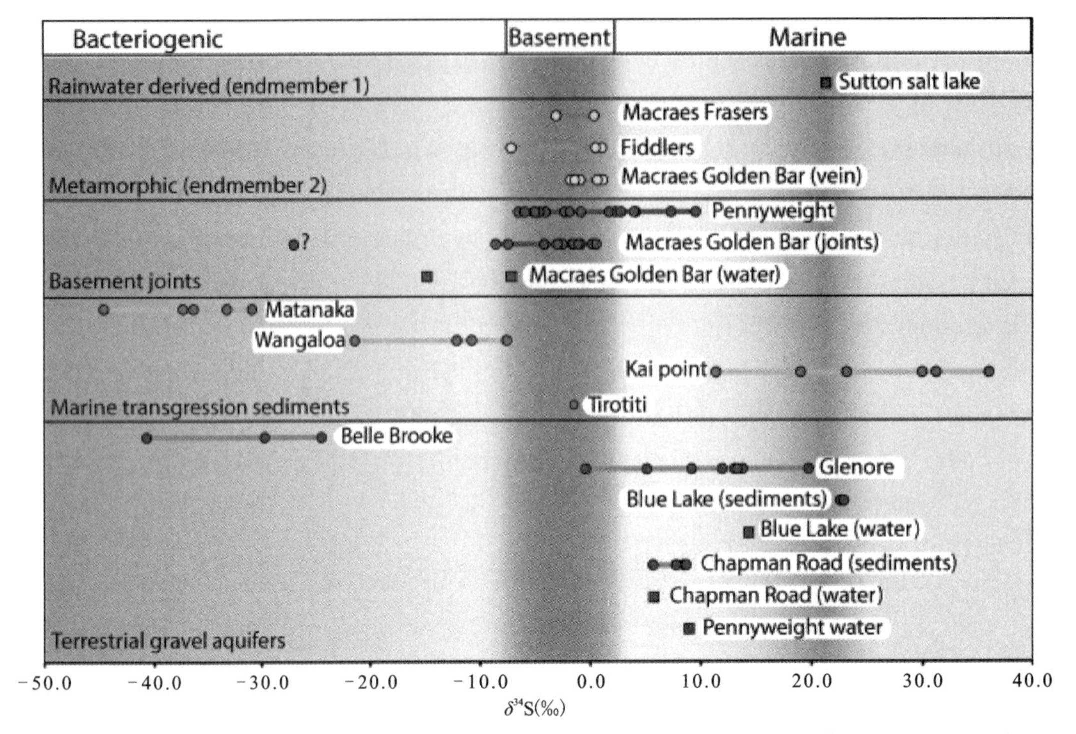

Figure 7.6 Summary of the distribution of $\delta^{34}S$ in different types of SO_4^{2-}. Rainwater-derived marine aerosols (blue square), and metamorphic basement (yellow circles). Joint pyrite (red) shows a mostly basement signal, with one outlying sample denoted by a question mark. Terrestrial gravel aquifers and Au-placers (purple) show a partly marine influenced signal. Marine transgression sediments (orange) are shown for comparison. Squares represent data from dissolved sulfate in surface water, and circles represent sulfide minerals.

The concentration and $\delta^{34}S$ of marine sulfate are globally homogeneous. The traditional accepted value for the $\delta^{34}S$ of marine sulfate is 20.3‰ ± 0.8‰ (2σ), but Rees (1978) demonstrated that there is a bias towards lighter values when measuring $\delta^{34}S$ via combustion to sulfur dioxide, and that $\delta^{34}S$ is closer to 21‰ (Paytan et al., 1998; Tostevin et al., 2016). Marine sulfate is the main source of atmospheric sulfur in coastal areas (Wadleigh et al., 1996), but organic sulfur compounds, such as dimethyl-sulfide (DMS), may constitute an additional minor source in regions of excess productivity, including saline mud flats and marine upwelling zones. DMS is typically depleted by as much as 3‰ compared with marine sulfate, but remains heavy compared with terrestrial sulfur sources (Amrani et al., 2013).

In anoxic environments, sulfate is reduced to sulfide, and in the presence of reactive Fe^{2+} this sulfide may be preserved as iron sulfide minerals. When sulfate reduction is biologically mediated, the light isotope is preferentially metabolized, resulting in a wide range of sulfur isotope signatures depleted by as much as 70‰ compared with starting sulfate (Canfield, 2001a, 2001b; Canfield et al., 2010; Detmers et al., 2001; Habicht et al., 2002; Leavitt et al., 2013; Sim et al., 2011). In contrast to these bacteriogenic signatures, inorganic thermochemical sulfate reduction (TSR) from groundwater below the oxygenated zone is accompanied by smaller isotope fractionations, 0～20‰, that may be positive or negative (Ohmoto and Goldhaber, 1997; Watanabe et al., 2009). TSR is limited to high temperature environments rarely found in groundwater reservoirs (Cross et al., 2004; Machel et al., 1995).

Anthropogenic signatures can originate from atmospheric sources, such as the burning of coal, and surface runoff from fertilizer or sewage. These two sources tend to have overlapped signatures around 0 (Samborska et al., 2013). In a case study from Pennsylvania, animal manure ranges from 3‰ to 5‰ and fertilizer between 2‰ and 9‰ (Cravotta, 2002). In England, fertilizer ranges from 0 to 6‰ (Moncaster et al., 2000), and in Spain, −2‰ to +7‰ (Otero and Soler, 2002). The $\delta^{34}S$ of fertilizer and sewage will vary depending on local farm practice.

7.5 Application

7.5.1 Geochemistry of sulfur isotopes in the high arsenic groundwater system (Xie et al., 2009)

The geochemical processes of sulfate residues have an important influence on the occurrence of redox environments and other biometabolic processes in groundwater. Sulfate reduction is often closely linked to the transport and transformation of contaminants (nitrate, organic matter and arsenic). The application of sulfur-oxygen isotopes of sulphate and their relationship with SO_4^{2-} concentrations can give a good indication of the occurrence of bacterial sulphate reduction and thus reveal the geochemical processes associated with it and their environmental significance.

The hydrologic and geologic background of Datong Basin was well described in the section of 4.4.1. For the present study, sixteen groundwater wells and two springs were selected for sampling in August 2007. The regions sampled include the recharge areas along the west and east margins of the basin and the high arsenic groundwater area close to the Huangshui River (Figure 7.7b and Figure 7.7c).

7.5.1.1 Range and distribution of As, sulfate and sulfur isotopes

Twelve out of the 16 groundwater samples contained arsenic exceeding the WHO guideline concentration of 10μg/L. The total dissolved concentration of arsenic in groundwater samples ranged from 3.4μg/L to 670μg/L. Two springs near the recharge areas

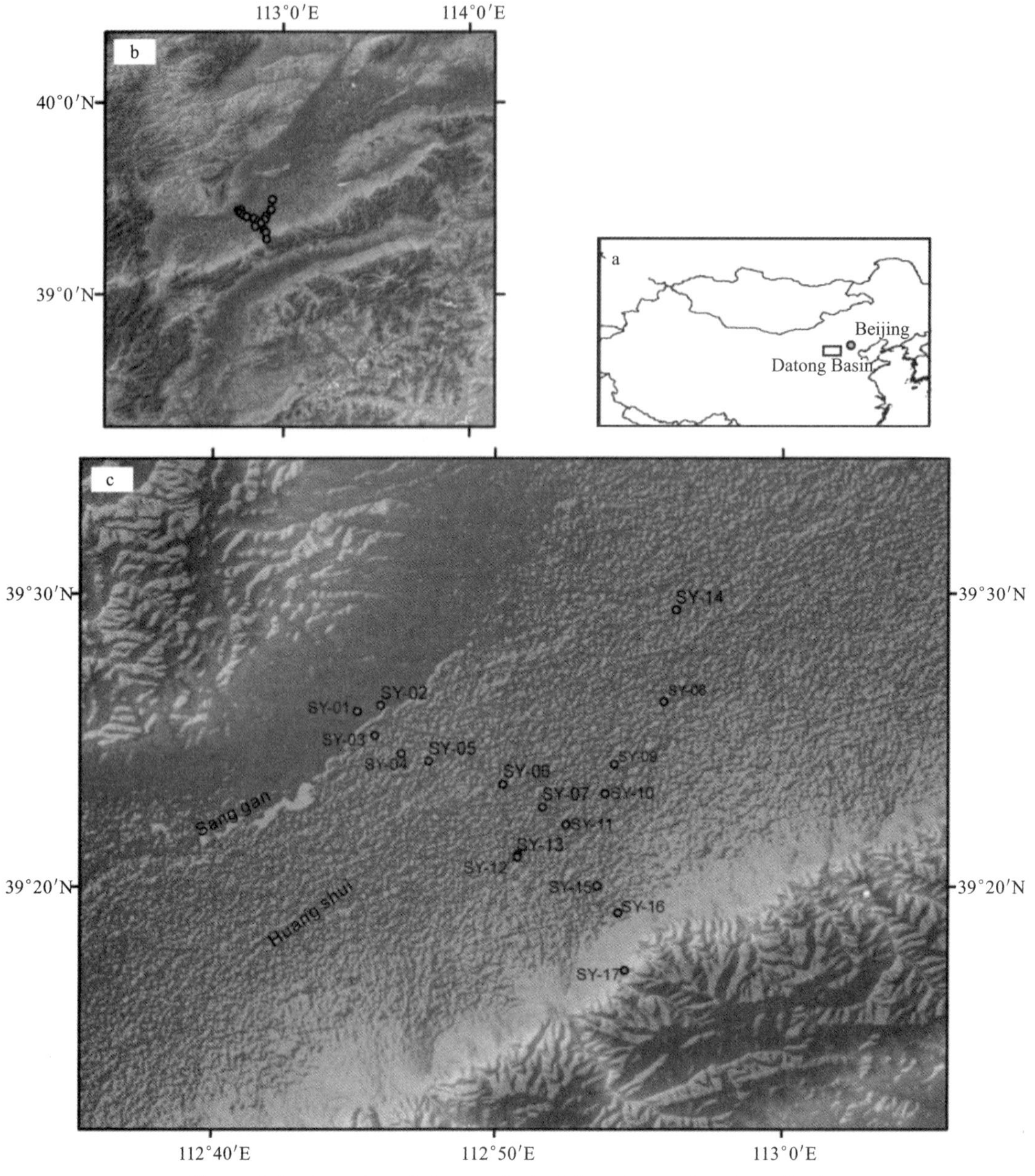

Figure 7.7 Location of Datong Basin (a) and sampling sites (b,c).

contained relatively low total dissolved arsenic (0.64μg/L and 3.1μg/L, respectively). As expected, arsenic concentrations are relatively low in the groundwater from the mountain front recharge areas. All the groundwater samples collected from the central basin have arsenic concentrations higher than 68μg/L.

The concentrations of SO_4^{2-} ranged from below detection limit in the basin to 365.7mg/L along the west recharge area (Table 7.2). With one exception, SO_4^{2-} values were above 100mg/L in the recharge areas and below 66mg/L in the basin. This distribution coincides with low arsenic in the recharge areas and high arsenic in the basin. The two springs had SO_4^{2-} concentrations of 107mg/L and 108.3mg/L respectively.

Table 7.2 Concentration of arsenic species and sulfur isotope composition of groundwater from Datong Basin.

Sample ID		Type	Depth (m)	*Eh* (mV)	$\delta^{34}S$ (‰)	SO_4^{2-} (mg/L)	As (Ⅲ) (μg/L)	As (Ⅴ) (μg/L)
Western margin	SY-01	Groundwater	20	75	10.4	111	<1	3.4
	SY-02	Spring		98	9.7	107	<1	3.1
	SY-03	Groundwater	30	−8	13.6	230	<1	3.9
	SY-04	Groundwater	33	90	18.7	169	24	3.8
	SY-05	Groundwater	40	−197	15.9	366	1.8	3.1
Center of the basin	SY-06	Groundwater	40	−212	N.D	<0.1	86	72
	SY-07	Groundwater	40	−214	33.4	173	400	120
	SY-08	Groundwater	50	−224	36.1	0.2	170	93
	SY-09	Groundwater	26	−289	23.5	66	61	5.1
	SY-10	Groundwater	120	−234	N.D	<0.1	92	40
	SY-11	Groundwater	30	−232	−2.5	0.9	230	120
	SY-12	Groundwater	23	−53	N.D	39.9	3.2	660
	SY-13	Groundwater	25	−126	8.5	0.7	200	100
Outlier	SY-14	Groundwater	40	−290	N.D	0.6	25	77
Eastern margin	SY-15	Groundwater	25	25	14.2	26.7	2.1	41
	SY-16	Groundwater	20	−170	N.D	108	10	23
	SY-17	Spring		N.D	8.7	114	N.D	N.D

N.D: not determined.

The $\delta^{34}S_{[SO_4]}$ values of aqueous sulfate from wells and springs exhibit wide isotopic variations (Table 7.2). The $\delta^{34}S_{[SO_4]}$ values for the west and east areas ranged from +8.8‰ to +15.9‰ while for the central basin had a wide range from −2.5‰ to +36.1‰. The wide ranges of the isotopic values probably reflect varying amounts of redox cycling in the basin and/or mixing of different water, though the highly enriched values are probably an indicator of sulfate reduction. No obvious correlation was observed between $\delta^{34}S_{[SO_4]}$ values SO_4^{2-} (Figure 7.8a). The $\delta^{34}S_{[SO_4]}$ values of samples in the center of basin (with the exception of SY-11; −2.5‰) are either at the same level or enriched relative to those from the margins. Considering the flow paths from the recharge areas to the basin center, there appears to be no correlation between $\delta^{34}S_{[SO_4]}$ and arsenic (Figure 7.8b). However, ignoring SY-11 and SY-13 with very low sulfate and depleted $\delta^{34}S_{[SO_4]}$, the samples in the basin with enriched $\delta^{34}S_{[SO_4]}$ values are associated with elevated contents of arsenic. Note that SY-11 and SY-13 are upgradient of the rest of the basin.

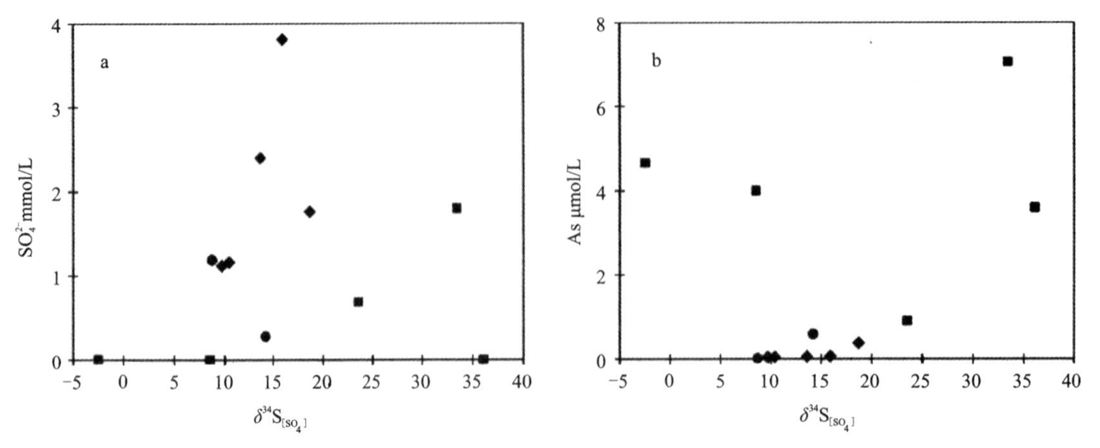

Figure 7.8 Concentration of SO_4^{2-} (a) and total arsenic (b) vs. $\delta^{34}S_{[SO_4]}$ in groundwater from Datong Basin.

7.5.1.2 As-S model

Sulfur isotopes could shed some light on the dominant process for sulfur cycling and in particular, sulfate reduction. The sulfur isotope signature of sulfate is a function of dissolution of sulfate minerals within aquifers or in surrounding bedrocks; atmospheric SO_4^{2-} via precipitation, and redox reactions within aquifers. There is no evidence for gypsum or other sulfate mineral in this area, although even small amounts of easily soluble sulfate salts that have been formed by evaporation in this semi-arid region will tend to dominate the dissolved load. The wide ranges of S isotope values in the basin indicate that the water are not well mixed, which is consistent with slow groundwater flow and solute transport. The $\delta^{34}S_{[SO_4]}$ values of two springs (+8.7‰ and +9.7‰, respectively) are close to and slightly higher than the $\delta^{34}S_{[SO_4]}$ values of atmospheric SO_4^{2-} in northern China (ranging from +3.7‰ to +8.1‰ with the average of +5.6‰, Mukai et al., 2001). The range of 8‰ to 15‰ of sulfate in the margins is within the range expected for evaporites (Clark and Fritz, 1997) and it is likely that 8‰~15‰ is the range for background sulfate from both the west and east margins. The $\delta^{34}S_{[SO_4]}$ of the most upgradient wells sampled is within this range. Though speculative at this point if we assume that the background $\delta^{34}S_{[SO_4]}$ values are around 8‰~15‰, the different trends towards enrichment and depletion of sulfur isotope ratios (about ±20‰) (Table 7.2, Figure 7.8) in the central parts of the basin indicate that both reduction of sulfate and oxidation of depleted sulfides may have taken place. Sulfate-reducing bacteria preferentially utilize ^{32}S and result in elevation of $\delta^{34}S$ value of SO_4^{2-} (Kaplan and Rittenberg, 1964). Except for the two samples (SY-11, SY-13) from the upgradient region of the basin, the high $\delta^{34}S_{[SO_4]}$ and high total arsenic content (Figure 7.8) and low sulfate concentrations and low *Eh* values are consistent with sulfate reduction accompanying arsenic reduction and mobilization. Interpreting the two upgradient wells in the central basin will need a more detailed study along that flow path. The low $\delta^{34}S_{[SO_4]}$ values observed (Figure 7.8), may be related to re-oxidization of depleted sulfide. Given that this is a sulfate-limited system, depleted sulfide formed due to sulfate reduction would have to occur early in the reduction reaction. As the amount of sulfate gets depleted, the unreduced sulfate will be enriched in the heavier isotope (Rayleigh distillation) and therefore the corresponding sulfide from this sulfate-enriched pool will be also be enriched relative to the sulfide formed in the earlier stage of the reaction.

In summary, $\delta^{34}S_{[SO_4]}$ values of the central basin have a wide range (from −2.5‰ to +36.1‰), possibly reflecting microbial reduction/oxidation of sulfate. The relation (albeit weak) between $\delta^{34}S_{[SO_4]}$ and total arsenic content indicates that sulfate reduction is probably occurring or occurred in the vicinity where arsenic is mobile. There are two samples with high total arsenic contents that have low $\delta^{34}S_{[SO_4]}$ values (−2.5‰ and +8.5‰, respectively), which could not be explained by microbial reduction alone but rather by reoxidation of sulfate (or some other sources).

7.5.2 Tracing nitrate and sulphate sources in groundwater using isotope mixing models (Torres-Martínez et al., 2020)

Sulfur isotopes provide a sensitive means to determine the origin of dissolved sulfate in surface water, and associated authigenic sulfide minerals precipitated from deep groundwater. Using an isotopic mixing model to quantify the groundwater nitrate and sulfate sources. The results are useful for decision-makers or water managers to better understand groundwater nitrate and sulfate pollution at the scale across which nitrate and sulfate is added to and accumulates within a waterhed and its underlying aquifer system and integrate this knowledge into the development of groundwater management plans.

7.5.2.1 Study Area

Monterrey is located north-northeast of the foothills of Sierra Madre Oriental (SMO) mountain range, this area is a hilly region with mountains rising in the west (Las Mitras) and southeast (La Silla) of the city and elevations varying from 260m to 3000m above sea level (m.a.s.l.) (Figure 7.9). The climate is semiarid with annual mean precipitation and temperature of 622mm and 22.3℃, respectively, with spatiotemporal variations for monthly average precipitation between 14.1mm and 150.6 mm, and monthly average temperature ranging from 14.5℃ to 28.4℃.

Groundwater provides approximately 40% of the urban water supply and is extracted mainly from three different aquifers or wellfields. These will be referred to herein as the Buenos Aires wellfield, Mina wellfield and Monterrey Metropolitan Area (MMA) aquifer. The Buenos Aires wellfield supplies approximately half of the total municipal water for Monterrey, covering the southwestern part of the city (Oesterreich and Medina Aleman, 2002). This wellfield consists of 42 deep production wells (usually >700m deep, and water level oscillating between 12m and 120m below ground level) delivering high-quality drinking water [twenty-six of these wells (62%) are used for drinking water supply], mainly from confined, Early Cretaceous limestone formations (Cupido, Aurora, Cuesta del Cura) from the Buenos Aires valley. The valley lies at a higher elevation than the city, so the cost of conveying the water is low. Horizontal infiltration galleries (Santiago system) also tap into this aquifer system (Figure 7.9c). Significantly, the land above the MMA aquifer is primarily urban, whereas the land overlying the Buenos Aires and Mina wellfields is piedmont and desert shrubland and mixed woodland (Figure 7.9b). Thirty-nine groundwater production wells, one spring, and the influent of 4 principle wastewater treatment plants (WWTP) were sampled (Figure 7.9b,c). Sites were selected using criteria such as their geographic distribution and representation of different well fields and aquifers, and relevance for Monterrey water supply system.

Samples were classified into five groups according to their statistical similarities and geographic correspondence to understand local geochemical trends. The obtained clusters were spatially defined as follow: Group 1 comprises samples collected from the Buenos Aires wellfield and La Estanzuela spring; Group 2 represents wells located in Mina wellfield; Groups 3, 4, and 5 are all located in the MMA aquifer and represent samples in the recharge area, transition zone and discharge area of this aquifer, respectively. Statistical summary of physicochemical parameters and isotopic ratios according to groups were shown in Table 7.3.

7.5.2.2 Pollution sources and attenuation processes of sulfate

The $\delta^{34}S_{[SO_4]}$ and $\delta^{18}O_{[SO_4]}$ signatures of samples from the study area show compositional ranges from +1.03‰ to +12.16‰, and from +5.79‰ to +12.12‰, respectively. Four sources can be related to SO_4^{2-} according to Figure 7.10a: atmospheric deposition from SO_2 emissions, the influence of soil-derived SO_4^{2-}, sewage and marine evaporites.

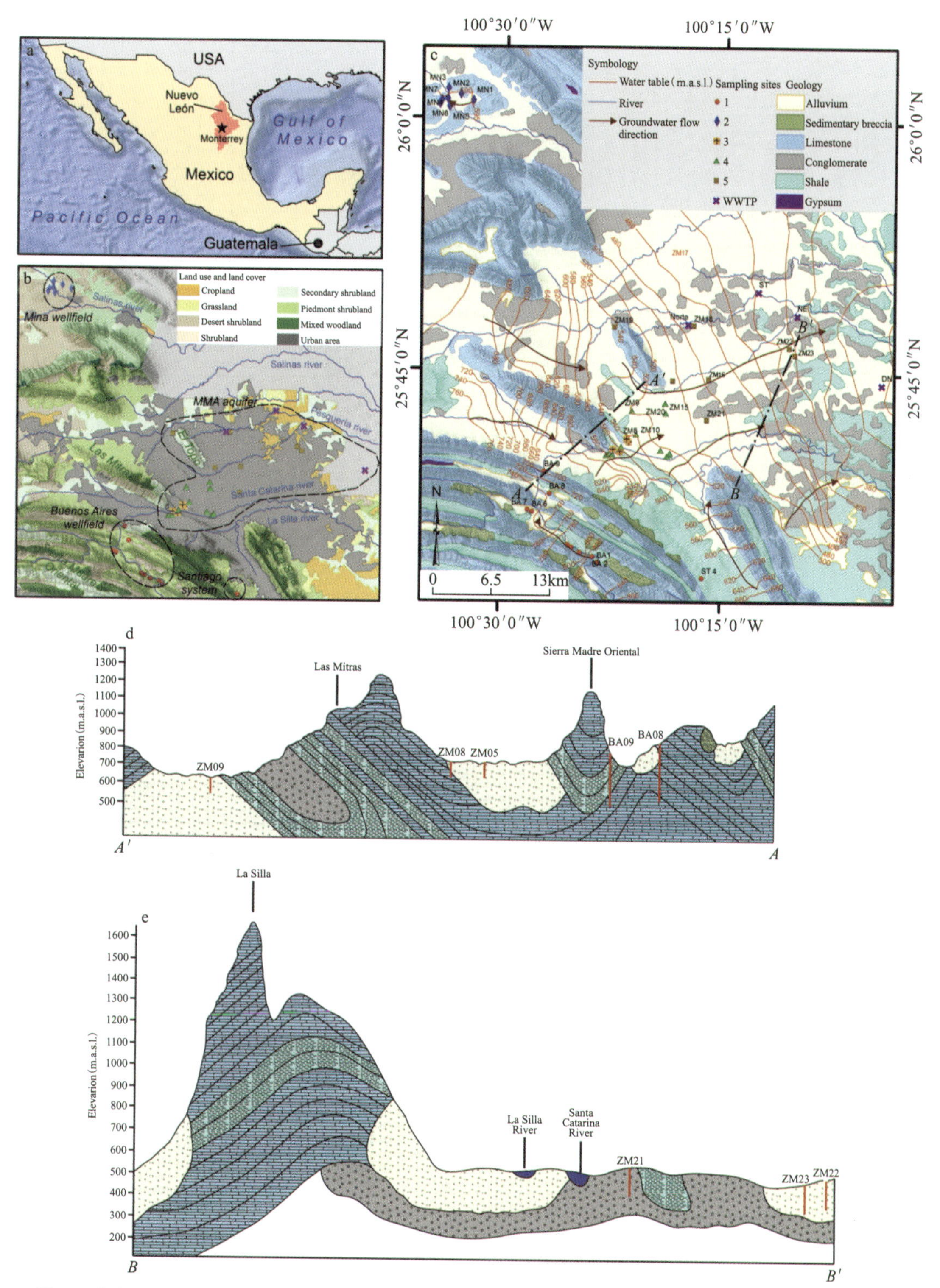

Figure 7.9 a. Location of the Monterrey Metropolitan Area (MMA) and the State of Nuevo Leon within Mexico; b. land use and land cover; c. surficial geology; d, e. hydrogeological cross-sections. Panels b and c include the locations of wells sampled for chemical and isotopic analysis.

Table 7.3 Statistical summary of physicochemical parameters and isotopic ratios according to groups.

Parameter	Unit	Group 1(n=10)		Group 2(n=7)		Group 3(n=4)		Group 4(n=12)		Group 5(n=7)	
		Mean	SD	Mean	SD	Mcan	SD	Mean	SD	Mcan	SD
Temp	℃	22.87	1.43	30.37	0.38	30.02	1.61	24.90	0.74	26.84	1.80
pH	—	7.35	0.31	7.20	0.10	7.37	0.09	7.14	0.10	6.98	0.17
EC	μS/cm	445.49	39.75	955.54	73.67	560.20	8.71	779.46	139.48	1 650.86	369.73
ORP	mV	362.42	8.14	373.00	13.89	328.25	17.56	368.33	33.18	343.14	30.79
TDS	mg/L	285.39	25.41	623.16	77.72	357.80	5.69	520.53	41.57	1 056.64	238.61
DO	mg/L	7.17	0.54	3.72	0.28	5.58	0.73	6.76	0.48	5.86	1.17
Ca	mg/L	74.47	3.66	110.04	7.76	93.53	1.01	111.10	9.11	181.89	36.95
Mg	mg/L	9.20	3.11	12.66	1.13	9.72	0.09	15.89	3.07	21.40	7.13
Na	mg/L	2.31	0.80	59.27	6.52	6.91	0.81	23.11	3.17	80.49	38.32
K	mg/L	0.46	0.10	1.17	0.25	0.41	0.27	1.08	0.42	0.77	0.49
HCO_3	mg/L	239.46	16.93	257.64	3.63	253.17	3.72	299.03	28.02	349.78	46.83
SO_4	mg/L	41.49	18.65	156.33	25.61	70.83	4.76	101.62	16.37	327.39	120.75
C	mg/L	3.73	1.30	119.57	19.45	18.33	1.47	40.62	5.75	151.74	52.96
NO_3	mg/L	1.18	0.43	1.46	0.36	3.58	0.42	11.22	2.69	17.02	7.57
F	mg/L	0.28	0.12	0.57	0.05	0.60	0.03	0.27	0.06	0.43	0.14
B	pg/L	14.60	5.76	62.43	23.34	21.00	4.69	75.67	28.29	115.57	146.28
Br	μg/L	23.10	5.86	85.00	25.07	21.25	3.10	110.83	33.46	217.57	280.77
I	μg/L	2.27	0.45	2.93	1.08	2.05	0.42	4.97	1.78	13.79	25.14
$\delta^{18}O-H_2O$	‰V-SMOW	−9.21	0.22	−7.54	0.16	−7.15	0.19	−7.56	0.80	6.91	1.28
$\delta^{2}H-H_2O$	‰V-SMOW	−61.63	1.84	−48.58	1.14	14.07	0.89	−50.08	5.57	−44.55	8.97
$\delta^{18}O-NO_3$	‰V-SMOW	1.68	1.14	5.20	0.68	4.04	2.89	2.39	0.79	5.58	2.60
$\delta^{15}N-NO_3$	‰Air	3.77	0.70	6.82	0.83	3.88.	0.38	9.64	0.60	12.55	2.02
$\delta^{18}O-SO_4$	‰V-SMOW	8.53	0.84	11.07	0.71	8.13	0.64	7.44	1.02.	9.38	0.37
$\delta^{34}S-SO_4$	‰V-CDT	4.89	2.58	9.56	1.97	6.65	1.05	6.65	0.79	7.67	1.32

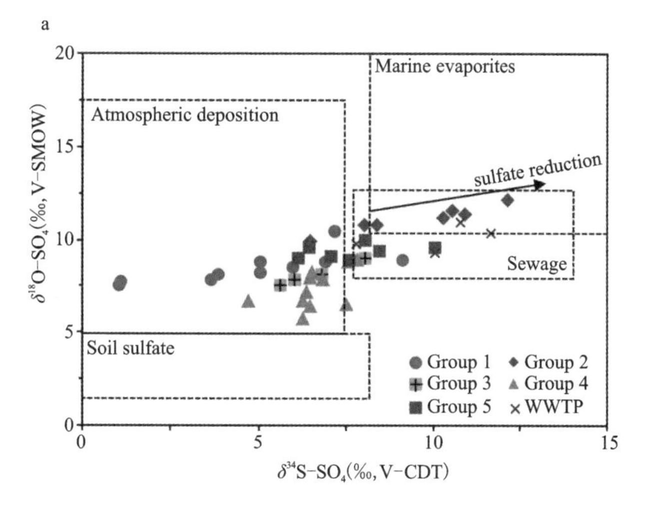

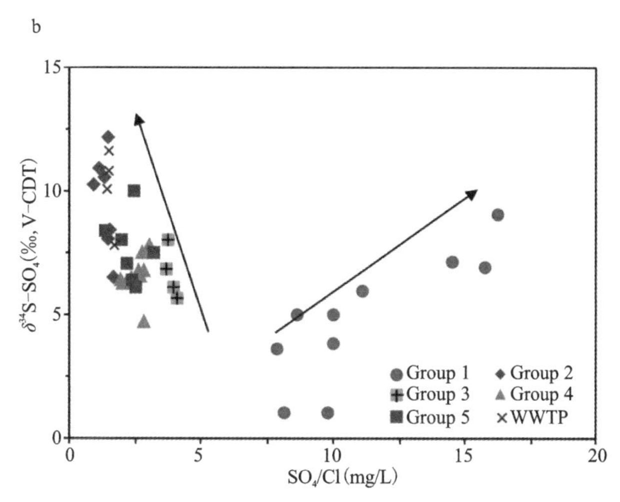

Figure 7.10 a. Dual-isotope plot of $\delta^{34}S_{[SO_4]}$ and $\delta^{18}O_{[SO_4]}$ in water samples collected in the study area. The isotopic composition of the primary SO_4 sources is represented by atmospheric deposition, soil sulfate, sewage and marine evaporites (Pittalis et al., 2018; Puig et al., 2017); b. $\delta^{34}S_{[SO_4]}$ vs. SO_4/Cl ratio in water samples collected in the study area.

Samples from Groups 1, 3 and 4 plot almost all within the range of atmospheric deposition, while samples from Group 5 lie between atmospheric deposition and sewage. For most samples of Group 2 it was not possible to identify the source of sulfate due to an overlapping of source fields (i. e. marine deposits, sewage) and sulfate reduction processes identified by the typical ratio trends between 1 : 4 and 1 : 2.5 (Mizutani and Rafter, 1973) (Figure 7.10a). This process is described by the following equation:

Furthermore, in all samples $\delta^{34}S_{[SO_4]}$ decreases with increasing SO_4/Cl ratio, except for Group 1 (Figure 7.10b), suggesting that isotopically light SO_4 was removed and the residual SO_4 became enriched in ^{34}S and ^{18}O during sulfate reduction in Groups 2, 3, 4, and

5 (Pittalis et al., 2018). For Group 1, an inverse trend is observed with smaller initial sulfate isotope values and larger SO_4/Cl ratios than the rest.

The dual-isotope diagram of $\delta^{15}N_{[NO_3]}$ vs. $\delta^{34}S_{[SO_4]}$ further constrains the variety of sources are responsible for each of these contaminants even within the same aquifer (Figure 7.11). It confirms that the main sources of pollution in the sites close to recharge areas (Groups 1, 2, and 3) are atmospheric deposition and mineralization of soil organic matter. In contrast, an additional source, the infiltration of sewage leaks, was identified for samples located in the urban area (Groups 4 and 5).

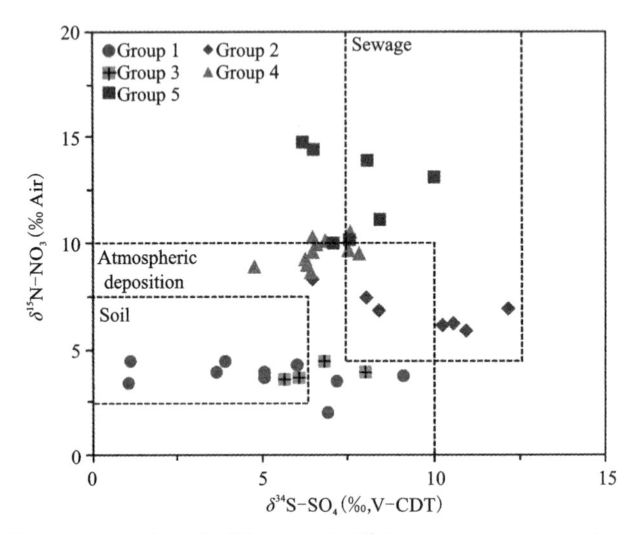

Figure 7.11 Dual isotope plot of $\delta^{34}S_{[SO_4]}$ and $\delta^{15}N_{[NO_3]}$ in water samples collected in the study area. The isotopic composition of the main pollution sources is represented by atmospheric deposition, soil, sewage and manure (Otero et al., 2009; Puig et al., 2017).

7.5.2.3 Apportionment of nitrate and sulfate using a Bayesian isotope mixing model

Estimation of the proportional contribution of nitrate, sulfate from three identified potential sources by Bayesian mixture models (MixSIAR). The results reveal that the nitrate source contribution in the study area generally followed soil organic nitrogen (SON) > manure and sewage (M&S) > atmospheric deposition (AD), with mean and standard deviation values of 60.6‰±12.4%, 37.8‰±12.5%, and 1.6‰±1.2%, respectively, as illustrated in Figure 7.12a, c. Specifically, Groups 1, 2, and 3 showed similar patterns, with the highest shares derived from SON with 96.8‰±4.2%, 68.4‰±15.9%, and 78.4‰±11.3%, respectively. In contrast, for Groups 4 and 5, the dominant nitrate source was M&S (54.5‰±18.6%, and 84.5‰±12.5%, respectively). Atmospheric deposition contributed little nitrate to the groundwater. Sulfate source contribution generally followed manure and sewage (M&S) > atmospheric deposition (AD) > marine evaporites (ME) >

soil sulfate (SS) with mean and standard deviation values of 40.6‰±18%, 24.9‰±10%, 18.4‰±5.5%, and 16.1‰±11.9%, respectively. The main sulfate source was M&S for Group 1 (38.9‰±29.8%), Group 3 (43.4‰±22.6%), Group 4 (57.0‰±22.7%), and 5 (41.1‰±21.6%), while for Group 2 it was marine evaporites (38.7‰±11.7%) (Figure 7.12b, c). This mixing model exercise helped to discern the contribution of distinct sulfate sources whose isotopic compositional ranges show overlaps in Figure 7.10a.

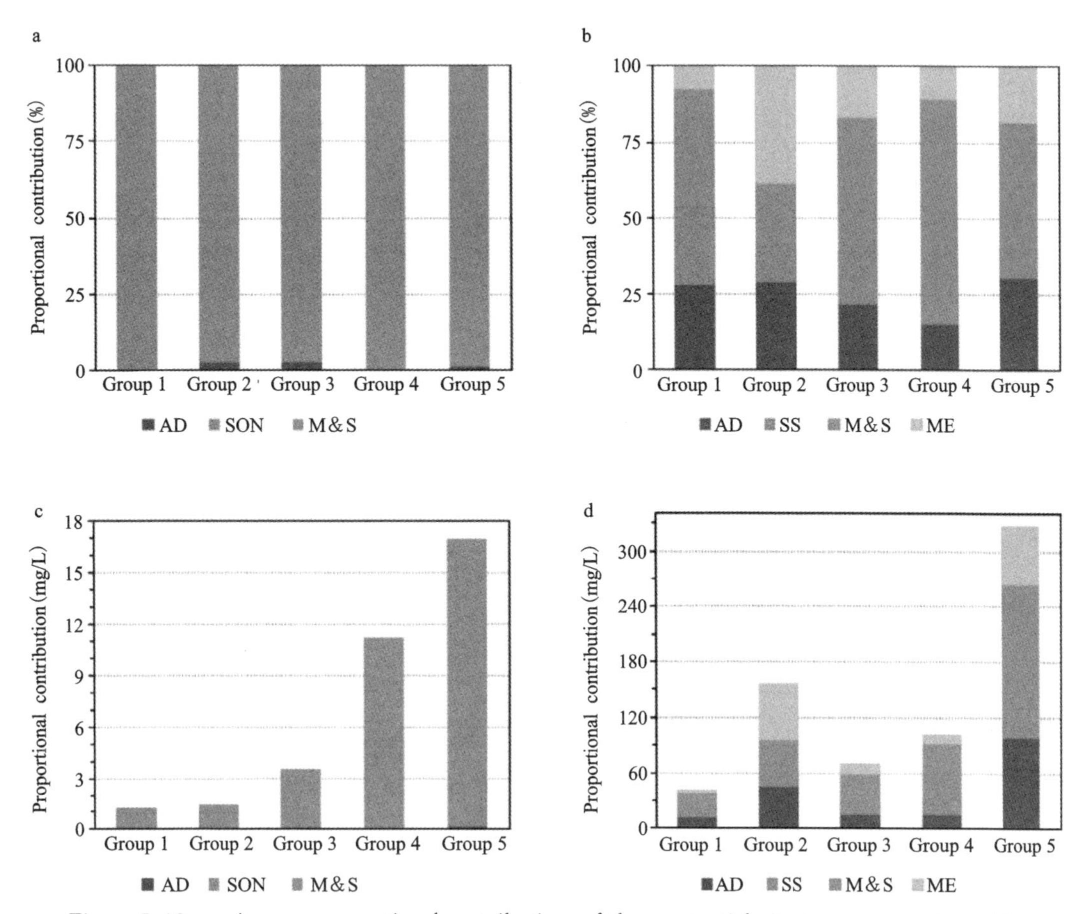

Figure 7.12 a. Average proportional contributions of three potential nitrate sources estimated by MixSIAR model for the different groups as percentage; b. Average proportional contributions of four potential sulfate sources as percentage estimated by MixSIAR model for the different groups as percentage; c. Average contribution of potential nitrate sources as concentration (mg/L); d. Average contribution of sulfate sources as concentration (mg/L).

References

AMRANI A, SAID-AHMAD W, SHAKED Y, et al., 2013. Sulfur isotope homogeneity of oceanic DMSP and DMS[J]. PNAS, 110:18 413-18 418.

ANTLER G, TURCHYN A V, RENNIE V, et al., 2013. Coupled sulphur and oxygen isotope insight into bacterial sulphate reduction in the natural environment[J]. Geochim. Cosmochim. Acta, 118: 98-117.

BACHINSKI D J, 1969. Bond strength and sulfur isotope fractionation in coexisting sulfides[J]. Econ. Geol., 64: 56-65.

BALCI N, SHANKS Ⅲ W C, MAYER B, et al., 2007. Oxygen and sulfur isotope systematics of sulfate produced by bacterial and abiotic oxidation of pyrite[J]. Geochim. Cosmochim. Acta, 71: 3796e3811.

BENDALL C, LAHAYE Y, FIEBIG J, et al., 2006. In-situ sulfur isotope analysis by laser-ablation MC-ICP-MS[J]. Appl. Geochem., 21: 782-787.

BOLLIGER C, SCHROTH M H, BERNASCONI S M, et al., 2001. Sulfur isotope fractionation during microbial reduction by toluene-degrading bacteria[J]. Geochim. Cosmochim. Acta, 65: 3289-3299.

BOTTRELL S, TELLAM J, BARTLETT R, et al., 2008. Isotopic composition of sulfate as a tracer of natural and anthropogenic influences on groundwater geochemistry in an urban sandstone aquifer, Birmingham, UK[J]. Appl. Geochem., 23: 2382-2394.

BöTTCHER M E, THAMDRUP B, VENNEMANN T W, 2001. Oxygen and sulfur isotope fractionation during anaerobic bacterial disproportionation of elemental sulfur[J]. Geochim. Cosmochim. Acta, 65: 1601-1609.

BRUNNER B, BERNASCONI S M, KLEIKEMPER J, et al., 2005. A model of oxygen and sulfur isotope fractionation in sulfate during bacterial sulfate reduction[J]. Geochim. Cosmochim. Acta, 69: 4773-4785.

BRÜCHERT V, KNOBLAUCH C, JÖRGENSEN B B, 2001. Controls on stable sulfur isotope fractionation during bacterial sulfate reduction in Arctic sediments[J]. Geochim. Cosmochim. Acta, 65: 763-776.

CALMELS D, GAILLARDET J, BRENOT A, et al., 2007. Sustained sulfide oxidation by physical erosion processes in the Mackenzie River Basin: climatic perspectives[J]. Geology, 35: 1003-1006.

CANFIELD D E, 2001a. Biogeochemistry of sulfur isotopes[J]. Rev. Mineral., 43: 607-636.

CANFIELD D E, TESKE A, 1996. Late Proterozoic rise in atmospheric oxygen concentration inferred from phylogenetic and sulphur-isotope studies[J]. Nature, 382: 127-132.

CANFIELD D E, THAMDRUP B, 1994. The production of ^{34}S depleted sulfide during bacterial disproportion to elemental sulfur[J]. Science, 266: 1973-1975.

CANFIELD D E, OLSEN C A, COX R P, 2006. Temperature and its control of isotope fractionation by a sulfate reducing bacterium[J]. Geochim. Cosmochim. Acta, 70: 548-561.

CANFIELD D E, 2004. The evolution of the Earth surface sulfur reservoir[J]. American Journal of Science,304 (10):839-861.

CANFIELD D E, FARQUHAR J, ZERKLE A L,2010. High isotope fractionations during sulfate reduction in a low-sulfate euxinic ocean analog[J]. Geology,38:415-418.

CHIVAS A R, ANDREWS A S, LYONS W B, et al. ,1991. Isotopic constraints on the origin of salts in Australian playas. 1. Sulphur[J]. Palaeogeog. Palaeoclimatol. Palaeoecol. , 84: 309-332.

CLARK I, FRITZ P,1997. Environmental isotopes in hydrogeology[M]. Boca Raton F L: Lewis Publishers:131-170.

CRADDOCK P R, ROUXEL O J, BALL L A,et al. ,2008. Sulfur isotope measurement of sulfate and sulfide by high-resolution MC-ICP-MS[J]. Chem. Geol. ,253:102-113.

CRAVOTTA C A, 2002. Use of Stable Isotopes of Carbon, Nitrogen, and Sulfur to Identify Sources of Nitrogen in Surface water in the Lower Susquehanna River Basin, Pennsylvania[J]. US Geological Survey (Water Supply Paper):2497.

CROSS M M, MANNING D A C, BOTTRELL S H. ,et al, 2004. Thermochemical sulphate reduction (TSR): experimental determination of reaction kinetics and implications of the observed reaction rates for petroleum reservoirs[J]. Org. Geochem. ,35:393-404.

CROWE D E, VALLEY J W, BAKER K L,1990. Micro-analysis of sulfur isotope ratios and zonation by laser microprobe[J]. Geochim. Cosmochim. Acta,54:2075-2092.

CZAMANSKE G K, RYE R O, 1974. Experimentally determined sulfur isotope fractionations between sphalerite and galena in the temperature range 600℃ to 275℃[J]. Econ. Geol. 69:17-25.

DE LAETER J R, BÖHLKE J K, DE BIèVRE P,et al. ,2003. Atomic weights of the elements: review 2000 (IUPAC technical report)[J]. Pure Appl. Chem. ,75:683-2000.

DETMERS J, BRUCHERT V, HABICHT K S, et al. , 2001. Diversity of sulfur isotope fractionations by sulfate-reducing prokaryotes[J]. Appl. Environ. Microbiol. ,67: 888-894.

ELDRIDGE C S, COMPSTON W, WILLIAMS I S, et al. , 1988. Sulfur isotope variability in sediment hosted massive sulfide deposits as determined using the ion microprobe SHRIMP. I. An example from the Rammelsberg ore body[J]. Econ. Geol. ,83: 443-449.

FRY B, RUF W, GEST H, et al. , 1988. Sulphur isotope effects associated with oxidation of sulfide by O_2 in aqueous solution[J]. Chem. Geol. ,73:205-210.

GARRELS R M, LERMAN A,1981. Phanerozoic cycles of sedimentary carbon and sulfur[J]. Proceedings of the National Academy of Sciences of the United States of America Physical Sciences,78 (8):4652-4656.

HABICHT K S, CANFIELD D E,1997. Sulfur isotope fractionation during bacterial sulfate reduction in organic-rich sediments[J]. Geochim. Cosmochim. Acta,61:5351-5361.

HABICHT K S, CANFIELD D E, RETHMEIER J C, 1998. Sulfur isotope fractionation during bacterial reduction and disproportionation of thiosulfate and sulfite[J]. Geochim. Cosmochim. Acta,62:2585-2595.

HEIDEL C, TICHOMIROWA M, 2011. The isotopic composition of sulfate from anaerobic and low oxygen pyrite oxidation experiments with ferric iron d new insights into oxidation mechanisms[J]. Chem. Geol. ,281:305-316.

HU G X, RUMBLE D, WANG P L,2003. An ultravioletlaser microprobe for the in-situ analysis of multisulfur isotopes and its use in measuring Archean sulphur isotope mass-independent anomalies[J]. Geochim. Cosmochim. Acta,67:3101-3118.

JOHNSTON D T, WOLFE-SIMON F, PEARSON A, et al. , 2009. Anoxygenic photosynthesis modulated Proterozoic oxygen and sustained Earth's middle age[J]. Proceedings of the National Academy of Sciences of the United States of America,106 (40): 16 925-16 929.

JǾRGENSEN B B, BÖTTCHER M A, LÜSCHEN H,et al. ,2004. Anaerobicmethane oxidation and a deep H_2S sink generate isotopically heavy sulfides in Black Sea sediments [J]. Geochim. Cosmochim. Acta,68:2095-2118.

KAPLAN I R, RITTENBERG S C, 1964. Microbiological fractionation of sulphur isotopes[J]. J. Gen. Microbiol. ,34:195-212.

KELLEY S P,FALLICK A E,1990. High precision spatially resolved analysis of $d^{34}S$ in sulphides using a laser extraction technique[J]. Geochim. Cosmochim. Acta,54:883-888.

KOZDON R, KITA R N, HUBERTY J M,et al. ,2010. In situ sulfur isotope analysis of sulfide minerals by SIMS: precision and accuracy with application to thermometry of similar to 3. 5Ga Pilbara cherts[J]. Chem. Geol. ,275:243-253.

KARIM A, VEIZER J, 2000. Weathering processes in the Indus River Basin: implications from riverine carbon, sulfur, oxygen, and strontium isotopes [J]. Chem. Geol. , 170 :153-177.

KIYOSU Y, KROUSE H R,1990. The role of organic acid in the abiogenic reduction of sulfate and the sulfur isotope effect[J]. Geochem. J. ,24:21-27.

KROUSE H R, VIAU C A, ELIUK L S,et al. ,1988. Chemical and isotopic evidence of thermochemical sulfate reduction by light hydrocarbon gases in deep carbonate reservoirs [J]. Nature,333:415-419.

LEAVITT W D, HALEVY I, BRADLEY A S, et al. , 2013. Influence of sulfate reduction rates on the Phanerozoic sulfur isotope record[J]. PNAS.

MACHEL H G, KROUSE H R, SASSEN P,1995. Products and distinguishing criteria of bacterial and thermochemical sulfate reduction[J]. Appl. Geochem. ,10:373-389.

MCCREADY R G L, 1975. Sulphur isotope fractionation by Desulfovibrioand Desulfotomaculum species[J]. Geochim. Cosmochim. Acta,39:1395-1401.

MONCASTER S J, BOTTRELL S H, TELLAM J H, et al., 2000. Migration and attenuation of agrochemical pollutants: insights from isotopic analysis of groundwater sulphate[J]. J. Contam. Hydrol., 43:147-163.

MUKAI H, TANAKA A, FUJII T, et al., 2001. Regional characteristics of sulfur and lead isotope ratios in the atmosphere at several Chinese urban sites[J]. Environ. Sci. Technol., 35:1064-1071.

NERETIN L N, BÖTTCHER M E, GRINENKO V A, 2003. Sulfur isotope geochemistry of the Black Sea water column[J]. Chem. Geol., 200:59-69.

OHMOTO H, GOLDHABER M B, 1997. Sulfur and carbon isotopes[J]// Barnes H L. Geochemistry of hydrothermal ore deposits, 3rd ed[M]. New York: Wiley Interscience, 435-486.

OHMOTO H, RYE R O, 1979. Isotopes of sulfur and carbon[J]// Geochemistry of hydrothermal ore deposits, 2nd ed[M]. New York: Holt Rinehart and Winston.

ONO S, SHANKS W C, ROUXEL O J, et al., 2007. S-33 constraints on the seawater sulphate contribution in modern seafloor hydrothermal vent sulfides[J]. Geochim. Cosmochim. Acta, 71:1170-1182.

OTERO N, SOLER A, 2002. Sulphur isotopes as tracers of the influence of potash mining in groundwater salinisation in the Lobregat Basin (NE Spain)[J]. Water Res., 36: 3989-4000.

OTERO N, SOLER A, CANALS A, 2008. Controls of $\delta d^{34}S$ and $d^{18}O$ in dissolved sulphate: learning from a detailed survey in the Llobregat River (Spain)[J]. Appl. Geochem., 23:1166e1185.

PARIS G, SESSIONS A, SUBHAS A V, et al., 2013. MC-ICP-MS measurement of $d^{34}S$ and $D^{33}S$ in small amounts of dissolved sulphate[J]. Chem. Geol., 345:50-61.

PAYTAN A, KASTNER M, CAMPBELL D, et al., 1998. Sulfur isotopic composition of Cenozoic seawater sulfate[J]. Science, 282:1459-1462.

PAWELLEK F, FRAUENSTEIN F, VEIZER J, 2002. Hydrochemistry and isotope geochemistry of the upper Danube River[J]. Geochim. Cosmochim. Acta, 66:3839-3853.

PITTALIS D, CARREY R, DA PELO S, et al., 2018. Hydrogeological and multi-isotopic approach to define nitrate pollution and denitrification processes in a coastal aquifer (Sardinia, Italy)[J]. Hydrogeol. J., 26:2021-2040.

PUIG R, SOLER A, WIDORY D, et al., 2017. Characterizing sources and natural attenuation of nitrate contamination in the Baix Ter aquifer system (NE Spain) using a multi-isotope approach[J]. Sci. Total Environ., 580:518-532.

REES C E, 1978. Sulphur isotope measurements using SO_2 and SF_6[J]. Geochim.

Cosmochim. Acta,42:383-389.

RUDNICKI M D, ELDERFIELD H, SPIRO B,2001. Fractionation of sulfur isotopes during bacterial sulfate reduction in deep ocean sediments at elevated temperatures[J]. Geochim. Cosmochim. Acta,65:777-789.

RYE R O, 1974. A comparison of sphalerite-galena sulfur isotope temperatures with filling-temperatures of fluid inclusions[J]. Econ. Geol., 69:26-32.

SAKAI H,1968. Isotopic properties of sulfur compounds in hydrothermal processes[J]. Geochem. J.,2:29-49.

SAMBORSKA K, HALAS S, 2010. ^{34}S and ^{18}O in dissolved sulfate as tracers of hydrogeochemical evolution of the Triassic carbonate aquifer exposed to intense groundwater exploitation (Olkusze Zawiercie region, southern Poland) [J]. Appl. Geochem., 25: 1397-1414.

SAMBORSKA K, HALAS S, BOTTRELL S H, 2013. Sources and impact of sulphate on groundwater of Triassic carbonate aquifers, Upper Silesia, Poland[J]. J. Hydrol., 486: 136-150.

SEAL R R, 2006. Sulfur isotope geochemistry of sulfide minerals[J]. Rev. Mineral Geochem.,61:633-677.

SEAL R R, ALPERS C N, RYE R O, 2000. Stable isotope systematics of sulfate minerals[J]. Rev. Mineral Geochem.,40:541-602.

SIM M S, BOSAK T, ONO S,2011. Large sulfur isotope fractionation does not require disproportionation[J]. Science, 333:74-77.

SOPHOCLEOUS M, 2002. Interactions between groundwater and surface water: the state of the science[J]. Hydrogeol. J., 10:52-67.

STALLARD R F, EDMOND J M, 1983. Geochemistry of the Amazon: 2. The influence of geology and weathering environment on the dissolved load[J]. J. Geophys. Res., 88: 9671-9688.

THODE H G, MACNAMARA J, COLLINS C B, 1949. Natural variations in the isotopic content of sulphur and their significance[J]. Can. J. Res.,27B:361.

TORRES-MARTíNEZ J A, MORA A, KNAPPETT P S K, et al., 2020. Tracking nitrate and sulfate sources in groundwater of an urbanized valley using a multi-tracer approach combined with a Bayesian isotope mixing model[J]. Water Res., 182: 115962.

TOSTEVIN R, CRAW D, VAN HALE R, et al., 2016. Sources of environmental sulfur in the groundwater system, southern New Zealand[J]. Appl. Geochem., 70: 1-16.

TROFIMOV A, 1949. Isotopic constitution of sulfur in meteorites and in terrestrial objects[J]. Dokl Akad Nauk SSSR, 66:181 (in Russian).

TUTTLE M L W, BREIT G N, COZZARELLI I M, 2009. Processes affecting d^{34}S and d^{18}O values of dissolved sulfate in alluvium along the Canadian River, central

Oklahoma, USA[J]. Chem. Geol., 265:455-467.

TURCHYN A V, TIPPER E T, GALY A, et al., 2013. Isotope evidence for secondary sulfide precipitation along the Marsyandi River, Nepal, Himalayas, Earth Planet [J]. Sci. Lett. http://dx.doi.org/10.1016/j.epsl.2013.04.033.

WADLEIGH M A, SCHWARCZ H P, KRAMER J R, 1996. Isotopic evidence for the origin of sulphate in coastal rain[J]. Tellus. B., 48:44-59.

WALKER J C G, 1986. Global geochemical cycles of carbon, sulfur and oxygen[J]. Marine Geology, 70 (1-2):159-174.

WATANABE Y, FARQUHAR J, OHMOTO H, 2009. Anomalous fractionations of sulfur isotopes during thermochemical sulfate reduction[J]. Science, 324:370-373.

WORTMANN U G, BERNASCONI S M, BÖTTCHER M E, 2001. Hypersulfidic deep biosphere indicates extreme sulfur isotope fractionation during single-step microbial sulfate reduction[J]. Geology, 29:647-650.

XIE X, ELLIS A, WANG Y, et al., 2009. Geochemistry of redox-sensitive elements and sulfur isotopes in the high arsenic groundwater system of Datong Basin, China[J]. Sci. Total Environ., 407(12): 3823-3835.

YUAN F, MAYER B, 2012. Chemical and isotopic evaluation of sulfur sources and cycling in the Pecos River, New Mexico, USA[J]. Chem. Geol., 291:13-22.

Chapter 8 Chlorine and Bromine Isotope

Both chlorine and bromine are the elements of the halogen group, which belong to the ⅦA group of elements in the periodic table. They have very similar atomic radii (181pm for Cl and 196pm for Br) and perform similar geochemical properties.

Chlorine exists predominantly as chloride ion, a trace component of all the Earth's geological compartments other than the oceans, its primary sink. Chlorine has two stable isotopes with the following abundances (Coplen et al. ,2002):

$^{35}Cl=75.78\%$

$^{37}Cl=24.22\%$

Natural isotope variations in chlorine isotope ratios might be expected due to the mass difference between ^{35}Cl and ^{37}Cl as well as to variations in coordination of chlorine in the vapor, aqueous and solid phases. Schauble et al. (2003) calculated equilibrium fractionation factors for some geochemically important species. They showed that the magnitude of fractionations systematically varies with the oxidation state of Cl, but also depends on the oxidation state of elements to which Cl is bound with larger fractionations for 2+cations than for 1+cations.

Bromine is widely distributed in nature but in relatively small concentrations compared to chlorine. Bromine has two stable isotopes with nearly equal abundances (Berglund and Wieser, 2011).

$^{79}Br=50.69\%$

$^{81}Br=49.31\%$

Bromine isotope fractionations between salts and brine are very small (Eggenkamp et al. ,2016). Although higher oxidation states of bromine exist in nature, little is known about the Br isotope composition of bromine oxyanions.

8.1 Chlorine and bromine species in groundwater

Chlorine and bromine reside in several major reservoirs(Figure 8.1): rock (the mantle and crust), soil (the pedosphere), freshwater (groundwater, lakes and rivers), salt-water (the oceans, saline lakes, inland seas, and subsurface crystalline and sedimentary brines), ice caps (the cryosphere), the lower atmosphere (the troposphere), and the middle atmosphere (the stratosphere).

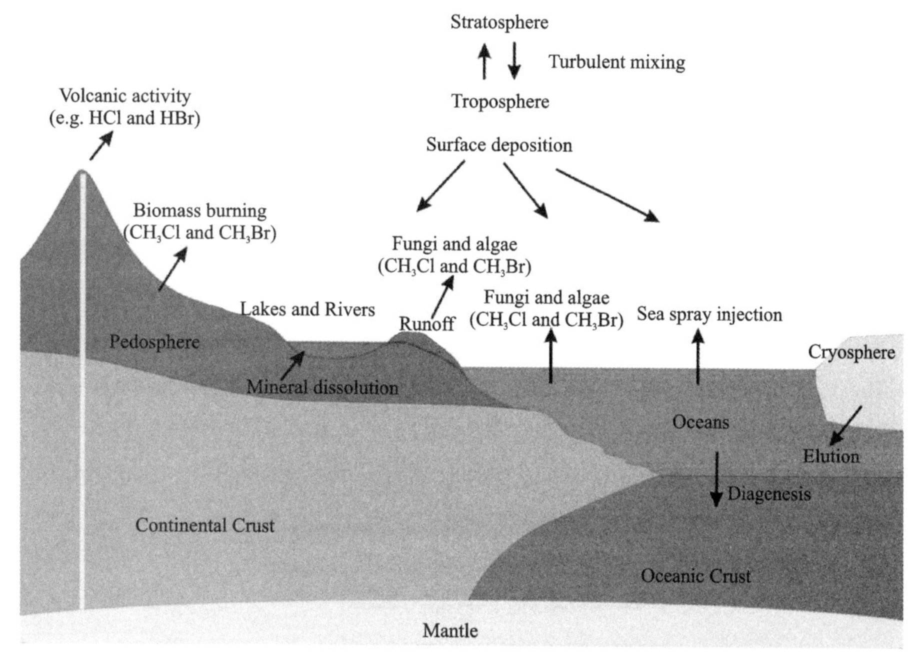

Figure 8.1 The Earth's major reservoirs of chlorine and bromine and some natural processes that transfer these two elements between these reservoirs (modified after Graedel and Keene, 1996).

8.1.1 Chlorine

The most common natural form of chlorine is the chlorine anion (Cl^-). Higher oxidation states for chlorine are very well known. In these higher oxidation states chlorine is present in the oxyanions hypochlorite (ClO^-), chlorite (ClO_2^-), chlorate (ClO_3^-) and perchlorate (ClO_4^-) where chlorine is in the +I, +Ⅲ, +V and +Ⅶ oxidation state respectively. As can be deducted from the Frost diagram (Frost, 1951) in acid solutions only chloride, chlorine and perchlorate are stable as the other forms tend to disproportionate, while in alkaline solutions chloride, hypochlorite, chlorite, chlorate and perchlorate can be stable. The hypochlorite ion is the strongest oxidiser of all chlorine oxyanions and it is a very unstable ion that only exists in solutions. When water is removed from a sodium hypochlorite solution it converts to a mixture of sodium chloride and sodium chlorate.

Chlorine isotopic composition is represented by the δ symbol, which is defined as the thousandth deviation of the $^{37}Cl/^{35}Cl$ ratio of the tested sample relative to the $^{37}Cl/^{35}Cl$ ratio of the standard sample. The calculation formula is as follows:

$$\delta^{37}Cl=\frac{R_{sam}-R_{sta}}{R_{sta}}\times 1000‰=\frac{(^{37}Cl/^{35}Cl)_{sam}-(^{37}Cl/^{35}Cl)_{sta}}{(^{37}Cl/^{35}Cl)_{sta}}\times 1000‰ \quad (8.1)$$

A summary of the chlorine stable isotope ranges of various materials is presented in Figure 8. 2. All ranges are reported relative to seawater "Standard Mean Ocean Chloride (SMOC)" as first proposed by Kaufmann (1984) and it has the value of 0. Recently, Godon et al. (2004) conducted a study on seawater chlorine isotopic composition where they analyzed 24 samples from different oceans and seas that confirmed the homogeneity and consistency of the 0 value assigned to SMOC by reporting that all values are within ±0. 08‰ (2σ).

In general, chlorine isotopes that exist in the chloride form in various natural materials and phases are found to range between ~−8‰ and ~+8‰. However, the vast majority of samples lie between −2‰ and +2‰. It seems that minerals and rocks are responsible, more than other sources, for extending the chlorine isotopic range to extreme positive values, while pore water in ocean and lake sediments are responsible for extending it to the other extreme.

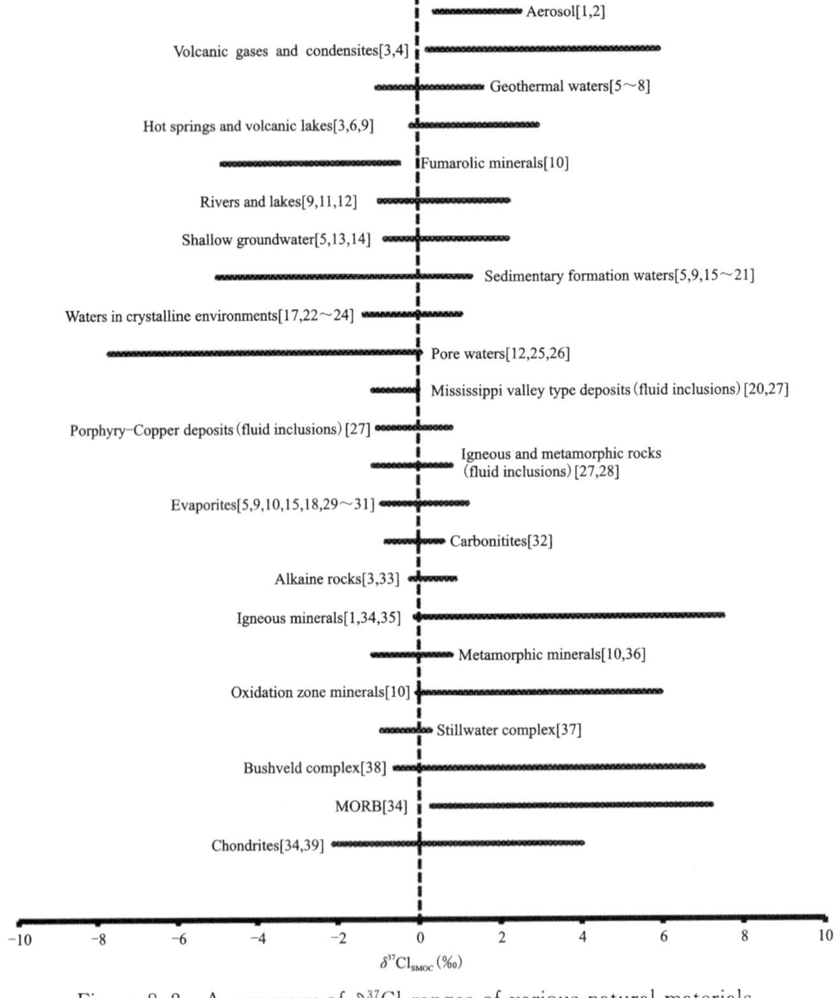

Figure 8. 2 A summary of δ^{37}Cl ranges of various natural materials.

8.1.2 Bromine

The most common natural form of bromine is also the bromide anion (Br^-). Higher oxidation states for bromine are comparable to those of chlorine and oxidation states +I and +V are well established. In acid solutions bromide, bromine and bromate are stable, while hypobromite will disproportionate, in alkaline solutions bromine will disproportionate to bromide and hypobromite. Bromine is also found in organic compounds. Compared to chlorine it has a much more important organic geochemistry and it is found in a considerable number of natural and artificial compounds.

Bromine isotopic composition is represented by the δ symbol, which is defined as the thousandth deviation of the $^{81}Br/^{79}Br$ ratio of the tested sample relative to the $^{81}Br/^{79}Br$ ratio of the standard sample. The calculation formula is as follows:

$$\delta^{81}Br=\frac{R_{sam}-R_{sta}}{R_{sta}}\times1000‰=\frac{(^{81}Br/^{79}Br)_{sam}-(^{81}Br/^{79}Br)_{sta}}{(^{81}Br/^{79}Br)_{sta}}\times1000‰ \quad (8.2)$$

Eggenkamp and Coleman (2000) were the first to propose the bromine isotopic composition of the ocean as standard and they called it Standard Mean Ocean Bromide (SMOB). They were also the first to report the bromine value in the standard delta notation ($\delta^{81}Br$) as the per mil deviation from the seawater composition that was assigned the value of 0. In their study, they analyzed bromine stable isotopes water samples from sedimentary and crystalline environments (Southern Ontario, Siberian Platform, Williston Basin, Canadian Shield, and the Fennoscandian Shield). The results revealed a total range of bromine stable isotopes between −1.43‰ and +3.35‰. The sedimentary formation water encompassed a larger range of isotopic variation (−1.43‰ to +3.35‰) in comparison to the water from the crystalline environments (+0.01‰ to +1.77‰). Figure 8.3 illustrates the $\delta^{81}Br$ ranges of different water from sedimentary and crystalline environments.

Eggenkamp (2014) collected the $\delta^{81}Br$ values of 128 groundwater samples from existing literature and found that: contrary to the $\delta^{37}Cl$ values, the $\delta^{81}Br$ values of 78.95% samples were positive($\delta^{81}Br \geqslant 0$) and 21.05% were negative ($\delta^{81}Br<0$).

8.2 Analytical techniques for the chlorine and bromine isotope

8.2.1 Chlorine

Measurements of chlorine isotope abundances have been made by different techniques. The first measurements by Hoering and Parker (1961) used gaseous chlorine in the form of HCl. The 81 samples measured exhibited no significant variations relative to the standard ocean chloride. In the early eighties a new technique has been developed by Kaufmann et al.

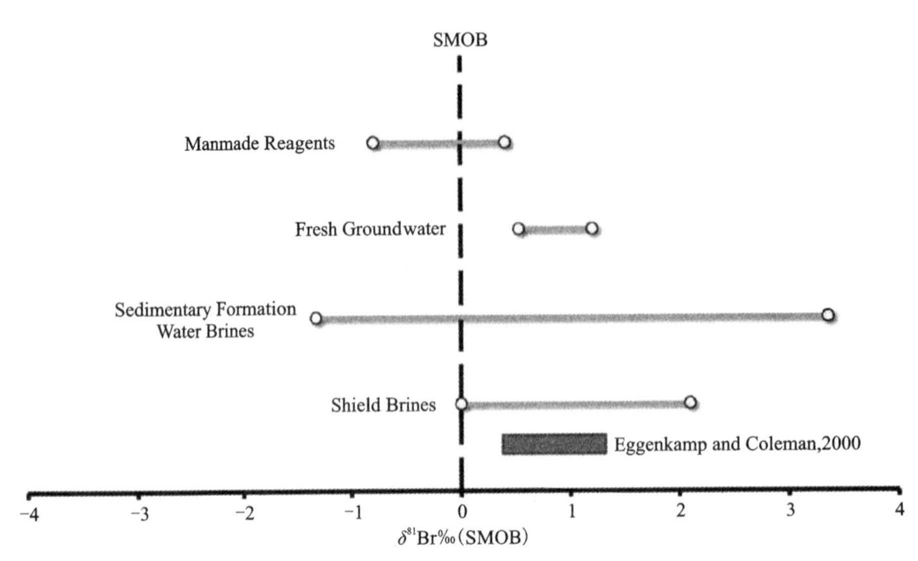

Figure 8.3 A summary of $\delta^{81}Br$ ranges of water from sedimentary and crystalline environments(Eggenkamp and Coleman, 2000; Shouakar-Stash et al., 2005).

(1984), that uses methylchloride (CH_3Cl). The chloride-containing sample is precipitated as AgCl, reacted with excess methyliodide, and separated by gas chromatography. The total analytical precision reported is near ±0.1‰ (Long et al.,1993; Eggenkamp,1994; Sharp et al.,2007). The technique requires relatively large quantities of chlorine (>1mg), which precludes the analysis of materials with low chlorine concentrations. Magenheim et al. (1994) described a method involving the thermal ionization of Cs_2Cl^+, which, as argued by Sharp et al. (2007), is very sensitive to analytical artefacts and therefore might lead to erroneous results. In any case both methods are laborintensive and rely on offline chemical conversion reactions. Recent attempts use continuous flow mass-spectrometry (Shouakar-Stash et al., 2005) or use MC-ICP-MS techniques (Van Acker et al., 2006). SIMS techniques have been described for glasses by Layne et al. (2004), Godon et al. (2004) and more recently by Manzini et al. (2017).

Currently, chlorine stable isotopes are determined by three different analytical techniques, two of which use thermal ionization mass spectrometry (TIMS). The first technique (NTIMS) is based on measuring [Cl^-] negative ions (Vengosh et al., 1989). This has reasonably high sensitivity; however, the technique is difficult and the precision is poor (>±0.4‰). The second technique (P-TIMS) is based on measuring positive ions [Cs_2Cl^+] (Xiao and Zhang, 1992; Magenheim et al., 1994; Xiao et al., 1995; Numata et al., 2001; Xiao et al., 2002). This technique is best for analyzing very small samples (1~50μg Cl) and has good precision (±0.2‰) for such small samples. The third technique is based on analyzing methyl chloride gas using dual inlet isotope ratio mass spectrometry (DI-IRMS) (Long et al., 1993). This last technique is reported to have the best precision

($\pm$0.09‰) available in analyzing chlorine stable isotopes; however, the technique is limited by the sample size requirement (300~1000μg of Cl^-) and also by the lengthy preparation process which limits the number of samples that can be analyzed in a prescribed period of time. Therefore, there is an interest in improving the methodology by lowering the required sample size without compromising the precision as well as shortening the analysis time in order to improve sample throughput.

Chlorine isotope samples in groundwater must be pre-treated to prepare Cl^- in groundwater into CH_3Cl gas suitable for instrument detection. During the pretreatment of samples, the amount of water sample taken is related to the concentration of Cl^-, and the general requirement is to ensure that the sample contains at least 3mg/L Cl^-. If the concentration of Cl^- in the water sample is less than 3mg/L, the water sample should be concentrated for easy operation. When the water sample is concentrated, the temperature should not be higher than 60℃. The sample pretreatment process refers to Long(1993), Kaufmann et al. (1984) and Taylor and Grimsrud sample preparation principle (1969). The pretreatment process mainly includes the preparation of Silver Chloride and CH_3Cl gas.

8.2.1.1 Silver chloride preparation

The first step in sample preparation for measurement of chlorine stable isotope ratios is precipitating inorganic chlorides in the form of silver chloride (AgCl). The method used in this technique follows that described in earlier studies (Eggenkamp, 1994; Holt et al., 1997). Briefly, the method aims at precipitating AgCl from solution at fixed [Cl^-] content, fixed ionic strength and fixed pH. The solution containing [Cl^-] is first diluted by ultra-pure water or evaporated gently to bring the solution to the desired concentration. Secondly, it is acidified to pH ~2 with ultra-pure nitric acid (HNO_3) and heated at 80℃ for a few minutes to drive off CO_2 (Long et al., 1993). Then, 0.4M potassium nitrate (KNO_3) solution is added to reach a high ionic strength which helps to form small crystals of AgCl. Anhydrous sodium phosphate dibasic (Na_2HPO_4) and citric acid monohydrate [$HOC(CH_2CO_2H)_2CO_2H \cdot H_2O$] (0.000 4mol and 0.009 8mol, respectively) are added to buffer pH ~2. This is important to remove small amounts of sulfide, phosphate and carbonate from the precipitate by keeping them in solution. Then 1mL silver nitrate ($AgNO_3$) solution (0.2M) is added to precipitate AgCl. The beakers are stored in a dark place overnight for the precipitation to come to completion. AgCl is stored in a dark place at all times, as it is sensitive to light in which it will photo-decompose. Once precipitation is completed, precipitates are transferred into amber vials, where they are left to settle. Then, they are rinsed a couple of times with 5% HNO_3. Samples are then placed into the oven overnight to dry. Dried samples are stored in a dark place until CH_3Cl preparation.

8.2.1.2 Methyl chloride preparation

The preparation procedure was first developed in 1954 to obtain the maximum yield of CH_3Cl (Langvard, 1954). Later, it was improved to achieve an even higher yield (98%) (Hill and Fry, 1962). However, the procedure was producing isotopic fractionation and, therefore, additional improvements were introduced during subsequent studies to solve previous problems (Taylor and Grimsrud, 1969; Kaufmann et al., 1984; Eggenkamp, 1994). AgCl is converted into CH_3Cl by reacting it with methyl iodide (CH_3I). Samples are weighed (0.2mg) into 20mL crimp amber vials where the reaction takes place. Vials are stored in a dark place if not to be prepared immediately. Once samples are ready to be prepared, they are placed in an inflatable glove bag connected to an ultra-pure helium tank. The glove bag should be placed in a fume hood during this procedure as CH_3I is toxic and prolonged exposure can damage lungs, kidneys, liver and the central nervous system. The helium tank is opened and helium is allowed to inflate the glove bag a few times by sealing the bag then opening it slightly to force the helium to flush out any air in the bag. Finally, the glove bag is sealed and the helium flow is reduced. Vials are tipped on an angle and flushed gently with a very low stream of helium to avoid losing sample. CH_3I (100mL) is added to the samples and then vials are sealed. It is recommended to add CH_3I to 3~4 vials at a time and then seal these vials, and so on, otherwise the CH_3I can evaporate from the early vials by the time addition of CH_3I is finished. Vials are checked and retightened once more at the end, because vials need to be sealed (crimped) very well to avoid any leakage, as the desired final product (CH_3Cl) is a gas. For the reaction to proceed to completion vials are placed in an oven for 48 hours at 80℃.

8.2.2 Bromine

Attempts to measure bromine stable isotopes were made as early as 1920. In 1936, more attempts were made using a Dempster-type mass spectrograph. The measurement was done by analyzing positive and negative ions (Br^+, Br^{2+} and Br^-), and a precision of ±25‰ was reported. Ten years later, another study used a mass spectrograph of the Nier type to analyze bromine stable isotopes. Results were obtained from measuring positive ions (Br^+, Br^{2+}) formed from electronic bombardment of bromine vapour. A precision of ±4‰ was achieved. In 1955, isotopic compositions of elemental bromine from various suppliers and diverse origins were determined by negative thermal ionization mass spectrometers (N-TIMS); however, no significant differences between these samples were found, and the reported precision was approximately ±4‰. In 1964, another study reported the use of TIMS in bromine stable isotope measurement with an improved precision of ±1.8‰. A much more precise technique was introduced in 1993 by means of positive-TIMS, based on measuring positive ions of Cs_2Br^+. This technique is useful for analyzing very small samples (4～32μg of Br) and has good precision(±1.2‰) for such small samples. Dual inlet isotope ratio mass spectrometry (DI-IRMS) was used for the first time to determine bromine stable isotope composition in 1978. The technique is based on analyzing methyl bromide gas. More recently, in 2000, a descriptive work was published on bromine separation and isotopic determination of bromine by DI-IRMS (Eggenkamp and Coleman, 2000). The study reported the use of 2～8mg of Br and a precision of ±1.8‰. Although the precision reported in the last technique is highly improved in comparison to previous techniques, there is an interest in improving the precision to better benefit from the small range of variation of the bromine stable isotope composition.

The pretreatment of bromine isotope samples includes three important steps: distillation separation of bromine and chlorine; Preparation of silver bromide; Preparation of bromomethane. Of these, the distillation separation of bromine and chlorine is particularly important because the concentration of bromine in natural samples is very low compared to chlorine.

8.2.2.1 Bromine separation

The methodology used for bromine separation follows closely the classical technique presented by earlier studies (Eggenkamp and Coleman, 2000). Briefly, the technique depends on the differences of oxidation-reduction behavior of different halogens. Halide ions with higher atomic masses are easier to oxidize than ones with lower atomic masses. Therefore, Br is oxidized more easily than Cl. The separation is conducted in the special distillation apparatus shown in Figure 8.4. The three-neck, round-bottom 500mL flask is filled with sample containing 1～10mg of Br^-. Ultra-pure water is added to bring the total volume in the flask to 100mL. Samples with low Br^- concentration are evaporated on a hot plate below boiling (～80℃) to concentrate the Br^- in solution. Then 10g of $K_2Cr_2O_7$ is added, and 6mm glass beads are added to facilitate gentle boiling. The graduated addition funnel is connected to the flask and filled with 20mL of 1∶1 H_2SO_4/H_2O (ultra-pure). A fritted gas dispersion tube is connected to the middle port of the flask to feed a steady flow of ultrapure helium to the flask to facilitate the movement of formed gases forward and to avoid back flushes. The two-neck, round-bottom 500mL flask is filled with 200mL of solution containing 2g of KOH. A cooling bath filled with crushed ice is placed under the two-neck flask. The two flasks are connected with a condenser, and the cooling ports are connected to the cooling water system.

A 125mL Erlenmeyer flask is filled with 100mL of solution containing 1g of KOH and connected to the two-neck flask. The power-controlled mantle, hemispherical heating round-bottom is placed below the three-neck flask. Once all connections are secured and everything is ready, the helium tank is opened, and a flow of 200mL/min is maintained during the entire separation period. Then the stopcock on the addition funnel is opened to allow the H_2SO_4/H_2O mixture to flow into the flask. After the addition of the mixture, 100mL of ultra-pure water is added via the addition funnel. The flask is then heated slowly to bring the solution to a boiling. It is at this stage that bromine gas (Br_2) (yellow-brown vapour) starts to form and flow, advancing to the second flask via the condenser. In the second flask, Br_2 gas reacts with the KOH solution to form KBr and KBrO. Any escaped Br_2 gas will be reduced in the Erlenmeyer flask with KOH solution. The distillation lasts for 20 minutes after boiling starts to ensure that all Br^- in solution has been oxidized and transferred. This has been tested by calculating the final yield of the Br^- from samples with known bromine concentration. Then the solutions from both the second flask and the Erlenmeyer are transferred to a beaker, 3g of zinc powder is added, and the mixture is boiled for 10 minutes to reduce all BrO^- ions in the solution to Br^-. The solution is then filtered through a 0.22μm Millipore Express Plus (PES).

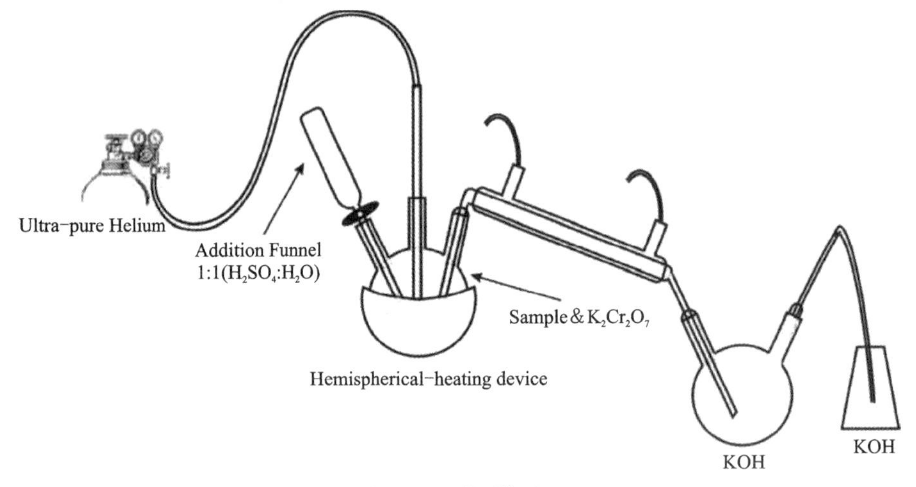

Figure 8.4 Bromine distillation apparatus.

8.2.2.2 Silver bromide preparation

After bromine is separated, it is precipitated as silver bromide (AgBr). The technique used is similar to that used in silver chloride precipitation and follows earlier studies (Taylor and Grimsrud, 1969; Eggenkamp, 1994). Briefly, the method aims at precipitating AgBr from solution at fixed Br-content, fixed ionic strength, and fixed pH. The solution containing Br is first acidified to pH ~2 by adding ultra-pure concentrated nitric acid (HNO_3). Then 18g of potassium nitrate (KNO_3) is added to the solution to increase the ionic strength, which helps to form small crystals of AgBr. Then 2mL of silver nitrate ($AgNO_3$) solution (0.2M) is added to precipitate AgBr. The beakers are stored in a dark place overnight for the precipitation to come to completion.

8.2.2.3 Methyl bromide preparation

Silver bromide is reacted with methyl iodide (CH_3I) to form CH_3Br gas. Samples are weighed (0.5mg) into 20mL amber crimp vials where the reaction takes place. Vials are stored under vacuum in a desiccator and in a dark place if the samples are not going to be prepared immediately.

8.3 Fractionation processes

8.3.1 Isotopic fractionation of diffusion processes

Diffusion probably is the most well known and most well understood process that is responsible for variations in chlorine and bromine isotope compositions. Molecular diffusion is the process in which matter is transported from one part of a system to another as a result of arbitrary molecular movements. The mass difference between different isotopes of elements will lead to isotope fractionation during diffusion. Diffusion, as a physical process, leads to similar isotopic fractionation effects for chlorine and bromine isotopes, but the degree of fractionation may be different. During pure diffusion, Cl (Br) diffused from the reservoir will enrich light isotopes ^{35}Cl (^{79}Br) due to the difference in activity between ^{35}Cl and ^{37}Cl (^{79}Br and ^{81}Br).

The definition of the fractionation as a result of diffusion ($^{35/37}\alpha$) for chlorine is:

$$^{35/37}\alpha=\frac{D_{35}}{D_{37}}=\sqrt{\frac{\mu_{37}}{\mu_{35}}}=\sqrt{\frac{m_{37}(m_{35}+M)}{m_{35}(m_{37}+M)}} \tag{8.3}$$

where D_{35} and D_{37} are the diffusion coefficients of the light and heavy isotopes of chlorine (Richter et al., 2006). Calculation of $^{35/37}\alpha$ requires knowledge of the hydration number n of the diffusing chloride ion [$Cl(H_2O)_n^-$], as would also be the case if Graham s Law was applicable. Based on the Debye-Hückel å parameter (Appelo and Postma, 2005) the hydration number n of Cl^- would be 3: [$Cl(H_2O)_3^-$].

8.3.1.1 Diffusion from a source with a constant concentration

Diffusion from a source with a constant concentration can be described as the situation where an infinite amount of matter with high chloride content diffuses into an infinite amount of matter with low chloride content. The chloride concentration at the boundary between these two parts will be constant during diffusion. The solution for the concentration as a function of x (distance from the boundary) and t (time) is:

$$c_{(x,t)} = c_0 \operatorname{erfc} \frac{x}{2\sqrt{Dt}} \tag{8.4}$$

where erfc is the complementary error function. Characteristic of this diffusion model is that in parts of the system with a low original chloride concentration $\delta^{37}Cl$ can be very low and also that it depends very much on the chloride concentration ratio between maximum and minimum original chloride concentrations in the system. Calculations from Eggenkamp (1994) showed that in extreme cases very large $\delta^{37}Cl$ variations could be obtained which, as shown later by experiments reported in Eggenkamp and Coleman (2009), can only be observed when the concentration in the low concentration part is so low that the analysis becomes non-trivial. In the part of the system with the higher original chloride concentration the $\delta^{37}Cl$ does increase due to the faster diffusion of the isotope ^{35}Cl. As the chloride concentration in this part of the system is relatively high the isotope effect is much less than in the low concentration part of the system (Figure 8.5).

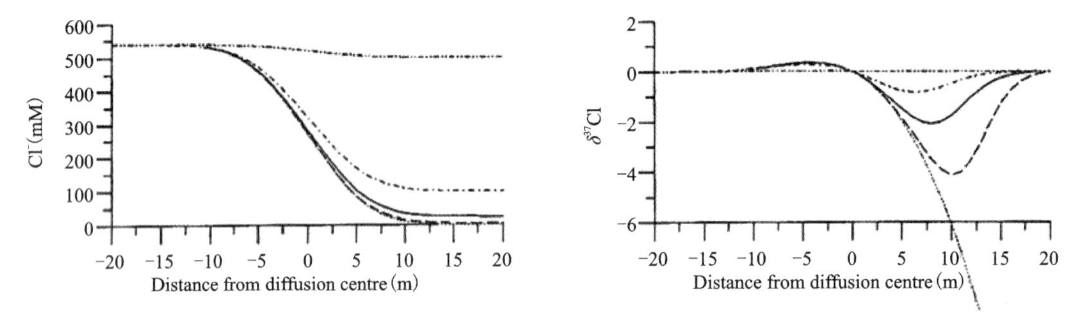

Figure 8.5 Chloride concentration (left) and chlorine isotope variation (right) profiles for a system with diffusion from a constant source for different chloride concentrations in the low concentration part of the system. Profiles are developed from Eq. 8.4, and drawn with a diffusion time of 250 years, a diffusion coefficient of 15×10^{-10} m^2/s, a diffusion coefficient ratio ($^{35/37}\alpha$) of 1.002 45, a chloride concentration in high concentration part of the system of 540mM, and in the low concentration part of 0mM, 5mM, 25mM, 100mM and 500mM (from Eggenkamp, 1994).

8.3.1.2 Diffusion after a momentary release of chloride or bromide

In this model, at time $t = 0$a fixed amount of chloride ions is released into the system. From the injection point (at location $x = 0$) the chloride/bromide will start to diffuse. In the following it is assumed that the original concentration in the region around the injection point is lower than the concentration after injection of the fixed amount of chloride/bromide at the injection point, the moment it is injected. As the release of chloride or bromide is momentarily the concentration at $x = 0$ will decrease from the start as t increases. The equation for linear diffusion after a momentary release of chloride or bromide is:

$$c_{(x,t)} = \frac{s}{\sqrt{4\pi Dt}} \exp \frac{x^2}{4Dt} \tag{8.5}$$

where s is the amount of the original momentary release of chloride/bromide. In this system the concentration of chloride/bromide in the diffusion center decreases rapidly, while the isotope ratio in the center increases due to the higher diffusion coefficient of the isotope ^{35}Cl or ^{79}Br relative to ^{37}Cl or ^{81}Br. The isotope ratio in the transported chloride/bromide can also reach very low values in this scenario, while, just as in the former model, the lowest values depend on the original chloride or bromide concentration in the environment where the momentary release diffuses to. The diffusion profile seems rather similar to the profile in the first model, however, as the chloride concentration in the diffusion center decreases the isotope effect also decreases as the chloride or bromide concentration near the diffusion center approaches the original chloride or bromide concentration in the environment. Figure 8. 6 shows the effect on the chloride concentration and the chlorine isotope composition for a system where after the initial release the chloride diffuses in one dimension (linear) away from the location of the initial momentary release.

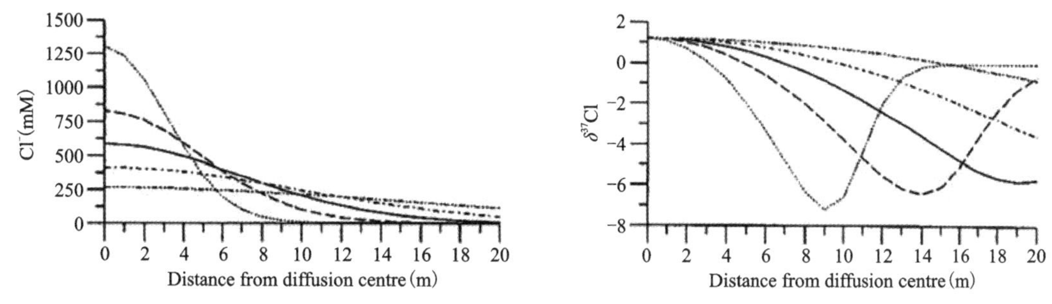

Figure 8. 6 Chloride concentration (left) and chlorine isotope variation (right) profiles for a system with diffusion after a momentary release of chlorine. The figure shows the effects for subsequent linear diffusion after Eq. 8. 5, and is drawn for an initial release of chloride of 10mol, a diffusion coefficient of $15\times10^{-10}m^2/s$, a diffusion coefficient ratio ($^{35/37}\alpha$) of 1. 002 45, and diffusion times of 100, 250, 500, 1000 and 2500 years. The initial chloride concentration in the area around the initial release was 5mM (from Eggenkamp, 1994).

8. 3. 1. 3 Diffusion from a source with a constant inflow

In this model, a constant inflow of chloride or bromide with an infinitely small volume is introduced from a source at a constant rate. It can be considered as an extension of the former model. The diffusion solution of this model can be found by integrating Eq. 8. 5 over time, replacing s by qdt,. In the mathematical solutions that describe the diffusion from a constant inflow q defines the amount of chloride added per unit of time.

For linear diffusion the solution is given by Carslaw and Jaeger (1959):

$$c_{(x,t)} = \frac{qx}{2D\sqrt{\pi}}\left[\frac{2\sqrt{Dt}}{x}\exp\left(\frac{-x}{4Dt} - \sqrt{\pi}\operatorname{erfc}\frac{x}{2\sqrt{Dt}}\right)\right] \tag{8.6}$$

In this system a continuously increasing chloride concentration at the location of the inflow is observed that decreases at larger distances from the inflow. The chlorine or bromine isotope ratios of the transported chloride or bromide reach a minimum which moves slowly away from the source as time progresses. As the chloride or bromide concentration at the source increases the isotope ratio of the minimum decreases due to the larger ratio between the chloride or bromide concentration between the source and the background. Figure 8.7 shows how the chloride or bromide concentration and chlorine or bromine isotope compositions evolve over time in the case of dimensional diffusion where chloride or bromide is continuously added at the source.

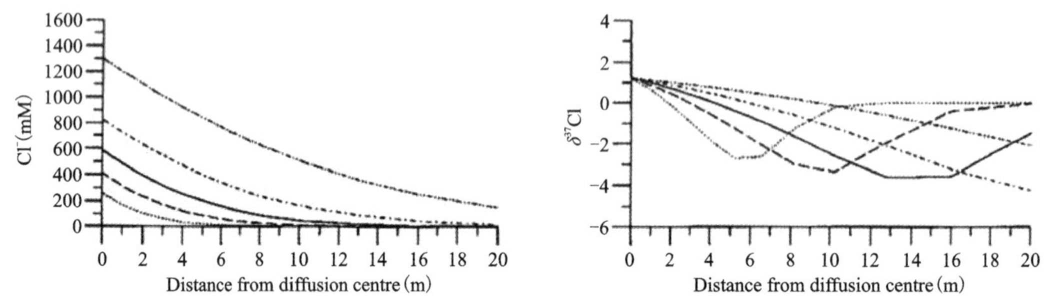

Figure 8.7 Chloride concentration (left) and chlorine isotope variation (right) profiles for a system with diffusion from a constant inflow of chloride. The figure shows the effects for subsequent linear diffusion after Eq. 8.6, and is drawn for an influx of chloride equal to 10mol/year, a diffusion coefficient of $15\times10^{-10}m^2/s$, a diffusion coefficient ratio ($^{35/37}\alpha$) of 1.002 45, and diffusion times of 100, 250, 500, 1000 and 2500 years. The initial chloride concentration in the area around the initial release was 5mM (from Eggenkamp, 1994).

8.3.2 Isotopic fractionation of salt precipitation process

In the evaporation salting process of brine, chlorine precipitates as salt minerals. During the crystallization of salt minerals, the fractionation of chlorine isotopes is influenced by the difference of bond energy of their compounds.

Hoering and Parker (1961) were the first to study the chlorine isotope fractionation between sodium chloride (halite) and a saturated sodium chloride solution. The fractionation factor obtained in their experiment was $1.000\,2\pm0.000\,3$, indicating that the chlorine isotope ^{37}Cl of the precipitated sodium chloride was probably higher than that of the saturated solution.

As part of their study on Zechstein evaporites Eggenkamp et al. (1995) determined equilibrium fractionation factors between sodium chloride, potassium chloride and magnesium chloride saturated solutions and their precipitating salts. They observed that the chlorine isotope fractionation is dependent on the cation that precipitates together with the chloride. Near-saturated solutions of reagent grade sodium chloride, potassium chloride, and magnesium chloride were allowed to evaporate at ambient temperature. After precipitating of the respective salts were formed, the solutions and the precipitates were separated by decanting the brine and filtering the precipitate. To remove any adhering solution the precipitate was rinsed with acetone. Both the brine and the salt were analyzed for chlorine isotopes composition and the fractionation was defined as the difference between the $\delta^{37}Cl$ of the precipitate and of the remaining brine. For sodium chloride they observed that the precipitate was 0.26‰±0.07‰ heavier than the brine, while for potassium chloride and magnesium chloride it was respectively 0.09‰±0.07‰ and 0.06‰±0.10‰ lighter. These fractionation factors were interpreted to explain the lack of variations in the studied Zechstein evaporites. As the sign of the isotope fractionation reverses from sodium chloride to potassium chloride this would mean that at a certain moment a minimum $\delta^{37}Cl$ would be reached, after which the $\delta^{37}Cl$ would increase again due to the domination of potassium and magnesium salt precipitate in the last stages of salt precipitation. The results obtained by Eggenkamp et al. (1995) were, at least qualitatively, confirmed in an experiment reported by Eastoe et al. (1999) who evaporated a batch of seawater and analyzed the $\delta^{37}Cl$ of the remaining water and the precipitated halite at certain stages. The $\delta^{37}Cl$ of the brine decreased during the early stages, but increased again during the later. The observation that in all stages the halite that was analyzed was considerably enriched in $\delta^{37}Cl$ compared to the brine was also in good agreement with earlier measurements.

Studies done in the Chinese Tarim and Western Qaidam Basins (Tan et al. ,2005, 2006) indicated that during the latest stages of salt precipitation in these basins $\delta^{37}Cl$ continued to decrease, and this effect was used locally as a proxy for the exploration of potassium deposits. It was shown however in an earlier study by Xiao et al. (2000) that equilibrium isotope fractionation in potassium and magnesium chlorides was not as large as chlorine isotope fractionation during halite precipitation. A recent study by Luo et al. (2012) confirmed these results. The effect that the chlorine isotope composition continues to decrease during precipitation of magnesium chlorides might however be the result of the specific character of these deposits, which are terrestrial rather than marine, or perhaps of kinetic effects.

Luo et al. (2014) presented newly determined fractionation factors for the precipitation of sodium, potassium and magnesium chloride from their respective saturated brines. The values they obtained appeared all to be higher than the data obtained by Eggenkamp et al. (1995), and more importantly in all three salts the precipitate appeared to have more positive $\delta^{37}Cl$ values than the brine. The error of the measurements seemed however to be rather large. The data obtained (with one standard deviation of the published measurements done per salt) were: NaCl 1.000 55±0.000 46, KCl 1.000 25±0.000 10 and $MgCl_2 \cdot 6H_2O$ 1.000 12±0.000 50. Except for KCl it is thus not obvious that these data are significantly different from the data presented by Eggenkamp et al. (1995).

Eggenkamp et al. (2011) presented determinations of chlorine isotope fractionation between several more chloride (and bromide) salts and their saturated solutions. These results have been presented at a conference only and the fractionation factors were only presented in a figure. It was clear however that different salts show very different isotope fractionation factors when they precipitate from a saturated brine, ranging from about +0.4‰ for barium chloride to about −0.2‰ for caesium chloride. For bromide salts the fractions factors appeared to be rather modest compared to the chloride salts.

8.3.3 Isotopic fractionation of ion osmosis process

In the process of solute passing through clay, due to the difference of isotope migration rate and the difference of negative charge repulsion on clay surface, isotope fractionation will occur.

Phillips and Bentley (1987) described the theory of ion-filtration with respect to chlorine isotopes in some detail. During the ion-filtration process the solvent is able to, while the dissolved ions are unable to move through the membrane. Ion-filtration is a process which is similar to reverse osmosis. Following the description by Kaufmann (1984) osmosis is the effect that appears when two reservoirs, one with water and one with an aqueous salt solution are in connection to reach other through a semi-permeable membrane. During this process water will be able to pass the membrane while the dissolved salt will not. As a result of this process the contents of the reservoirs will interact through the membrane and will try to balance the chemical potential between the two reservoirs. That means that water flows through the semi-permeable membrane in the direction of the reservoir with the higher chemical potential (concentration of salts) in an attempt to equal the potentials, effectively diluting the contents of the reservoir with the higher concentration. In the case of reverse osmosis, the process is comparable (two reservoirs with aqueous solutions of different chemical potential), but now a pressure is applied on the reservoir with the higher concentration. Because of this pressure water is forced through the membrane in the direction of the freshwater side. As a result, the salinity increases on the saline side of the membrane. This effect may have a consequence on the isotope composition of the chlorine in the water. In many cases semipermeable membranes are consisting of clay-minerals. These minerals have negative charges at their surfaces, which means that negative ions are repelled from the membrane. The effect is comparable to diffusion as the lighter ion will be repelled more efficiently than the heavier ion, so that isotope fractionation will take place.

Phillips and Bentley (1987) described a different model to explain chlorine isotope fractionation occurring from ion-filtration. In their model actually ions move through the membrane, but because the lighter isotope moves away from the membrane more efficiently than the heavier isotope the result was that more of the heavier isotopes are actually transferred through the membrane, increasing the $\delta^{37}Cl$ in the sedimentary layer behind the membrane. In their simplified model development of the chloride concentration and chlorine isotopes is described as the result of compaction of three sedimentary layers. At the start the chloride concentration is highest in the lower and lowest in the shallowest sedimentary layer, and the chlorine isotope composition is chosen to be zero in the upper two layers and +0.25‰ in the lowest. When pressure is applied it is assumed that the compaction starts in the lower and progresses to the shallower layers. In step 1 the lowest layer is compacted. As a result of this both water as well as chloride with an increased $\delta^{37}Cl$ moves to the second layer, and in this layer both the chloride and the chlorine isotope composition increase. When the lowest layer is fully compacted the second layer starts to compact. During this process, again water and isotropous heavy chloride move to the third layer. As a result, the chloride concentration increases and the isotope ratio decreases in this second layer, while in the upper layer both the concentration and the isotope composition increase. When the second layer is fully compacted the third starts to compact resulting in higher concentrations and a lower isotope composition in this layer. The effect of this process is that considerable positive $\delta^{37}Cl$ values can be theoretically be found in transported chloride as opposed to diffusion which tents to lead to more negative $\delta^{37}Cl$ values in transported chloride.

8.3.4 Isotopic fractionation induced by microbially activity

Several microbes are able to either produce or decompose organohalogen compounds. During bacterial dehalogenation of organochlorine compounds bacteria obtain energy through reduction of organically bound chlorine to chloride ions which are then released into environment. Potentially this process can cause isotope fractionation.

One of the first studies in which both chlorine and carbon isotope fractionation as a result of microbial degradation was determined was the experimental study by Heraty et al. (1998) on dichloromethane. They determinedboth the chlorine and carbon isotope composition in the remaining dichloromethane, and compared it to the original isotope compositions, after it was aerobically degraded by MC8b, a gram-negative methylotrophic bacteria related to *Methylobacterium* or *Ochrobactrum*. The isotope fractionation factor that was determined from this reaction was 0.996 2±0.000 3 for chlorine and 0.957 6±0.001 5 for carbon, indicating that both $\delta^{37}Cl$ and $\delta^{13}C$ of the remaining dichloromethane decreased. These fractionation factors correspond to kinetic isotope effects of 1.003 8 and 1.042 4 respectively for chlorine and carbon.

Chlorine isotope fractionation during bacterial dehalogenation of dichloromethane was also shown by Zyakun et al. (2007). They determined the isotope fractionation of dichloromethane during microbial dehalogenation by *Methylobacterium dichloromethanicum* and *Albibacter methylovorans*. Contrary to the study by Heraty et al. (1998) they determined the isotope com-position of the reacted dichloromethane directly on the dichloromethane molecule that was introduced into the mass spectrometer. To determine isotope fractionation masses 84 ($^{35}Cl_2CH_2$), 86 ($^{35}Cl^{37}ClCH_2$) and 88 ($^{37}Cl_2CH_2$) were compared. The dichloromethane that remained after the reaction had more positive isotope ratios than the original dichloromethane, which is quite the opposite when compared to the experimental observations from Heraty et al. (1998). The increased isotope ratio was explained by the process of diffusion of isotopically lighter dichloromethane through the cell walls of the bacteria so that the residual dichloromethane ends up being more positive.

The experimental results that are obtained in these studies suggest that chlorine isotope fractionation during microbial degradation of organochlorine compounds is very much dependent of the reaction path. The data obtained so far show that a reaction can either induce isotope fractionation, or show no fractionation at al. The processes that take place during degradation of chlorinated organic compounds are slowly better understood and it is important that the chlorine isotope effects are dependent on the type and the pathway of the reactions that take place.

8.4 Variations of chlorine and bromine isotopes in groundwater

Groundwater is present in virtually all sedimentary basins. This groundwater is, especially at greater depths, mostly saline, which containing large amounts of chloride and bromide. Recent estimates are that the amount of chloride in groundwater may be even larger than the total amount of chloride in the oceans and more than double the amount of chloride in evaporites at about 330×10^{20} g (Land, 1995). The origin of groundwater ultimately is from the oceans as it was suggested by Knauth (1998) that before the development of the continental crust, all salt and brine currently found on the continents would have been entirely in the oceans. The composition of groundwater is changed by chemical processes such as water-rock inter-action, salt dissolution, albitization as well as physical processes such as diffusion, ion exchange and ion-filtration and mixing of water bodies with different compositions. During these processes not only the chemical but also the chlorine and bromine isotopic composition of the water can alter considerably.

8.4.1 Chlorine isotope variations in groundwater

Figure 8.8 shows a histogram with the $\delta^{37}Cl$ values from 324 sediment pore water samples taken from the above publications. This histogram shows very clearly that most measurements have negative $\delta^{37}Cl$ values, with as largest sample class $-0.2‰$. Some basic statistics from these data can be found in Table 8.1. The data in this figure confirm the fact that negative $\delta^{37}Cl$ data are more common than positive data with a median value of −0.23‰. What is well known regarding chlorine isotopes is the narrow range of the data. 50% of the measurements are between −0.55‰ and +0.04‰ versus SMOC. Virtually all samples have values that are not more than 2‰ from SMOC, the only real deviating samples are a few samples of formation water sampled from the Oseberg and Forties Fields in the North See (Eggenkamp, 1994; Ziegler et al., 2001). These observations are also nicely illustrated in Figure 8.9, showing the chloride concentrations and the $\delta^{37}Cl$ data from the 274 samples taken from the above publications from which also chloride concentrations were reported. In this figure it is clearly visible that the total $\delta^{37}Cl$ variation is larger at lower chloride concentrations than at higher chloride concentrations. The reason for this is that physical processes such as diffusion and ion-filtration cause larger variations at lower concentrations than at higher concentrations.

Table 8.1 Basic statistical characteristics of 324 chlorine and 128 bromine isotope ratio measurements of deep formation water.

	$\delta^{37}Cl$ value	$\delta^{81}Br$ value
Average	−0.37	+0.57
Standard deviation	0.85	0.63
1% percentile	−3.60	−0.34
10% percentile	−1.21	−0.06
1st quartile	−0.55	+0.16
Median	−0.23	+0.37
3rd quartile	+0.04	+0.68
90% percentile	+0.40	+1.48
99% percentile	+1.38	+2.24

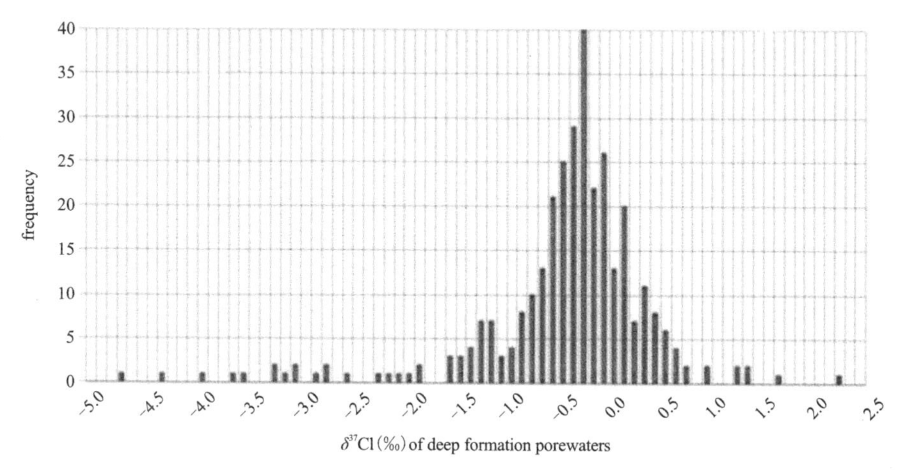

Figure 8.8 Histogram of 324 $\delta^{37}Cl$ values of deep saline formation water.

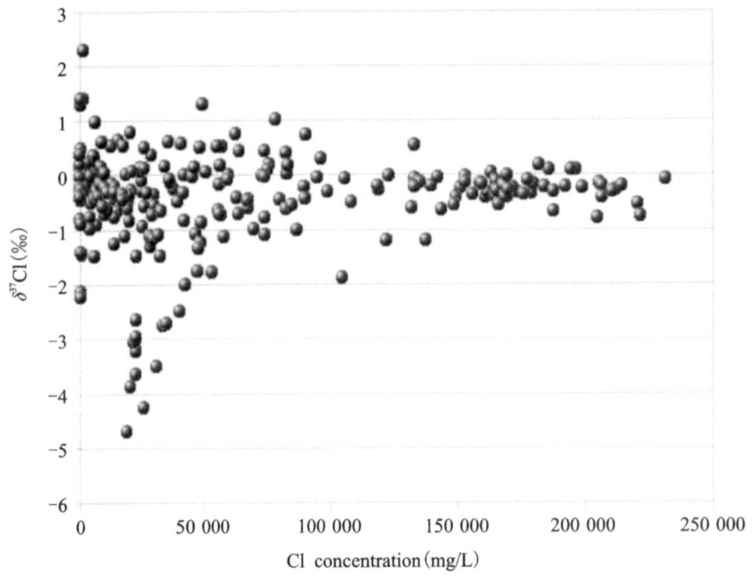

Figure 8.9 Relationship between the chloride concentrations and chlorine isotope composition of 274 formation water analyses from which both data are reported. From this figure it is visible that at higher chloride concentrations the variation in δ^{37}Cl values is smaller, but that the average δ^{37}Cl value still is less than 0.

8.4.2 Bromine isotope variations in groundwater

Figure 8.10 shows a histogram of the 128 bromine isotope measurements published in Eggenkamp and Coleman (2000), Shouakar-Stash et al. (2007), Stotler et al. (2010) and Boschetti et al. (2011). In this histogram it is clearly shown that bromine isotopes tend to have mostly positive values in pore water. The median value for example is +0.37‰ relative to SMOB, compared to −0.23‰ relative to SMOC for chlorine. Other statistical characteristics can be found in Table 8.1 where they can be compared to chlorine isotope characteristics.

This effect is also seen in Figure 8.11 where the relationship between the bromide concentrations and the bromine isotope composition in these samples is shown. In this figure it is easily visibly that in samples with lower bromide concentrations the bromine isotope variations are much larger than in samples with higher concentrations, and also that in samples with high bromide concentrations ($>2000\times10^{-6}$) the isotope compositions tend to be relatively close to SMOB. Considering the fact that it can be expected that, just as for chloride, all bromide in sediment pore water is ultimately of ocean water origin (with a δ^{81}Br of 0‰) these data confirm again that the fractionation processes that define the bromine isotope ratios in sedimentary basins are quite different than the processes that define the chlorine isotope ratios.

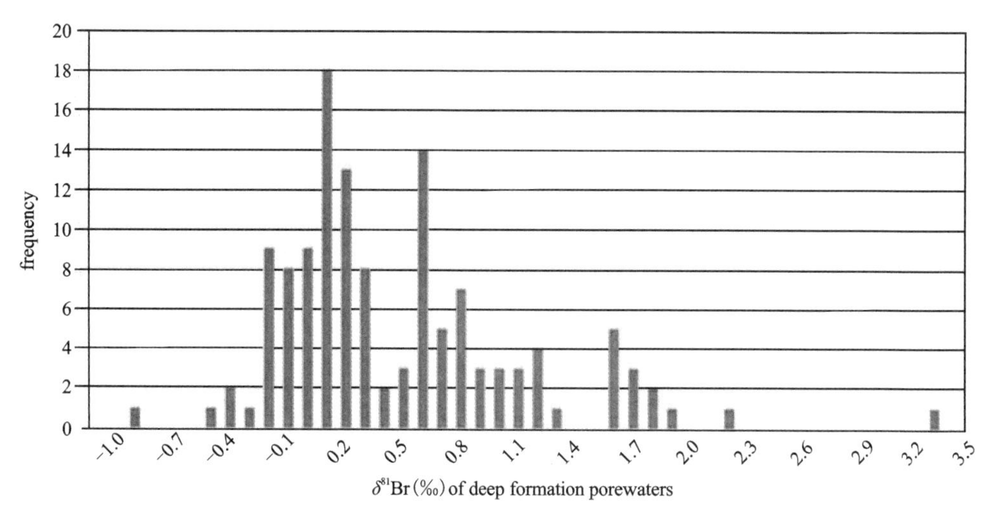

Figure 8.10 Histogram of 128 δ^{81}Br values of deep saline formation water.

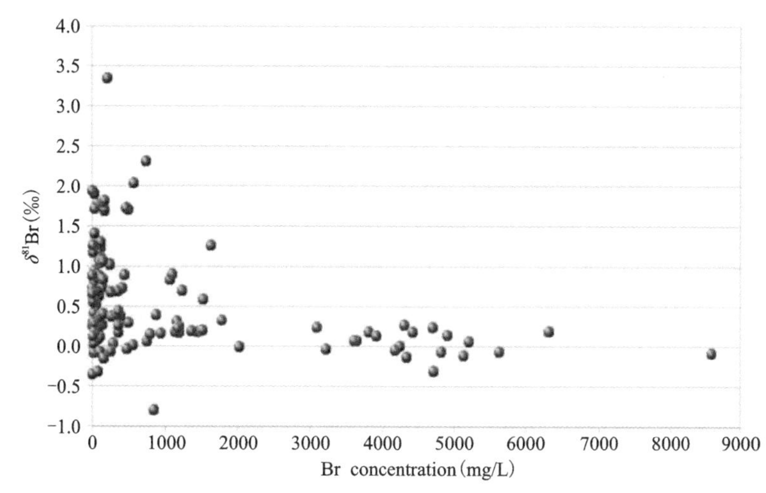

Figure 8.11 Relationship between the bromide concentrations and bromine isotope composition of 128 formation water analyses from which both data are reported. From this figure it is visible that at higher bromide concentrations the variation in δ^{81}Br values is smaller, and that the average δ^{81}Br value is fairly close to 0.

8.4.3 The relationships between bromine and chlorine isotope ratios in groundwater

Chloride and bromide concentrations in formation water and salt deposits have been studied extensively as proxies for their origin. Looking at the chloride and bromide concentrations Rittenhouse (1967) was able to determine five groups of formation water. These five groups were defined by him as follows:

Ⅰ. Formation water in which the bromine and total mineralization approximate that what would be expected from simple evaporation of seawater ("Seawater evaporation" on

Figure 8.12) or dilution of seawater with water of low mineralization and bromide content ("Dilution with fresh water" on the figure).

Ⅱ. Formation water which are similar to formation water of group I except for having about twice as much bromine for a specified total mineralization. The additional bromine may have been added during early diagenesis.

Ⅲ. Formation water in which the total-dissolved-solids content is greater than that of seawater but the bromine content is less than would be expected from simple concentration of seawater. These water appear to have dissolved some halite.

Ⅳ. Formation water in which the total dissolved solids content is less than that of seawater but the bromine content is less than would be expected from simple dilution of seawater with water of low total mineralization and bromine content. These water may have been diluted with water having fairly high dissolved solids concentrations but low bromine content, or may have been group Ⅲ water diluted with water of low total mineralization and bromine content.

Ⅴ. These formation water have a high salinity and higher bromide contents than water from group Ⅱ. Some of these water are associated with thick salt deposits and appear to be the residual brine which is left after salt deposition ("Salt precipitation" on Figure 8.12).

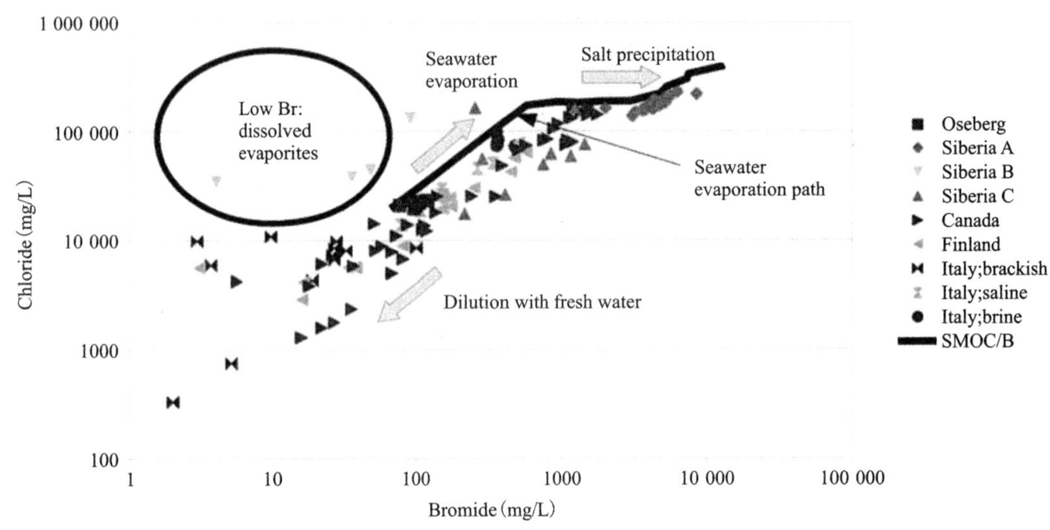

Figure 8.12 Chloride and bromide concentrations from formation water in those studies that reported both concentration and isotope data.

Figure 8.12 shows the chloride and bromide concentrations of the samples that are reported in Eggenkamp and Coleman (2000), Ziegler et al. (2001), Shouakar-Stash et al. (2007), Stotler et al. (2010) and Boschetti et al. (2011). Samples from Siberia are subdivided into the three groups as defined by Shouakar-Stash et al. (2007) and the samples from Italy are subdivided according to the subdivisions from Boschetti et al. (2011). The other samples are plotted according to the country they come from. In this figure the seawater evaporation path is also shown. The groups as they were defined by Rittenhouse (1967) are indicated in this figure and fairly easy to recognize, although it is also clear that there are no easily recognizable borders visible between the different groups, so that not all samples can be placed easily into a certain group. When the distribution of the chlorine and bromine isotopes of the same samples are plotted on a δ^{37}Cl versus δ^{81}Br diagram the groups as observed by Rittenhouse (1967) are much less clear. Figure 8. 13 shows the chlorine and bromine isotope data of all samples plotted together. As most samples fall into the fairly restricted, area with the values for chlorine between −1‰ and +1‰ and for bromine between −0. 5‰ and +1. 5‰ this area is blown up in Figure 8. 14. Looking at this figure it is clear that the origin of the samples is probably not as simple as could be deducted from the chloride and bromide concentrations only.

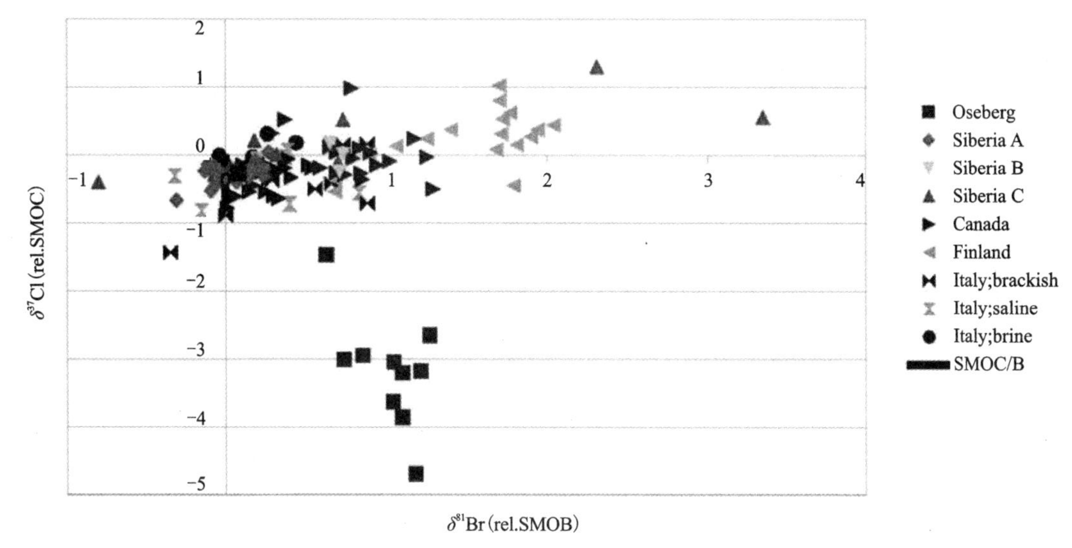

Figure 8. 13 Chloride and bromide isotope data from formation water in those studies that reported both concentration and isotope data.

The most extreme samples with regard to chlorine isotopes, the samples from the Norwegian Oseberg field with δ^{37}Cl values as low as −4. 7‰ have chloride and bromide concentrations which are very close to seawater. The low δ^{37}Cl values indicate however very clearly that these water are not simply seawater. Ziegler et al. (2001) explained the low δ^{37}Cl values as the result of a combination of strong diffusion and membrane-filtration effects. Unfortunately, the bromine isotope composition in this study have not been interpreted together with the chlorine isotope compositions as they were not published together.

Samples from later studies were interpreted in tandem. Shouakar-Stash et al. (2007) analyzed a large set of formation brines from Siberia. They were able to divide their dataset

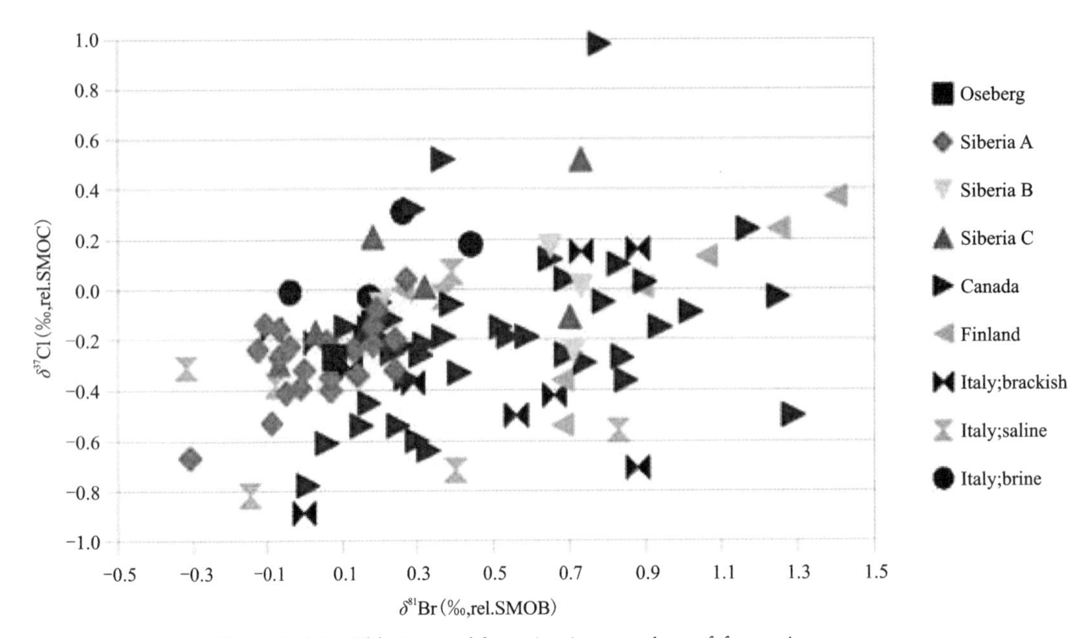

Figure 8. 14 Chlorine and bromine isotope data of formation water.

into four groups, one of which contained of fresh water with bromide concentrations too low to be analyzed for bromine isotopes. The other groups however could be distinguished based on chlorine and bromine concentrations and isotopes. The first group, Siberia A in Figure 8. 12～8. 14, consists of calcium-chloride brines that are believed to be the residual of an evaporated paleoseawater, which is indicated by Br/Cl ratios in these samples which are much higher than the Br/Cl ratio in seawater, and isotopic compositions which are fairly close to 0 for both isotopic systems. The second group, Siberia B consists of sodium chloride brines that are derived mainly from halite dissolution, as is shown by very low Br/Cl ratios. These samples are characterized by relatively positive δ^{81}Br data, and from this set it was concluded for the first time that evaporites might be characterized by generally positive δ^{81}Br data. The third grouping consisted of mostly very saline water (although not as saline as group Siberia A) from which the origin could not be clearly determined, although the chemical and isotopic characteristics of this group suggested that they were the product of various complex scenarios such as metamorphism, water-rock interaction, permafrost freezing and mixing. This group showed very large variations in both chlorine and bromine isotope compositions, among which the most positive bromine isotope data from which also chlorine isotope data were measured. Because of the large variation in chlorine and bromine isotope compositions it was impossible to obtain a clear explanation of the origin of these samples and for that reason a complex origin was suggested.

Stotler et al. (2010) analyzed the chlorine and bromine isotope compositions of a large group of 42 samples from the Canadian shield and 18 samples from the Fennoscandian shield. Based on the chloride and bromide concentrations these samples would mostly fall into Group 2 of Rittenhouses (1967) classification. The Br/Cl ratio of the Canadian samples averages around 0.009 and of the Fennoscandian samples around 0.007, which is between twice and three times the ratio in seawater. The $\delta^{37}Cl$ values of the Canadian samples vary between −0.78‰ and +0.98‰ and the $\delta^{81}Br$ between +0.01‰ and +1.29‰. For the Fennoscandian samples the chlorine isotopes vary between −0.54‰ and +1.52‰ and the bromine between +0.26‰ and +2.04‰. A weak positive correlation between chlorine and bromine isotopes (with an r^2 of 0.36) was observed in this set. At one site with serpentinite rocks present it was observed that the $\delta^{37}Cl$ isotope values showed a large range while these samples showed only very small variations in $\delta^{81}Br$ values and this was attributed to ion filtration through serpentinite, which apparently affected the chloride but not the bromide ions. Comparisons with other isotopic systems, such as $^{87}Sr/^{86}Sr$, indicated that water-rock interactions at some sites were likely to influence both halogen isotope systems. The large variation of $\delta^{37}Cl$ and $\delta^{81}Br$ values observed in the samples measured by Stotler et al. (2010) did not support a marine origin for these brines. This indicated that, if a seawater origin were to be considered for the fluids, a process or combination of processes significantly altered both the chlorine and the bromine isotope signatures. These unidentified processes probably explain the shift of the Br/Cl ratios from 0.035 to higher values showing that the samples in Group 2 from Rittenhouse (1967) in reality do have a no seawater origin.

A large set of 23 formation water from the Emilia-Romagna region of the Northern Apen-nine Foredeep in Northern Italy have been ana-lysed for (amongst others) chloride and bromide as well as chlorine and bromine isotopes (Boschetti et al., 2011). Based on the total mineralization of the samples the set is divided into three groups; brackish, saline and brine samples. The Br/Cl ratios indicate that the brackish samples are mostly diluted seawater (average Br/Cl is 0.004 1), although a few of them have very low Br/Cl ratios indicating evaporite dissolution followed with subsequent dilution with fresher water. These samples appear to have the most positive $\delta^{81}Br$ values, showing a further independent indication for positive $\delta^{81}Br$ values in evaporite deposits. The saline samples seem to be altered samples (Group 2, Br/Cl is on average 0.006 1) while the brine samples show Br/Cl ratios with are in between the brackish and saline samples (average Br/Cl is 0.004 9). On the $\delta^{81}Br$ versus $\delta^{37}Cl$ plot the other samples show a considerable scatter. This is explained in part as a possible contribution of marine organogenic bromine. Most samples however seem to describe a rough positive trend with a ~0.5 slope. This is explained by probable halogen (both chloride and bromide) diffusion from brines toward aquifers with lower salinities. This slope seems to agree with the fact that the isotope fraction factor for chlorine diffusion is about double the isotope fractionation for bromide diffusion as shown by Eggenkamp and Coleman (2009).

Three of the four samples analyzed by Bagheri et al. (2014b) show Br/Cl ratios that are between this ratio in seawater and double this amount and their isotopes indicate some moderate secondary processes. One of the samples has a low Br/Cl ratio indicating an evaporite origin, which again is confirmed by a rather positive δ^{81}Br value.

8.5 Application

8.5.1 Isotope evidences for groundwater salinization(Li et al., 2016)

Groundwater Cl chemical and isotopic signatures can help to provide useful information on salt transport in groundwater system. Here, we take the shallow groundwater located in the Datong Basin in the northern of China as an example to present the application of chlorine and bromine isotope for groundwater salinization. The hydrologic and geologic background of Datong Basin was well described in the section of 4.4.1. The states of saline soil and sampling location was presented in Figure 8.15 and Figure 8.16, respectively.

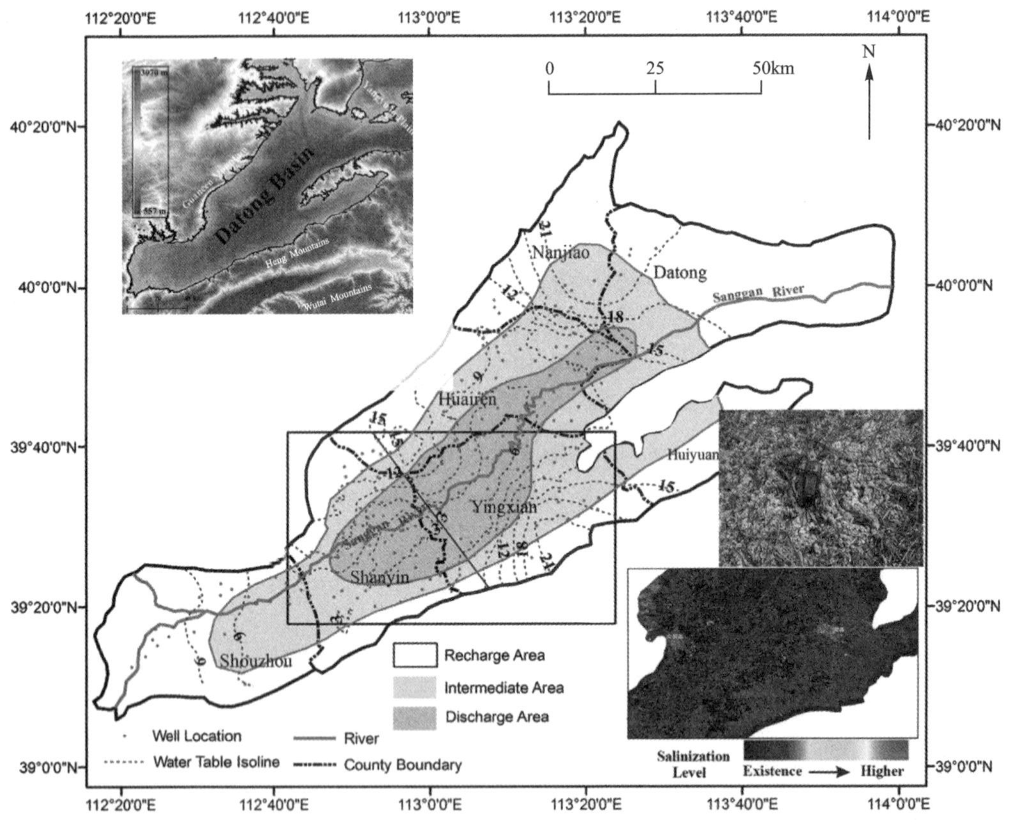

Figure 8.15 Well locations for water table survey, maps of basin-scale shallow groundwater table contour and the spatial distribution of saline soil in the central area of Datong Basin. The photo in the lower central part is a picture of the white saline soil.

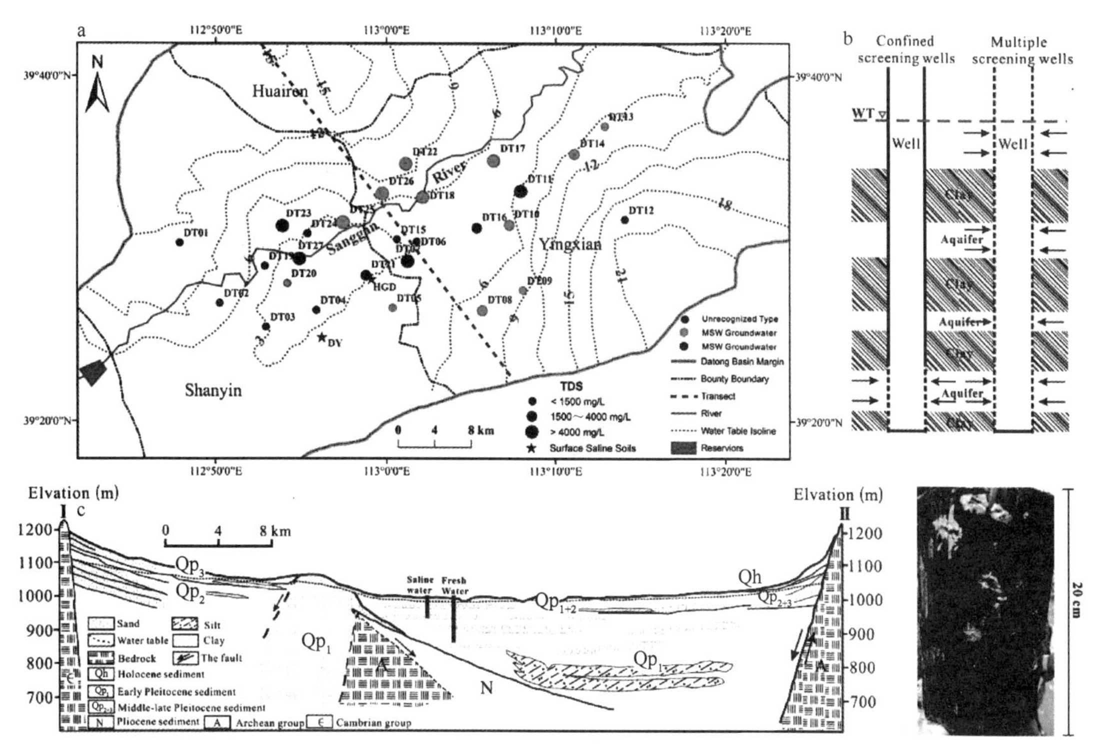

Figure 8. 16 Sampling sites of groundwater in the central area of Datong Basin (a), well structures for MSW and CSW (b), hydrogeological cross-section along the transect line labeled in Figure 8. 16, which is modified from Guo and Wang (2005) (c). The photo in the lower right corner is a core sample from the dark clay layer at depth of around 20m below the land surface.

8.5.2 Endogenous source of salinity: chlorine stable isotope and inverse modeling

As shown in Table 8. 2, the chlorine stable isotope composition of groundwater samples ranged from 0. 70‰ to 2. 39‰, the sample DT12 located in east margin of the Datong Basin had the highest δ^{37}Cl value. There is no clear correlation between groundwater Cl isotope vs. well depth and structure. For MSW groundwater, the δ^{37}Cl varied from 0. 70‰ to 1. 9‰, while the range for CSW groundwater was 0. 78‰～2. 39‰.

The correlation between groundwater Cl concentration and δ^{37}Cl value (Figure 8. 17a), indicates that the δ^{37}Cl value tends to decrease with the increase of Cl concentration. For CSW groundwater, except the DT02, δ^{37}Cl values for the rest of samples displayed a slight but regular decrease with gradually increase of groundwater Cl concentrations, and the regression equation is as follows (r^2=0. 933) (Figure 8. 17b):

$$\delta^{37}\text{Cl}=7.6\times10^{-6}\text{Cl}^2-0.00625\text{Cl}+2.367 \quad (8.7)$$

The similar trend for Cl concentration and δ^{37}Cl value has also been recorded by Bonifacie et al. (2007b) for the sediment pore water, which was attributed to the potential effects from fluid circulation in the magmatic sediment of oceanic crust. The Quaternary

Table 8.2 Chemical and isotopic compositions of groundwater samples from Datong Basin.

ID	Well type	D[a]	pH	T	Eh	WT[b]	TDS	Water type	$\delta^{37}Cl$	σ	Cl/Br	Cl	alite SI[c]	Calcite SI	Dolomite	As	I	F
		m		℃	mV	m	mg/L		‰	±	Molar ratio	mg/L			SI	μg/L	μg/L	mg/L
DT02	CSW[d]	48	8.17	13.4	−113	3	1148	Na-Cl	1.7	0.3	1340	296	−5.805 9	0.399 5	1.220 7	53	51	0.79
DTO3	CSW	55	8.16	13.1	−141	2.2	1085	Na-HCO_3	1.95	0.17	688	109	−6.185 5	0.311 3	1.383 8	255	162	0.46
DT04	CSW	58	7.97	14.4	−81	4.36	510	Na-HCO_3	2.13	0.09	625	16.9	−7.395 6	0.029 6	0.561 0	302	29	0.72
DT06	CSW	70	9.05	14.5	10	1.69	1090	Na-HCO_3	0.78	0.18	240	47.5	−6.481 6	0.785 4	2.279 5	206	207	1.03
DT12	CSW	75	7.68	—[e]	160	34.32	451	Ca-HCO_3	2.39	0.33	788	14	−8.031 6	0.316 7	0.391 1	2	12	0.20
DT15	CSW	83	8.32	12.6	−13	2.3	1465	Na-HCO_3	1.6	0.22	224	122	−5.925 9	0.258 3	1.133 5	166	792	2.23
DT16	CSW	13	8	13.4	129	—	3789	Na-HCO_3	1.12	0.33	518	497	−4.950 3	0.312 7	1.501 6	26	1187	7.93
DT19	CSW	95	8.52	—	128	1.96	947	Na-HCO_3	1.39	0.28	560	172	−5.972 9	0.413 6	1.318 9	94	318	1.32
DT21	CSW	17	8.29	13.1	97	2.67	1592	Na-HCO_3	1.75	0.2	595	129	−5.923 2	0.461 0	1.359 5	16	214	4.55
DT24	CSW	80	8.4	—	136	6.68	1377	Na-HCO_3	1.16	0.45	351	348	−4.404 5	0.307 4	1.254 1	200	1286	2.26
DTO5	MSW[f]	20	7.66	12	101	2.5	770	Na-HCO	1.61	0.18	1979	115	−6.441 2	0.144 7	0.385 5	1	61	0.60
DT08	MSW	27	7.37	14.3	139	7.7	1897	Na-Cl	0.7	0.26	2156	417	−5.650 8	0.239 7	0.648 0	3	38	0.31
DT09	MSW	17	7.92	13.2	54	11.84	1156	Na-HCO_3	1.49	0.19	2436	144	−6.128 9	0.238 0	1.282 8	367	231	1.76
DT10	MSW	50	8.2	11.8	106	—	2515	Na-HCO_3	1.9	0.18	1035	239	−5.382 4	0.349 2	1.318 7	6	471	2.17
DT13	MSW	34	8.54	12.5	125	9.9	1182	Na-HCO_3	1.46	0.16	1295	164	−5.905 7	0.584 9	1.446 5	14	77	3.91
DT14	MSW	17	8.4	12.6	179	12.72	3183	Na-HCO	1.66	0.18	1531	500	−4.301 4	1.048 2	2.623 9	2	1079	3.31
DT17	MSW	10	7.2	11.9	206	3.62	8250	Na-Cl	1.31	0.1	2656	2884	−4.090 7	0.374 3	1.192 9	5	244	0.35
DT18	MSW	18	7.76	12.2	157	3.9	5998	Na-SO_4	0.98	0.28	1717	1125	−4.516 8	0.486 0	1.626 1	3	314	1.85
DT20	MSW	100	8.54	13.1	52	1.53	1282	Na-HCO	1.65	0.22	171	51.9	−6.397 9	0.610 6	1.697 2	342	663	1.80
DT22	MSW	6	7.52	—	196	2.87	6414	Na-Cl	1.34	0.22	2152	1822	−4.321 9	0.421 9	0.940 6	13	69	0.16
DT25	MSW	12	7.76	—	41	5.65	6569	Na-Cl	0.97	0.1	1310	1885	−4.301 4	0.732 7	2.088 7	6	1041	0.01
DT26	MSW	10	7.73	12.9	27	3.23	4088	Na-Cl	1.88	0.23	1736	1158	−4.650 3	0.512 2	1.595 2	3	309	1.90

Continue

ID	Well type	D^a	pH	T	Eh	WT^b	TDS	Water type	$\delta^{37}Cl$	σ	Cl/Br	Cl	alite SI^c	Calcite SI	Dolomite	As	I	F
		m		℃	mV	m	mg/L		‰	±	Molar ratio	mg/L			SI	μg/L	μg/L	mg/L
DT01	—	55	7.82	—	224	11.3	738	Na-HCO	1.11	0.25	670	51.5	−6.783 6	0.100 9	0.555 3	2	18	1.54
DT07	—	27	7.87	13.1	96	5	4336	Na-Cl	1.36	0.16	602	965	−4.672 7	0.556 4	1.896 8	64	381	1.82
DT11	—	12	7.98	13.2	154	2.18	5493	Na-Cl	1.05	0.07	1370	1060	−4.492 5	0.334 1	1.534 2	7	1043	1.93
DT23	—	10	7.38	11.7	156	1.55	4531	$Na-SO_4$	1.6	0.12	1525	1053	−4.785 9	0.301 3	1.033 9	2	462	1.59
DT27	—	16	7.68	10.9	−100	3.5	6821	Na-Cl	1.04	0.11	1483	2463	−4.404 5	0.503 7	1.808 7	81	409	0.37

a. Depth; b. Water table; c. Saturation index; d. Confined screening well; e. No measurement; f. Multiple screening well.

sediment at Datong is mainly composed of terrestrial alumino-silicate minerals (Li et al., 2013; Wang et al., 2009). Due to the extremely low Cl contents, it is difficult to precisely measure the Cl stable isotope composition in silicate minerals (Bonifacie et al., 2007a; Eggenkampet al., 1995). However, during water-sediment interaction along groundwater flow path, the preferential movement of lighter isotope resulted in the decrease of δ^{37}Cl with increasing groundwater Cl (Schauble et al., 2003). Therefore, considering the well structure of CSW, the observed depletion of ^{37}Cl with increasing Cl concentrations of groundwater might suggest the contribution of water-sediment interaction to groundwater Cl concentration. The deviation of DT02 might be related to the potentially extra Cl contribution from surface halite dissolution because of its Cl/Br molar ratio (1340) evidently higher than those of CSW samples. Additionally, two outliers (DT06 and DT08 in Figure 8.17a) with lower δ^{37}Cl values can be attributed to local-scale diffusion process in the central area, where fine clay/silt prevails and sufficiently high gradient of groundwater Cl concentrations exists, which provides favorable condition for the occurrence of diffusion process (Eggenkamp, 2014; Krooss et al., 1992).

As can be seen in Figure 8.18, evapotranspiration could elevate Cl concentration but without introducing new Cl sources into groundwater systems. Meanwhile, the negative SI of halite in groundwater indicates that it is not expect for halite to be precipitated, therefore, the ground water system of Datong is closed in terms of Cl flux under the effect of evapotranspiration. It also means that any distinctive Cl isotope signature is hard to be left under the effects of evapotranspiration (Eggenkamp, 2014). Accordingly, the relatively stable variation of δ^{37}Cl values with the evident increase of groundwater Cl in Figure 8.18 further demonstrate the dominant effects of evapotranspiration on groundwater Cl and chemistry in Datong.

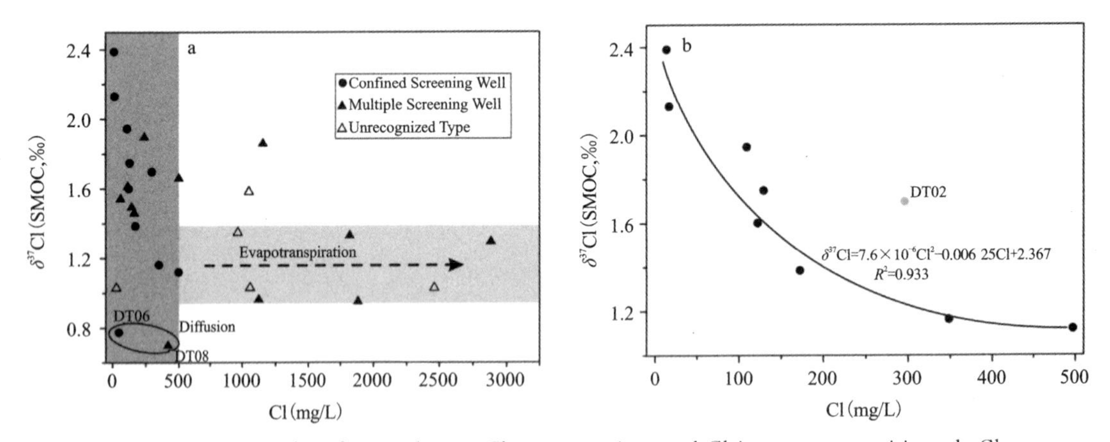

Figure 8.17 a. Plot of groundwater Cl concentrations and Cl isotope compositions; b. Cl isotope profiles of groundwater under the dominant effects of water-rock interaction.

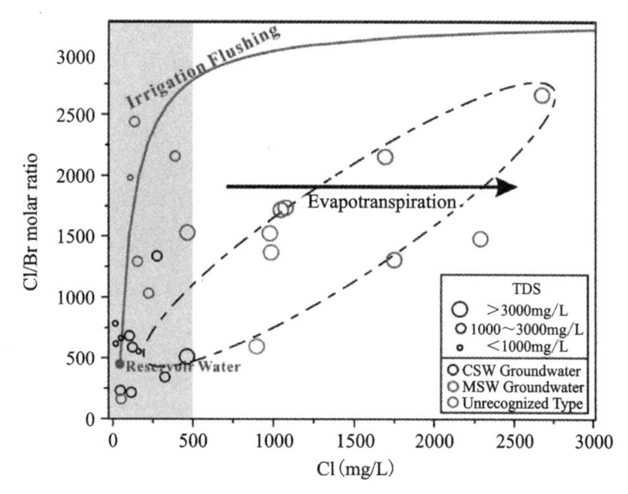

Figure 8.18 Plot of groundwater Cl concentration and Cl/Br molar ratio. The Cl concentration and Cl/Br molar ratio of upstream reservoir end-member are 40mg/L and 436, respectively.

To quantitatively assess the evapotranspiration influence, inverse geochemical modeling of PHREEQC can be used to provide the important clues (Parkhurst and Appelo, 1999). Based on the sampling location in central area, two shallow samples of DT18 and DT27 and two deep samples of DT15 and DT19 were selected to represent initial and final solutions, respectively. Because of the evident differences in groundwater nitrate between CSW and MSW groups, which clearly indicate the effect of contamination of agricultural activities, two types of fertilizers [$(NH_4)_2SO_4$ and NH_4HCO_3] were assumed to be available throughout the simulation. It indicates that more than 66% loss caused by evapotranspiration can be predicted to explain the hydrochemical evolution of shallow groundwater. This further confirms that climate-driven evapotranspiration has major impact on shallow groundwater chemistry and groundwater salinization at Datong.

8.5.3 Origin and evolution of formation water in North China Plain based on hydrochemistry and stable isotopes (^2H, ^{18}O, ^{37}Cl and ^{81}Br) (Chen et al., 2014)

The application of bromine stable isotope to evaluate the origin and evolution of formation water is still in its early stage, and only a few researches exist on this topic, though considerable work has been done about natural variation of $\delta^{81}Br$. The groundwater located in the North China Plain (NCP) was selected as an example to present the application of chlorine and bromine isotope to better assess the origin and evolution of formation water.

8.5.3.1 Study area

The oilfield in NCP located in Hebei Province, is one of the largest oilfields in China and has been explored for over 50 years. The formation water in NCP have a complicated origin and result from complex evolution processes. The NCP is a typical large-scale Meso-Cenozoic sedimentary basin and consists of Achaean and Paleoproterozoic basement composed of a set of complex metamorphic rock, overlain by marine carbonate ranging from Mesoproterozoic to early Palaeozoic and continental clastic rock of Cenozoic. Lacuna exists for the strata from Upper Ordovician to Lower Carboniferous. Cenozoic strata are widely distributed in NCP with general thickness ranging from 1000m to 3500m and the thickest at 5000 m, most of which are Tertiary deposit. For Quaternary sediments, the properties are controlled by basement tectonics and geography. The oil reservoir in NCP developed from bedrock to overlying Neocene sediments, while Paleogene strata are serving as the primary source rock. In order to study the origin and evolution of formation water in NCP, thirteen formation water were sampled from motor-pumped well. In addition, two formation water from South China Sea and Jianghan oilfield, one Quaternary brine in the southern coast of Laizhou Bay, three salt lake water (Qinghai salt lake, Gahai salt lake and Xiaochi salt lake) and four seawater samples were also collected.

8.5.3.2 Origin and evolution of groundwater in North China Plain based on hydrochemistry and stable isotopes ^{37}Cl and ^{81}Br

The TDS versus δ^{37}Cl and δ^{81}Br values are shown in Figure 8.19. The negative correlation between TDS and δ^{81}Br is significant ($r^2=0.68$) for the formation water, while this correlation between TDS and δ^{37}Cl is relatively slight. Nonetheless, the highest TDS with the lowest δ^{37}Cl and δ^{81}Br values is observed in the 30th sample. For samples from Jizhong Depression, δ^{81}Br values are larger than +0.6‰ except the 30th sample with the value of +0.28‰, and those are less than +0.6‰ for the samples from Huanghua Depression.

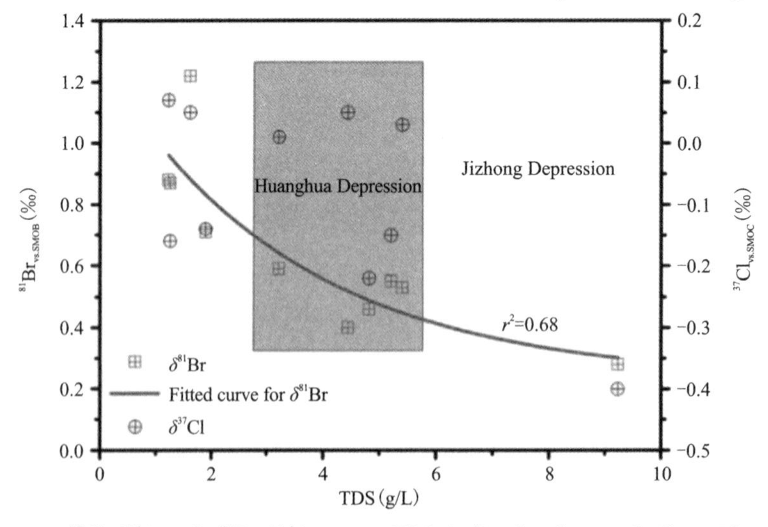

Figure 8.19 δ^{37}Cl (‰) and δ^{81}Br (‰) versus TDS (g/L) for the North China Plain samples.

Lots of research (Eggenkamp et al., 1995; Luo et al., 2012) reported that chlorine stable isotope can be fractionated during evaporation process, in which precipitation is enriched with ^{37}Cl and residual brine is enriched with ^{35}Cl. In other words, chlorine isotopic fractionation will occur after the precipitation of halite, which is surely true of bromine isotope. However, it is worth noting that all the samples in NCP are far away from halite precipitation point, so it is not possible that isotope values were affected by the same process. The most likely interpretation for the negative correlations between TDS and both $\delta^{37}Cl$ and $\delta^{81}Br$ is that the long evaporation process enhances the original formation water to be enriched with heavy isotopes (^{37}Cl and ^{81}Br), and this process resulted in increased TDS. For $\delta^{37}Cl$ and $\delta^{81}Br$ in the present study, large δ values are probably owing to evaporation, while low δ values may be attributed to mixing with seawater. Formation water of Jizhong Depression mainly occur in fluvio-lacustrine deposits and are influenced by evaporation extensively. Meanwhile, formation water of Huanghua Depression also occur in fluvio-lacustrine deposits. Though they also underwent evaporation, they are affected more by transgression, resulted in mixing with seawater and lowering $\delta^{37}Cl$ and $\delta^{81}Br$ values.

The explanation above can be further revealed in the relationship of $\delta^{37}Cl$ versus $\delta^{81}Br$ (Figure 8.20). Different water groups are easily distinguished from each other based on $\delta^{37}Cl$ and $\delta^{81}Br$ characteristics. It can be seen that both modern salty lake water and formation water in NCP show the positive relationships between $\delta^{37}Cl$ and $\delta^{81}Br$. The $\delta^{37}Cl$ and $\delta^{81}Br$ of salty lake water are richer than those of formation water, which is more vivid for $\delta^{81}Br$. The relationship of $\delta^{37}Cl$ versus $\delta^{81}Br$ exhibits a hypothetical continental effect, which shows a regular variation from coast to inland. The 8th and the 9th samples collected from Qinghai salty lake and Gahaisaltlake, located near inland Qaidam Basin, are the product of continental salinization with relatively higher $\delta^{37}Cl$ and $\delta^{81}Br$ values. Salty lake water (the 10th sample) sampled from Xiaochi, Shanxi province is more depleted in $\delta^{37}Cl$ and $\delta^{81}Br$, which is probably owing to other evolution processes except continental salinization. The location of formation water in NCP on the $\delta^{37}Cl$ versus $\delta^{81}Br$ plot is intermediate between seawater and salty lake water, on which formation water in Jizhong Depression show a tendency to be closer to salty lake water and those in Huanghua Depression are closer to seawater. This information is consistent with the discussion presented above, and further supports the assumption that formation water in NCP are probably the product of mixing between evaporated river water (lake water) and seawater, and inaddition, formation water in Huanghua Depression are influenced more intensively by seawater, while formation water in Jizhong Depression are affected more by evaporation.

In these diagrams (Figures 8.19 and Figures 8.20), the signature is quite obvious that $\delta^{37}Cl$ exhibits disperse values compared to $\delta^{81}Br$, which is speculated that diffusion and ion filtration may be responsible for partial scatter $\delta^{37}Cl$ values in sedimentary rock.

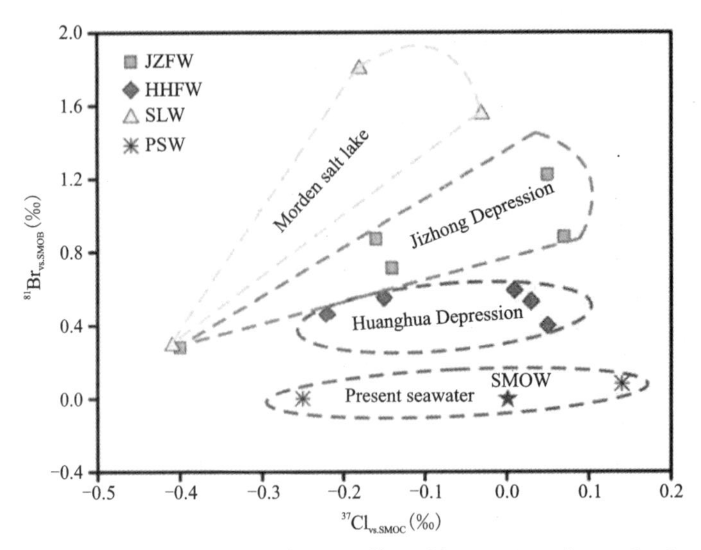

Figure 8.20 Relationship of $\delta^{37}Cl$ (‰) and $\delta^{81}Br$ (‰) for the different kinds of samples [JZFW means formation water from Jizhong Depression, HHFW means formation water from Huanghua Depression, SLW means salty lake water, PSW means present seawater, the values of both $\delta^{37}Cl$ and $\delta^{81}Br$ for Standard Mean Ocean Water (SMOW) are 0].

Experiments and geological processes have been conducted to interpret that diffusion of chlorine often occurs under condition of concentration gradients and results in effluent chloride enriched in ^{35}Cl (Deasaulniers et al., 1986; Eastoe et al., 2001; Eggenkamp, 1994; Eggenkamp and Coleman, 1997). For ion filtration, it generally takes place when formation water is forced to get through clay sediments. The calculation of Phillips and Bentley (1987) and experiments of Campbell (1985) indicated that effluent chloride is enriched in ^{37}Cl. Li et al. (2012) also reported that ion filtration might occur resulting in chlorine isotope fraction between aquifers. Therefore, scattered $\delta^{37}Cl$ in NCP formation water could be the result of potential natural processes such as diffusion and ion filtration.

In summary, the evolution history of formation water in NCP could be described using a concept model in Figure 8.21. Formation water in the two areas both occur in fluvio-lacustrine deposit, and mainly originated from ancient meteoric water (river water or lake water). The two tectonic units both experienced evaporation processes and were affected by transgression for several times. The present formation water are the mixture of evaporated meteoric water (river water and/or lake water) and seawater, which was subsequently altered by water rock interaction. The formation water in Jizhong Depression are located far away from Bohai Sea and were influenced by evaporation intensively, which resembles the formation process of the present salty lake, while formation water in Huanghua Depression are near Bohai Sea, and underwent more transgressions in geological history, showing more signatures affected by seawater than those of Jizhong Depression. This geochemical difference between Jizhong Depression and Huanghua Depression also illustrates the tectonic discrepancy of two depressions. Besides, from the present case, it can be concluded that ^{81}Br behaves more conservatively during groundwater evolution under certain circumstances, in which ^{37}Cl would be changed to some degree during process like diffusion and ion filtration.

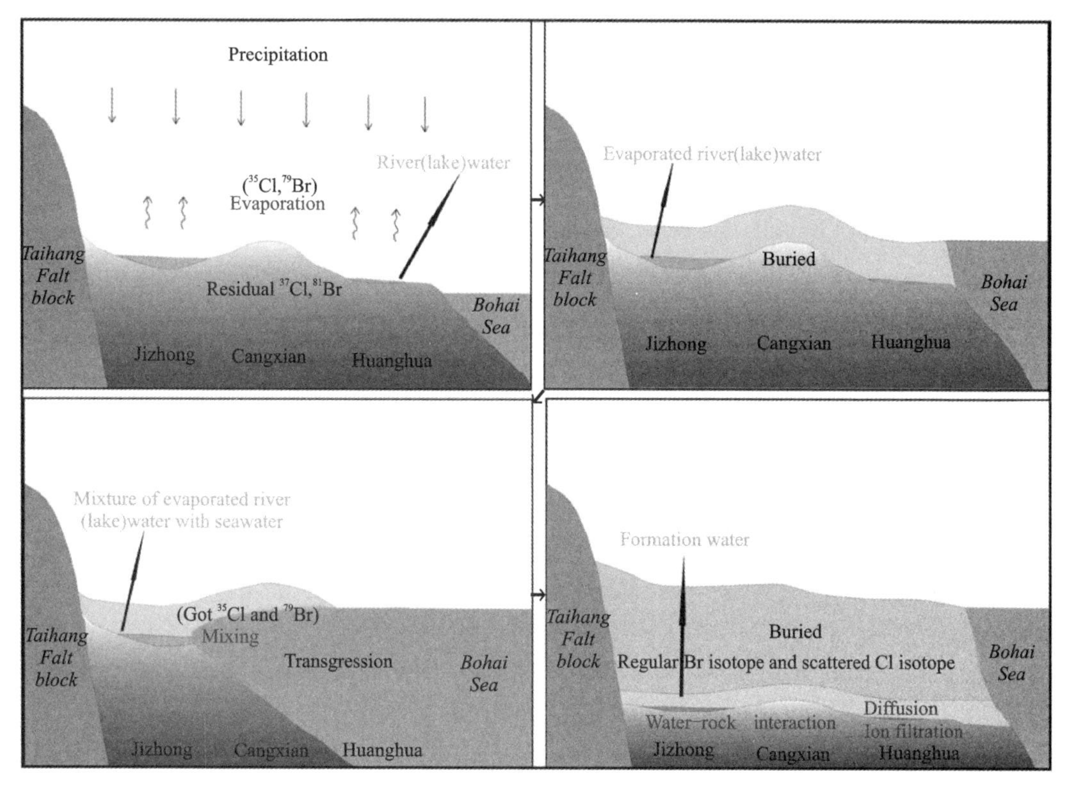

Figure 8. 21 Concept model for evolution of formation water in NCP.

References

APPELO C A J, POSTMA D, 2005. Geochemistry, groundwater and pollution (2nd ed)[M]. Leiden: Balkema.

BERGLUND M, WIESER M E, 2011. Isotopic compositions of the elements 2009 (IUPAC technical report)[J]. Pure. Appl. Chem., 83: 397-410.

BOSCHETTI T, TOSCANI L, SHOUAKAR-STASH O, et al., 2011. Salt water of the northern apennine foredeep basin (Italy): origin and evolution[J]. Aquat. Geochem., 17: 71-108.

CAMPBELL D J, 1985. Fraction of stable chlorine isotopes during transport through semipermeable membranes (MS Thesis)[D]. Univ. of Arizona.

COPLEN T B, et al., 2002. Isotope abundance variations of selected elements[J]. Pure. Appl. Chem., 74: 1987-2017.

CARSLAW H S, JAEGER J C, 1959. Conduction of heat in solids (2nd ed) [M]. London: Oxford University Press.

CHEN L, MA T, DU Y, et al., 2014. Origin and evolution of formation water in North China Plain based on hydrochemistry and stable isotopes (^{2}H, ^{18}O, ^{37}Cl and ^{81}Br) [J]. J. Geochem. Explor., 145: 250-259.

EASTOE C J, LONG A, KNAUTH L P, 1999. Stable chlorine isotopes in the Palo Duro Basin, Texas: evidence for preservation of Permian evaporite brines[J]. Geochim. Cosmochim. Acta, 63: 1375-1382.

EGGENKAMP H G M, COLEMAN M L, 2009. The effect of aqueous diffusion on the fractionation of chlorine and bromine stable isotopes[J]. Geochim. Cosmochim. Acta, 73: 3539-3548.

EGGENKAMP H G M, 1994. $d^{37}Cl$: the geochemistry of chlorine isotopes [D]. Utrecht: University of Utrecht.

EGGENKAMP H G M, KREULEN R, KOSTER VAN GROOS A F, 1995. Chlorine stable isotope fractionation in evaporites[J]. Geochim. Cosmochim. Acta, 59: 5169-5175.

EGGENKAMP H G M, BONIFACIE M, ADER M, et al., 2011. Fractionation of Cl and Br isotopes during precipitation of salts from their saturated solutions[J]. 21th Annual V. M. Goldschmidt Conference. Prague, Czech Republic. Mineral Mag., 75: 798.

EGGENKAMP H G M, BONIFACIE M, ADER M, et al., 2016. Experimental determination of stable chlorine and bromine isotope fractionation during precipitation of salt from a saturated solution[J]. Chem. Geol., 433: 46-56.

EGGENKAMP H G M, COLEMAN M, 2000. Rediscovery of classical methods and their application to the measurement of stable bromine isotopes in natural samples [J]. Chem. Geol., 167: 393-402.

EGGENKAMP H G M, 2014. The geochemistry of stable chlorine and bromine isotopes [M]. New York: Springer.

GODON A, WEBSTER J D, LAYNE G D, et al., 2004. Secondary ion mass spectrometry for the determination of $d^{37}Cl$. Part Ⅱ: intercalibration of SIMS and IRMS for alumino-silicate glasses[J]. Chem. Geol., 207: 291-303.

GRAEDEL T E, KEENE W C, 1996. The budget and cycle of earth's natural chlorine [J]. Pure. Appl. Chem., 68: 1689-1697.

GUO H, WANG Y, 2005. Geochemical characteristics of shallow groundwater in Datong Basin, northwestern China[J]. J. Geochem. Explor., 87(3): 109-120.

HERATY L J, FULLER M E, HUANG L, et al., 1998. Isotope fractionation of carbon and chlorine by microbial degradation of dichloromethane[J]. Org. Geochem., 30: 793-799.

HILL J W, FRY A, 1962. Chlorine isotope effects in the reactions of benzyl and substituted benzyl chlorides with various nucleophiles[J]. J. Amer. Chem. Soc. , 84: 2763-2769.

HOERING T, PARKER P L, 1961. The geochemistry of the stable isotopes of chlorine [J]. Geochim. Cosmochim. Acta, 23: 186-199.

HOLT B D, STURCHIO N C, ABRAJANO T A, et al. , 1997. Conversion of chlorinated volatile organic compounds to carbon dioxide and methyl chloride for isotopic analysis of carbon and chlorine[J]. Anal. Chem. , 69: 2727-2733.

KNAUTH L P, 1998. Salinity history of the Earth's early ocean[J]. Nature, 395: 554-555.

KAUFMANN R S, LONG A, BENTLEY H, et al. , 1984. Natural chlorine isotope variations[J]. Nature, 309: 338-340.

LANGVAD T, 1954. Separation of chlorine isotopes by ionexchange chromatography [J]. Acta Chem. Scand. , 8: 526-527.

Land L S, 1995. The role of saline formation water in crustal cycling[J]. Aquat. Chem. , 1: 137-145.

LAYNE G, GODON A, WEBSTER J, et al. , 2004. Secondary ion mass spectrometry for the determination of $d^{37}Cl$. Part I: ion microprobe analyses of glasses and fluids[J]. Chem. Geol. , 207: 277-289.

LI J, WANG Y, XIE X, 2016. Cl/Br ratios and chlorine isotope evidences for groundwater salinization and its impact on groundwater arsenic, fluoride and iodine enrichment in the Datong Basin, China[J]. Sci. Total Environ. , 544: 158-167.

LI X Q, ZHOU A G, LIU Y D, et al. , 2012. Stable isotope geochemistry of dissolve chorine in relation to hydrogeology of the strongly exploited Quaternary aquifers, North China Plain[J]. Appl. Geochem. , 27: 2031-2041.

LONG A, EASTOE C J, KAUFMANN R S, et al. , 1993. High precision measurement of chlorine stable isotope ratios[J]. Geochim. Cosmochim. Acta, 57: 2907-2912.

LUO C G, XIAO Y K, WEN H J, et al. , 2014. Stable isotope fractionation of chloride during the precipitation of single chloride minerals[J]. Appl. Geochem. , 47: 141-149.

LUO C G, XIAO Y K, MA H Z, et al. , 2012. Stable isotope fractionation of chlorine during evaporation of brine from a saline lake[J]. Chin. Sci. Bull. , 57: 1833-1843.

MAGENHEIM A J, SPIVACK A J, VOLPE C, et al. , 1994. Precise determination of stablechlorine isotope ratios in low-concentration natural samples[J]. Geochim. Cosmochim. Acta, 58: 3117-3121.

MANZINI M, BOUVIER A S, et al. , 2017. SIMS chlorine isotope analyses in melt inclusions from arc settings[J]. Chem. Geol. , 449: 112-122.

NUMATA M, NAKAMURA N, GAMO T, 2001. Precise measurement of chlorine stable isotopic ratios by thermal ionization mass spectrometry[J]. Geochem. J. , 35: 89-100.

PARKHURST D L, APPELO C A J, 1999. User's Guide to PHREEQC (ver. 2)—A computer program for speciation, reaction-path, one-dimensional transport, and inverse geochemical calculations[C]. U. S. Geological Survey Water-Resources Investigations.

PHILLIPS F M, BENTLEY H W, 1987. Isotopic fractionation during ion filtration: I. Theory[J]. Geochim. Cosmochim. Acta, 51: 683-695.

RICHTER F M, MENDYBAEV R A, CHRISTENSEN J N, et al., 2006. Kinetic isotopic fractionation during diffusion of ionic species in water[J]. Geochim. Cosmochim. Acta, 70: 277-289.

RITTENHOUSE G, 1967. Bromine in oil field water and its use in determining possibilities of origin of these water[J]. AAPG Bull., 17: 1213-1228.

SCHAUBLE E S, ROSSMAN G R, TAYLOR H P, 2003. Theoretical estimates of equilibrium chlorine-isotope fractionations[J]. Geochim. Cosmochim. Acta, 67: 3267-3281.

SHARP Z D, BARNES J D, BREARLEY A J, et al., 2007. Chlorine isotope homogeneity of the mantle, crust and carbonaceous chondrites[J]. Nature, 446: 1062-1065.

SHOUAKAR-STASH O, DRIMMIE R J, FRAPE S K, 2005. Determination ofinorganic chlorine stable isotopes by continuous flow isotope mass spectrometry[J]. Rapid Commun. Mass Spectr., 19: 121-127.

STOTLER R L, FRAPE S K, SHOUAKAR-STASH O, 2010. An isotopic survey of δ^{81}Br and δ^{37}Cl of dissolved halides in the Canadian and Fennoscandianshields[J]. Chem. Geol., 274: 38-55.

TAN H B, MA H Z, XIAO Y K, et al., 2005. Characteristics of chlorine isotope distribution and analysis on sylvinite deposit formation based on ancient salt rock in western Tarim basin[J]. Sci. China (ser. D), 48: 1913-1920.

TAN H B, MA H Z, WEI H Z, et al., 2006. Chlorine, sulfur and oxygen isotopic constraints on ancient evaporite deposit in the Western Tarim Basin, China[J]. Geochem. J., 40: 569-577.

Taylor J W, Grimsrud E P, 1969. Chlorine isotopic ratios by negative ion mass spectrometry[J]. Anal. Chem., 41: 805-810.

VAN ACKER M, SHAHAR A, YOUNG E D, et al., 2006. GC/Multiple collector-ICPMS method for chlorine stable isotope analysis of chlorinated aliphatic hydrocarbons[J]. Anal. Chem., 78: 4663-4667.

VENGOSH A, CHIVAS A R, MCCULLOCH M T, 1989. Direct determination of boron and chlorine isotopic compositions in geological materials by negative thermalionization mass spectrometry[J]. Chem. Geol. (Isot. Geosci. Sect.), 79: 333-343.

XIAO Y K, ZHANG C G, 1992. High precision isotopic measurement of chlorine by thermal ionization mass spectrometry of the Cs_2Cl+ion[J]. Intl. J. Mass Spectrom. Ion Proc., 116: 183-192.

XIAO Y K, ZHOU Y M, LIU WG, 1995. Precise measurement of chlorine isotopes based on Cs_2Cl^+ by thermal ionization mass-spectrometry[J]. Anal. Lett., 28:1295-1304.

ZIEGLER K, COLEMAN M L, HOWARTH R J, 2001. Palaeohydrodynamics of fuids in the Brent Group (Oseberg Field, Norwegian North Sea) from chemical and isotopic compositions of formation water[J]. Appl. Geochem., 16:609-632.

ZYAKUN A M, FIRSOVA Y E, TORGONSKAYA M L, et al., 2007, Changes of chlorine isotope composition characterize bacterial dehalogenation of dichloromethane[J]. Appl. Biochem. Microbiol., 43:593-597.

Chapter 9 Calcium and Strontium Isotope

Study of the isotopic variations of calcium is of interest because Ca is important in geochemical and biochemical processes, and is one of few major cations in rocks and minerals with demonstrated isotopic variability. Calcium is critical to life and a major component of the global geochemical cycles that control climate. Studies to date show that biological processing of calcium produces significant isotopic fractionation (4‰ to 5‰ variation of the $^{44}Ca/^{40}Ca$ ratio has been observed). Calcium isotopic fractionation due to inorganic processing at high (e. g. magmatic) temperatures is small. There are few studies of calcium isotope fractionation behavior for low-temperature inorganic processes. Calcium has six stable isotopes in the mass range of 40～48 being the largest relative mass difference except hydrogen and helium:

$^{40}Ca=96.94\%$

$^{42}Ca=0.647\%$

$^{43}Ca=0.135\%$

$^{44}Ca=2.087\%$

$^{46}Ca=0.004\%$

$^{48}Ca=0.187\%$

The strontium Sr isotope method can be a powerful tool in studies of chemical weathering and soil genesis, cation provenance and mobility, and the chronostratigraphic correlation of marine sediments. It is a sensitive geochemical tracer, applicable to large-scale ecosystem studies as well as to centimeter-scaled examination of cation mobility within a soil profile. The $^{87}Sr/^{86}Sr$ ratios of natural materials reflect the sources of strontium available during their formation. Isotopically distinct inputs from precipitation, dryfall, soil parent material, and surface or groundwater allow determination of the relative proportions of those materials entering or leaving an ecosystem. The isotopic compositions of labile soil exchange complex and soil solution strontium and Sr in vegetation reflects the sources of cations available to plants. Strontium isotopes can be used to track the biogeochemical cycling of nutrient cations such as calcium. The extent of cation contributions from in situ weathering and external additions to soil from dust and rain can also be resolved with this method. Sr has 4 stable isotopes.

$^{84}Sr = 0.56\%$

$^{86}Sr = 9.86\%$

$^{87}Sr = 7.00\%$

$^{88}Sr = 82.58\%$

9.1 Analytical techniques for the Calcium and Strontium Isotope

9.1.1 Analysis of calcium isotopes

Ca isotopes were measured using the double-spike method via Thermo-Finnigan Triton thermal ionization multi-collector mass spectrometer (TIMS)(Figure 9.1). The procedures for chemical separation and mass spectrometric analysis of Ca are modified by Russell et al. (1978) and Marshall and DePaolo (1989). A total of 20μg Ca for water samples were spiked with a ^{42}Ca-^{48}Ca double spike for Ca analysis. The spiked aliquots were dried down, dissolved in 2~3 drops 3mol/L nitric acid and then loaded onto Ca separation column filled with calcium-specific cation exchange resin (DGA resin, Eichrom Technologies, normal variety). The final Ca fraction was eluted in 1mL DI water and dried down. These separated and spiked samples were loaded onto degassed Re double filaments using 20% phosphoric acid as activator and measured by TIMS. NIST SRM 915a was used as a laboratory standard and yielded an average value of -0.97 ± 0.1 ($n=23$) during the analysis. Ca isotopic compositions

Figure 9.1 NETL's Thermo Scientific TIMS at the University of Pittsburgh, Department of Geology and Planetary Science.

are expressed as $\delta^{44/40}Ca$ values and reported relative to Bulk Silicate Earth (BSE) value, which is 1.00‰ higher than the standard:

$$\delta^{44/40}Ca(‰)=\left[\frac{(^{44}Ca/^{40}Ca)_{sample}}{(^{44}Ca/^{40}Ca)_{standard}}-1\right]\times 1000 \qquad (9.1)$$

The procedures of measurement are analogous to those used for other radiogenic isotope measurements. Mass discrimination in the spectrometer must be corrected for using a natural isotope ratio that is not affected by radioactive decay. Because the amount of isotopic discrimination is roughly proportional to the mass difference between isotopes, it makes most sense to choose either the isotope ratio $^{40}Ca/^{42}Ca$ or $^{40}Ca/^{44}Ca$ to monitor radiogenic enrichments of ^{40}Ca. To measure the mass discrimination, it can be advantageous to use an isotope ratio that encompasses a large mass difference between isotopes (such as $^{48}Ca/^{42}Ca$ or $^{48}Ca/^{44}Ca$). However, there are other considerations that make these ratios unattractive for this purpose. The accuracy of the mass discrimination correction depends on the precision with which the isotope ratio can be measured, which is poorer for lower abundance isotopes. Hence ^{48}Ca is a poor choice, and the combination of ^{48}Ca and ^{42}Ca is particularly unattractive. Also, because the mass discrimination is mass dependent as discussed below, it is advantageous to use an isotope ratio that is similar in average mass to that of the target isotope ratio being measured.

The result of the various considerations is that it is generally best to use either the $^{40}Ca/^{42}Ca$ or $^{40}Ca/^{44}Ca$ ratios to monitor the radiogenic enrichments of ^{40}Ca, and to use $^{42}Ca/^{44}Ca$ to measure mass discrimination. Marshall and DePaolo (1982) chose to use $^{40}Ca/^{42}Ca$ as the target ratio and to use $^{42}Ca/^{44}Ca$ for the mass discrimination correction. They found that when using the value of $^{42}Ca/^{44}Ca=0.312\ 21$ reported by Russell et al. (1978) for the mass discrimination correction, the initial solar system value of $^{40}Ca/^{42}Ca$ is 151.016 ± 0.008.

Precise measurement of Ca isotope ratios is challenging because the range of isotopic abundances is large. The value of the ratio $^{40}Ca/^{42}Ca$ of ca. 150 means that achieving a typical strong beam intensity of 5×10^{-11} amp for the $^{40}Ca^{2+}$ ion beam leaves the $^{42}Ca^{2+}$ ion beam at an intensity of 3.3×10^{-13} amp. At these levels, counting statistics are a limitation on the measurement precision. Marshall and DePaolo (1982, 1989) used a single collector TIMS instrument to make their measurements, with one-second integration of the $^{40}Ca^{+}$ beam and four-second integration of the $^{42}Ca^{2+}$ beam. At the ion beam intensities quoted above, this yields $3.1\times 10^{8}\ ^{40}Ca^{2+}$ ions per measurement and $8.4\times 10^{6}\ ^{42}Ca^{2+}$ ions per measurement. Ignoring any complications from background and electronic noise, this means that the accuracy of a single ratio measurement as determined by counting statistics is approximately the uncertainty associated with the measurement of ^{42}Ca, which is about $\pm(8.4\times 10^{6})^{-1/2}\approx\pm 0.3$‰. The theoretical accuracy of the ratio after correction for mass discrimination includes the uncertainty on the $^{42}Ca/^{44}Ca$ measurement, which is about ±0.5‰, so the overall limit imposed by counting statistics is about 0.6‰ at the 1σ level. By making ca. 200 measurements, this uncertainty can theoretically be reduced to about ±0.1‰ (2σ, or about 95% confidence limits) if there are no other sources of noise in the data. In practice, the precision that is achieved with the quoted beam intensities is about ±0.2‰ to 0.3‰. Although one might think it possible to improve the measurement by integrating the ion beams longer, this does not work for single collector measurements because it increases the amount of time between measurements of the reference isotope, and hence degrades the corrections for the changing beam intensity with time. Counting statistics can also theoretically be greatly improved by use of a multicollection mass spectrometer, where all of the isotopes can be collected at the same time. However, in practice it has been found that while multicollection does in fact improve the precision of individual measurements, there is unaccountable drift that worsens the reproducibility between mass spectrometer runs. Russell et al. (1977, 1978) and Marshall and DePaolo (1982, 1989) achieved a ca. 50% improvement in measurement precision by increasing the beam intensities by 10× relative to those quoted above (i. e. $^{40}Ca^{2+}$ beam intensity of 4×10^{-10} amp). Hence the best that has so far been done is a reproducibility of about ±0.15‰, or ±1.5 units of ε_{Ca} at the 2σ level.

9.1.2 Analysis of strontium isotopes

Sr isotope analysis can be achieved by either using thermal ionization mass spectrometry (TIMS) or multi-collector-inductively coupled plasma-mass spectrometer (MC-ICP-MS) (Figure 9.2), which is a newly established method that allows for higher sample throughput (Fortunato et al., 2004; Vanhaecke et al., 2009). However, accurate and precise Sr isotope measurements by either technique require the separation of isobaric interferences (mainly Rb) and the removal of matrix elements (Balcaen et al., 2005; Becker, 2005; Pin et al., 2003). Strontium is routinely separated from sample matrices using Sr specific extraction resins loaded onto chromatographic columns (Horwitz et al., 1991, 1992). While these gravity flow methods are extremely effective at purifying aqueous samples for Sr, they can be time consuming.

In order for Sr isotopes to serve as a useful tool for long-term monitoring of fluid migration related to oil and natural gas production or for monitoring, verification, and accounting (MVA) of CO_2 storage, rapid Sr separations methods are required. Improvements to gravity flow separation methods have come with the development of coupled chromatographic/MC-ICP-MS techniques and vacuum assisted methods (EPA, 2010; Galler et al., 2008; Garcia-Ruiz et al., 2008; Maxwell, 2006; Maxwell et al., 2010). Coupled methods use ion chromatographs attached directly to mass spectrometers (Galler et al., 2008; Garcia-Ruiz et al., 2008). These methods result in high sample throughput, but their benefit can only be realized in the presence of a MC-ICP-MS, eliminating the option for TIMS analysis or offsite sample processing. Alternatively, vacuum assisted Sr separation can be carried out anywhere there is a clean laminar flow hood. However, these vacuum extraction techniques require large, expensive resin cartridges, and current vacuum-based techniques are designed specifically for evaluation of radioactive Sr in environmental samples and have not been optimized specifically for isotope ratio measurement of the eluted Sr in terms of procedural blanks and proper concentrations for analysis by MC-ICP-MS (Maxwell, 2006; Maxwell et al., 2010).

Figure 9.2 NETL's Thermo Scientific NEPTUNE PLUS MC-ICP-MS at the University of Pittsburgh, Department of Geology and Planetary Science.

Sr isotope measurements performed by MC-ICP-MS were compared to TIMS on a variety of samples with a wide range of $^{87}Sr/^{86}Sr$. All of these samples had been previously processed through gravity-fed Sr-Resin® columns and analyzed by TIMS (Brubaker et al., 2013; Chapman et al., 2012, 2013; Sharma et al., 2013). Unseparated aliquots of the same samples were obtained and processed through the entire chemical separation and MC-ICP-MS procedure described in this paper. Even when measured by TIMS, the $^{87}Sr/^{86}Sr$ ratio of the NIST SRM 987 standard on different instruments can vary significantly, depending on the instrument geometry and measurement algorithm. The measured value of this standard is routinely reported to allow inter-laboratory comparison of $^{87}Sr/^{86}Sr$ ratios. Therefore, all measured ratios were normalized to a common value of 0.710 240 for the NIST SRM 987 standard by applying the offset between the standard measurements and this value to all measured ratios. To assess the agreement of the $^{87}Sr/^{86}Sr$ ratios measured by TIMS and MC-ICP-MS, a Bland-Altman plot was used (Bland and Altman, 1986). The average difference of the $^{87}Sr/^{86}Sr$ ratios measured by TIMS and MC-ICP-MS is 1.7×10^{-6}. While this indicates that, for the ten samples that were measured, $^{87}Sr/^{86}Sr$ ratios were measured to be slightly higher using TIMS, this value is smaller than the average standard error of sample measurements by TIMS (8.0×10^{-6}). Therefore, this can conclude that any bias related to the average difference $^{87}Sr/^{86}Sr$ ratios measured by TIMS and MC-ICP-MS is within the error of the measurement. Furthermore, difference between $^{87}Sr/^{86}Sr$ ratios measured TIMS and MC-ICP-MS is no greater than 1.8×10^{-5} as indicated by the 95% confidence levels (2σ, dark dashes). This variability is close to the long-term precision (2σ) of standard measurements by TIMS (1.7×10^{-5}). These results suggest that Sr isotopic measurements with a precision and accuracy comparable to routine measurements made by TIMS are achievable using this method.

9.2 Fractionation processes

9.2.1 Calcium isotope fractionation

9.2.1.1 Igneous and metamorphic rocks and petrogenetic processes

The measurements that are available to assess high temperature fractionation of Ca isotopes in nature do not constitute a representative sampling of crystalline rocks; most of the samples measured are volcanic and are relatively common rock types (Table 9.1, Figure 9.3). There are few or no measurements available for clastic sedimentary rocks, soils, high-K granites, and many other rock types that might be of interest. All of the measured $\delta^{44}Ca$ values of typical volcanic rocks are between $-0.3‰$ and $+0.3‰$. The mean and standard deviation of the values from the table is $-0.11‰\pm0.18‰$. Since virtually all of the samples listed in Table 9.1 are mantle-derived rocks with no significant involvement of old continental rocks, the $\delta^{44}Ca$ values do not reflect any radiogenic ^{40}Ca enrichment. Hence the slightly negative average value may in fact be confirmation that the mantle $^{40}Ca/^{44}Ca$ ratio is slightly lower than the value for the Skulan et al. (1997) standard. The variation of the $\delta^{44}Ca$ values is larger than analytical uncertainty, which suggests that there are small variations in igneous rocks that are barely resolvable with current techniques. There are some hints that the $\delta^{44}Ca$ values may correlate with other isotopic parameters (Figure 9.3), but there are too few data to be definitive.

Richter et al. (2003) have shown experimentally that there are physical processes that can fractionate Ca isotopes significantly (Figure 9.4). A diffusion couple consisting of basaltic liquid (10.38% CaO) and rhyolitic liquid (0.5% CaO) was allowed to evolve for 12 hours at a temperature of 1450℃. Isotopic effects are generated during multi-component diffusion where Ca (as well as Mg, Fe, etc.) diffuses from the basalt liquid into the rhyolite liquid. The basalt and rhyolite start out with identical $\delta^{44}Ca$ values of $-0.2‰$. The lighter isotopes of Ca diffuse slightly faster than the heavier isotopes and consequently the basalt becomes enriched in heavy Ca while the rhyolite develops a negative δ^{44}Ca value. As shown in the figure, the experiment generates an isotopic contrast of about 6.5‰ in δ^{44}Ca where none existed at the outset. This type of effect has not yet been observed in natural samples, but the experiments confirm that diffusion within silicate melts can fractionate Ca isotopes. Richter et al. (2003) modeled the results and found that the ratios of the diffusivities of the two isotopes is described approximately by:

$$\frac{D_{44}}{D_{40}} = \left(\frac{m_{40}}{m_{44}}\right)^{0.075} \tag{9.2}$$

where D_{44}/D_{40} is the ratio of the diffusivities of the diffusing species containing ^{44}Ca and ^{40}Ca respectively, m_{40} and m_{44} are the nuclidic masses of the two calcium isotopes. This result indicates that the masses of the diffusing species are considerably larger than the masses of the individual Ca atoms. Richter et al. (2003) did not comment on the nature of the diffusing species in the silicate liquid. However, it may be noteworthy that, for the ratio of the diffusivities to equal the square root of the mass of the diffusing species, the diffusing species containing ^{40}Ca needs to have a mass of about 278, exactly the mass of the anorthite formula unit ($^{40}CaAl_2Si_2O_8$). Hence, a model using the anorthite formula unit as the diffusing species would be consistent with the observations. This approach could be useful for estimating the polymer sizes associated with cations in silicate melts.

Table 9.1 Calcium isotopic compositions of terrestrial volcanic rocks, with other related isotopic parameters.

Sample Number	CaO (wt.%)	δ^{18}O	δ^{44}Ca	±2s	ε_{Nd}	^{87}Sr/^{86}Sr	Sample Description (Ref.)
D54G	11.32	5.22	−0.33	0.07			Basalt, Marianas dredge (1)
KOO-10	8.13	5.71	−0.17	0.22			Tholeiitic Basalt, Koolau (1)
KOO-21	8.39	5.99	−0.10	0.19		0.704 26	Tholeiitic Basalt, Koolau (1)
KOO-55	8.64	5.75	−0.09	0.14		0.704 16	Tholeiitic Basalt, Koolau (1)
GUG-6	10.82	5.03	0.21	0.22	7.9		Basalt, Marianas (1)
ALV-1833	11.68	5.10	−0.17	0.23	8.3	0.702 93	Basalt, Marianas (1)
HK-02	10.18		−0.16	0.26	7.7	0.703 25	Alkali basalt, Haleakala (2)
HU-24	9.42		−0.32	0.23	5.7	0.703 57	Alkali basalt, Hualalai (2)
HK-11	9.67		−0.37	0.27	8.2	0.703 10	Alkali basalt, Haleakala (2)
HU-05	10.09		0.14	0.16	5.2	0.703 60	Alkali basalt, Hualalai (2)
HSDP 452	10.46	4.80	−0.16	0.12	6.9	0.703 58	Tholeiitic basalt, Mauna Kea (1)
HSDP 160	10.45	4.87	0.34	0.25	7.3	0.703 50	Alkali basalt, Mauna Kea (1)

Continue

Sample Number	CaO (wt. %)	$\delta^{18}O$	$\delta^{44}Ca$	±2s	ε_{Nd}	$^{87}Sr/^{86}Sr$	Sample Description (Ref.)
92-12-29	4.89		−0.15	0.22	1.6	0.704 46	Unzen Dacite (1992 eruption) (3)
94-02-05	4.91		0.02	0.24	1.4	0.704 46	Unzen Dacite (1994 eruption) (3)
76DSH-8	5.35		0.12	0.21		0.703 09	Shasta Dacite (Black Butte) (4)
76DSH-8	5.35		−0.11	0.21		0.703 09	Shasta Dacite, Hornblende (4)
76DSH-8	5.35		−0.27	0.06		0.703 09	Shasta Dacite, Plagioclase (4)
92DLV-113	0.53		−0.09	0.20	−0.3	0.706 11	Inyo rhyolite
IO-14	11.82		−0.16	0.19			Basalt, Indian Ocean MORB
IO-38	11.30		−0.23	0.10			Basalt, Indian Ocean MORB
SUNY MORB	10.38		−0.22	0.03			MORB (5)
SUNY MORB	10.38		−0.27	0.12			MORB (5)
Lake Co. Obsidian	0.53		−0.23	0.05			Rhyolite (5)

References: (1)Eiler et al. (1996, 1997); (2) Sims et al. (1999); (3) Chen et al. (1999); (4) Getty and DePaolo (1995); (5) Richter et al. (2003).

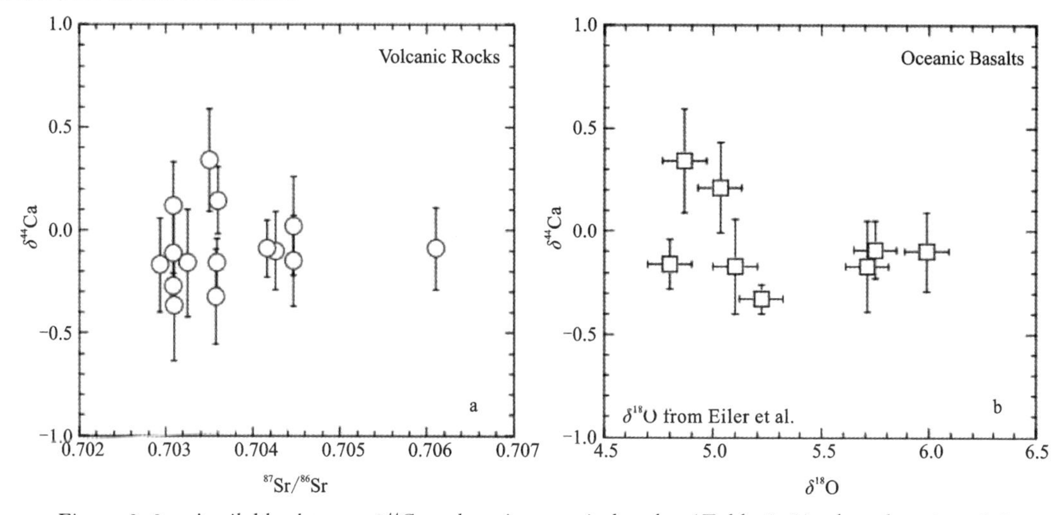

Figure 9.3 Available data on $\delta^{44}Ca$ values in oceanic basalts (Table 9.1) plotted against (a) $^{87}Sr/^{86}Sr$ and (b) $\delta^{18}O$. The average $\delta^{44}Ca$ value for all igneous rocks measured so far is $-0.05‰ \pm 0.2‰$. Small variations in mantle-derived igneous rocks could be expected as a result of recycling (subduction) of seawater-altered rocks, which would have relatively high $\delta^{44}Ca$, recycling of weathering products such as clay-rich sediment, which are also expected to have high $\delta^{44}Ca$ (not confirmed by measurements), recycling of oceanic carbonates, which have relatively low $\delta^{44}Ca$, and recycling of materials enriched in radiogenic ^{40}Ca.

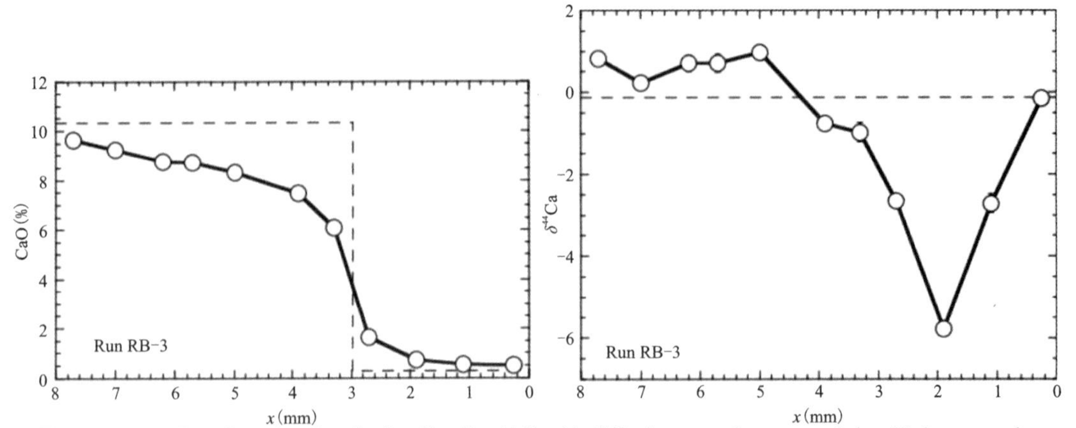

Figure 9. 4 Results of a basalt-rhyolite liquid-liquid diffusion couple reported by Richter et al. (2003). The top figure shows the initial CaO concentration profile as a dashed line, and the final concentration profile as measured by electron microprobe. The lower figure shows the initial profile of δ^{44}Ca as a dashed line (the basalt value is −0. 24‰ ±0. 05‰ and the rhyolite value −0. 23‰ ± 0. 05‰, so they are identical) and the final values as measured on the run products. The ca. 6. 5‰ variation was generated by preferential diffusion of light Ca isotopes from the basalt to the rhyolite during the run.

9. 2. 1. 2 Biological fractionation through food chains

The observation suggesting that biological processes fractionate Ca isotopes is the systematic lowering of δ^{44}Ca values through food chains (Skulan et al. , 1997). In both terrestrial and marine environments, carnivores at the end of the food chain have significantly lighter skeletal δ^{44}Ca values (Figure 9. 5). The starting point for the marine systems is seawater, which is about 1‰ higher in δ^{44}Ca than average igneous rocks (the reason for this is discussed further below). Organisms that obtain their Ca directly from seawater, such as foraminifera, mollusks, and fishes, have δ^{44}Ca values of about +0. 5‰ to −1. 0‰. Seals, which get their Ca from smaller organisms rather than seawater, have an average δ^{44}Ca value of about −1. 3‰. Bone from an orca is the lightest material so far measured in a higher marine organism at −2. 3‰. The total difference between orca bone and seawater is 3. 2‰. For terrestrial organisms the range of δ^{44}Ca is similar to that found in the marine environment, but the δ^{44}Ca values are lower. In a small sampling of materials in Northern California and in New York (Skulan, 1999; Skulan and DePaolo, 1999) it is found that whole soils are close to or slightly higher than rock values for δ^{44}Ca, plants are about 1‰ lower, horse and deer have bone δ^{44}Ca values of about −2‰ and a cougar has a bone δ^{44}Ca value of −3. 2‰. The total range of δ^{44}Ca values in the terrestrial environment is about 3. 5‰.

The Ca isotope data from food chains indicate that the Ca fixed in mineral matter in organisms is typically lighter than the Ca available to them from their surroundings or through their diets. In succeeding levels of a food chain, the δ^{44}Ca values decrease by about

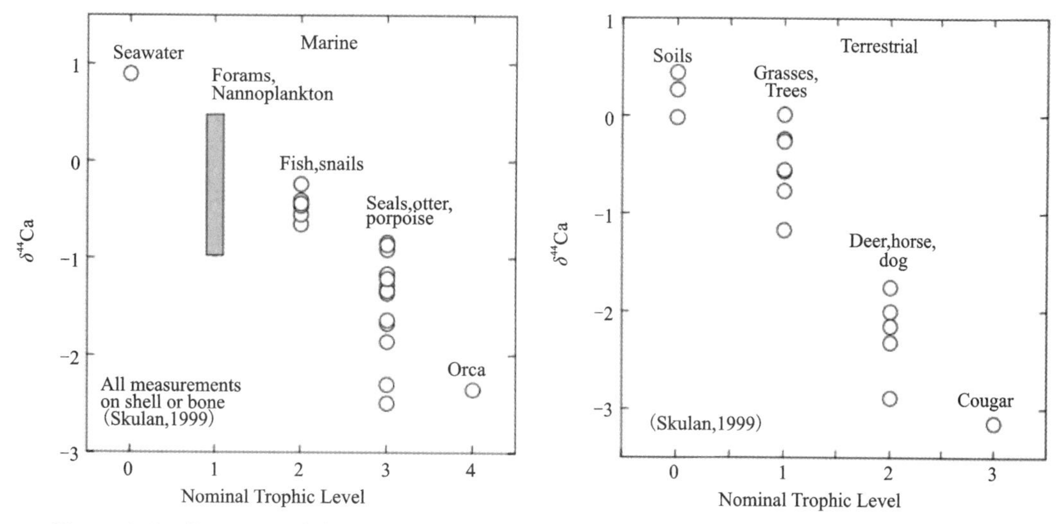

Figure 9.5 Summary of data suggesting that Ca isotopes are fractionated as calcium proceeds through food chains. Data are from Skulan et al. (1997), Skulan and DePaolo (1999) and Skulan (1999). Additional data are reported by Clementz et al. (2003).

1 unit per level. Terrestrial animals at analogous levels of the food chain tend to have δ^{44}Ca values that are about 1 unit lower than their marine counterparts. A recent report by Clementz et al. (2003) confirms the systematic relationship between δ^{44}Ca and trophic level in marine organisms, and suggests that the relationship is preserved in 15-million-year-old bone specimens as well. Skulan (1999) also analyzed a large number of fossil bone and shell samples and suggests that Ca isotopes can be used for paleodiet reconstruction.

9.2.2 Strontium isotope fractionation

9.2.2.1 Strontium isotope systematics

Table 9.2 shows the average concentrations of rubidium, strontium, potassium, and calcium in common natural materials. Rb averages 90×10^{-6} in crustal rocks, but its concentration ranges from $<1 \times 10^{-6}$ in most carbonates to $\sim 200 \times 10^{-6}$ in some granites. The average Rb content of soils is 70×10^{-6}, with a general range of $(20 \sim 500) \times 10^{-6}$ (Bohn et al., 1979). The average concentration of Sr in crustal rocks is 370×10^{-6}, but can vary from 1×10^{-6} in some ultramafic rocks to several percent in some aragonitic corals. The average concentration of Sr in soils is 240×10^{-6}, but can fall below 10×10^{-6} or exceed 1000×10^{-6} in some cases (Bohn et al., 1979). Soil retention of divalent cations is generally greater than retention of monovalent cations; K^+ is generally more readily exchangeable than Ca^{2+}, and Sr^{2+} is relatively strongly adsorbed to phyllosilicates and organic matter in soil. Soil retention of monovalent alkali ions tends to increase with atomic weight: Cs$>$Rb$>$K$>$Na$>$Li. For divalent alkaline earth ions, soil retention increases in the order Ba$>$

Sr>Ca>Mg. This can result in some fractionation of Rb from K and Sr from Ca under certain conditions. Seawater has ~8×10^{-6} Sr and ~0.1×10^{-6} Rb. River water and precipitation generally have <0.1×10^{-6} Sr and an order of magnitude less Rb (Bohn et al., 1979; Sposito, 1989; Faure, 1986).

If Rb and Sr are incorporated into a mineral or rock at its formation and the system remains closed with respect to those elements, then the amount of ^{87}Sr increases over time as radioactive ^{87}Rb decays; the amounts of ^{84}Sr, ^{86}Sr, and ^{88}Sr remain constant. Therefore, older rocks will in general have higher $^{87}Sr/^{86}Sr$ ratios than younger ones with the same initial Rb/Sr ratio. Over geologic time, rocks of a given age composed of minerals with a high Rb/Sr (e. g. granites in the continental crust), will develop a higher $^{87}Sr/^{86}Sr$ than rocks with a lower Rb/Sr (e. g. oceanic basalt) (Figure 9. 6). Thus, $^{87}Sr/^{86}Sr$ ratios in geologic materials are indicators of both age and geochemical origin.

Table 9. 2 Average or ranges of concentrations of strontium, calcium, rubidium, and potassium in earth materials.

		Sr	Ca	Rb	K	Reference[a]
Geologic ($\times10^{-6}$)	Average crust	370	41 000	90	21 000	1
	Exposed upper crust	337		95		2
Soil	Soil minerals	240	24 000	67	15 000	1
	Soil (labile)	0. 2~20				3, 4
Individual rock types	Ultramafic rock	1	25 000	0. 2	40	5
	Sandstone	20	39 100	60	10 700	5
	Low-Ca granite	100	5100	170	42 000	5
	Deep-sea clay	180	29 000	110	25 000	5
	Syenite	200	18 000	110	48 000	5
	Shale	300	22 100	140	26 600	5
	High-Ca granite	440	25 300	110	25 200	5
	Basalt	465	76 000	30	8300	5
	Carbonate	610	302 300	3	2700	5
	Deep-sea carbonate	2000	312 400	fs10	2900	
Biologic ($\times10^{-6}$)	Wood	8~2500				3, 6~9
	Roots (spruce)	19				9
	Conifer needles	2~20				3, 8

Continue

		Sr	Ca	Rb	K	Reference[a]
Hydrologic (μg/L)	Seawater	7620	414 000	110	425 000	10
	Rivers	6～800	15 000	1.3	2300	10～13
	Rain	0.7～383	800～56 000		55～1340	3, 6～8, 14
	Snow[b]	0.01～0.76	8～75		5～20	15, 16

[a] 1. Sposito (1989); 2. Goldstein and Jacobsen (1988); 3. Miller et al. (1993); 4. Bullen et al. (1997); 5. Faure (1986); 6. Graustein and Armstrong (1983); 7. Gosz et al. (1983); 8. Åberg et al. (1989); 9. Åberg et al. (1990); 10. Holland (1984); 11. Wadleigh et al. (1985); 12. Goldstein and Jacobsen (1987); 13. Yang et al. (1996); 14. Herut et al. (1993); 15. Andersson et al. (1990); 16. Baisden et al. (1995).

[b] Determined on volume of melted snow.

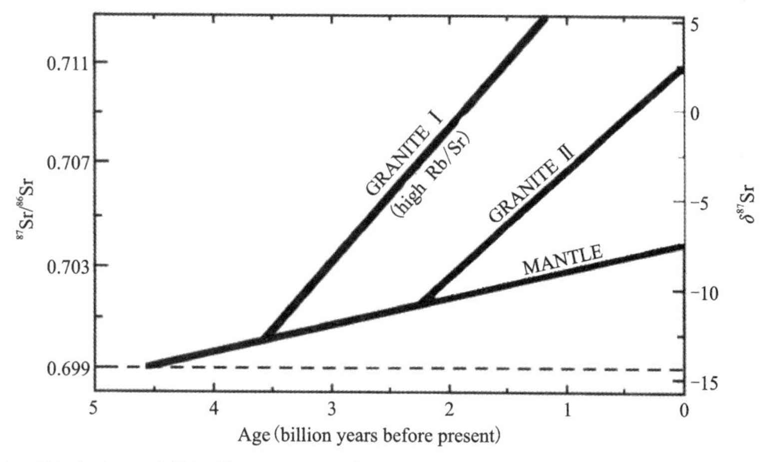

Figure 9.6 Evolution of $^{87}Sr/^{86}Sr$ over geologic time. A hypothetical granite that crystallized around 3.6 Ga ago from a mantle-derived melt (Granite Ⅰ) evolves along a steep trajectory toward high $^{87}Sr/^{86}Sr$ values because of its high Rb/Sr ratio; Granite Ⅱ also evolves along a steep trajectory, but its $^{87}Sr/^{86}Sr$ measured today is lower than that of Granite Ⅰ because it crystallized later. Two rocks of the same age but with different Rb/Sr ratios will evolve along different trajectories, with the lower Rb/Sr rock yielding a lower $^{87}Sr/^{86}Sr$.

Strontium is a relatively high-mass element, and therefore fractionation from geologic and biological processes is very small compared to that in low-mass isotopic systems (e.g., H, C, O, S). Moreover, mass-dependent fractionation of Sr isotopes (whether natural or instrument-induced) is corrected for during mass spectrometric measurement by normalization of the non-radiogenic isotopes to known values. Therefore, the measured $^{87}Sr/^{86}Sr$ ratio reflects only variations in the amount of radiogenic ^{87}Sr present in the sample, which is a function of its source. The net result is that the strontium isotopic composition of a sample yields information about provenance or geologic interactions, unobscured by local temperature variations or internal biologic processes. Rb-Sr systematics are well understood because of their extensive application to geochronological studies, and a large database exists for the estimation of $^{87}Sr/^{86}Sr$ ratios of potential source materials to ecological systems (Faure and Powell, 1972; Graustein, 1989). Strontium is ubiquitous in nature, and is one of the most abundant of the trace elements in surficial deposits and rocks (Sposito, 1989; Table 9.2). Although Sr is of little significance itself as a nutrient, it can be used as a chemical proxy for Ca because of its chemical similarity to that element.

9.2.2.2 Strontium isotope tracers of terrestrial processes

Once the isotopic compositions of the sources of strontium to the soil-atmosphere-biosphere system are known, $^{87}Sr/^{86}Sr$ ratios in soil, precipitation, dust, and vegetation can track the geochemical cycling of Sr within and between components. This provides insight into the terrestrial processes that affect the budget of Sr and other elements. The range of Sr isotopic variation in each component determines the resolution of the method.

Mass balance techniques (Chadwick et al., 1990; Brimhall et al., 1991) are used to describe the movement of elements into and out of the soil column. This approach can be extended to quantify the proportion of material added to a soil by eolian processes or lost by leaching. Eolian material in a soil can be identified by comparison with dust collected above ground or by the chemistry of labile cations in the soil and abundances of silt and clay at the top of the soil profile. However, because of the complexity of soil-forming processes, quantification of eolian input to a soil, identification of its provenance, and determination of its contribution to pedogenesis relative to *in situ* parent material weathering can be difficult. For example, dust is the dominant source of calcium involved in the formation of calcrete in some soils of the southwestern U. S. (Gile, 1979; Reheis and Kihl, 1995; Reheis et al., 1995). However, Rabenhorst et al. (1984) and West et al. (1988) found that eolian input was not a significant factor in the development of calcareous soils developed on limestone in central and west Texas.

Mixing equations are used in conjunction with strontium isotope measurements to determine the relative contribution of individual inputs to the reservoir of interest. To extrapolate the Sr isotope results to other components in a system, the concentration and the isotopic composition of Sr in each component must be determined. The isotopic composition, δ_{mix}, of a mixture of n components is given by:

$$\delta_{mix} = \frac{M_1^{Sr}\delta_1 + M_2^{Sr}\delta_2 + \cdots + M_n^{Sr}\delta_n}{M_1^{Sr} + M_2^{Sr} + \cdots + M_n^{Sr}} \tag{9.3}$$

where M_n^{Sr} represents the mass of Sr in component n.

This equation is valid for all notations of $^{87}Sr/^{86}Sr$. For a two-component system, the contribution of Sr from component 1 to a mixture is calculated from the isotopic ratios of the mixture and the endmembers:

$$\frac{M_1^{Sr}}{M_1^{Sr}+M_2^{Sr}}=\frac{\delta_{mix}-\delta_2}{\delta_1-\delta_2} \tag{9.4}$$

This simple relationship can be used to calculate the relative contributions of atmospheric input and soil mineral weathering to vegetation (Graustein and Armstrong, 1983; Gosz and Moore, 1989; Wickman and Jacks, 1993), labile Sr in soil (Graustein, 1989; Åberg et al., 1989; Miller et al., 1993), and soil carbonate (Capo and Chadwick, 1993). In dynamic natural systems, it is often useful to frame problems in terms of fluxes, rather than static mixing of two endmembers. For a system in steady state (i. e. constant source fluxes mixing together into a reservoir in which the output rate is identical to the input rate), an expression similar to Eq. 9.3 is obtained for combining material from n sources into a single reservoir:

$$\delta_{mix}=\frac{J_1^{Sr}\delta_1+J_2^{Sr}\delta_2+\cdots+J_n^{Sr}\delta_n}{J_1^{Sr}+J_2^{Sr}+\cdots+J_n^{Sr}} \tag{9.5}$$

where J_n^{Sr} is the flux of strontium (in μg/year) from source n. As an example of the application of Eq. 9.5, the relative fluxes of strontium to a soil profile from precipitation (J_P^{Sr}) and soil mineral weathering (J_M^{Sr}) can be calculated from the isotopic composition of labile soil Sr. The Sr flux from precipitation would be equivalent to AJ_PC_A, where A is waterhed area, J_P is precipitation rate in $g/cm^2 \cdot a$, and C_A is concentration of dissolved atmospheric Sr in rainwater; the flux of strontium from dissolution or alteration of primary soil minerals would be equivalent to AJ_MC_M, where J_M is the weathering flux density in $g/cm^2 \cdot a$ and C_M is the concentration of Sr in soil minerals. In principle, Eq. 9.5 can be rearranged to calculate a weathering rate, provided the exchange factor for Sr between rainwater and the labile fraction of the soil can be determined.

The amount of calcium (or any other element which does not fractionate from Sr) contributed by an endmember in a two-component system can be calculated from the Sr isotope data, provided the Sr/Ca concentration ratio is known for each component. The fraction of Ca contributed to a mixture by component 1 is given by:

$$\frac{M_1^{Ca}}{M_1^{Ca}+M_2^{Ca}}=\frac{(\delta_{mix}-\delta_2)K_2}{(\delta_{mix}-\delta_2)K_2+(\delta_1-\delta_{mix})K_1} \tag{9.6}$$

where K_1 and K_2 are the Sr/Ca concentration ratios for components 1 and 2. Mixing curves for different Sr/Ca ratios of a hypothetical two component system are shown in Figure 9.7.

Åberg et al. (1990) used Sr as a Ca proxy to estimate Ca budgets in different media: precipitation, throughfall, runoff, soil water, soil, trees, and mussel shells. However, they found that budgeting Ca inputs and outputs in the soil-vegetation ecosystem is difficult, because of the complications involved in quantifying the weathering of primary minerals. Although pedogenic processes may decouple the behavior of Sr from other elements (even geochemically similar elements like Ca) these complexities can in principle be accounted for by determining the Sr concentration or Sr/element ratio in each of the system endmembers. Quantitative modeling of the soil-vegetation-atmosphere system using Sr isotopes, including non-steady state models, is explored in greater depth in a companion paper (Stewart et al., 1998).

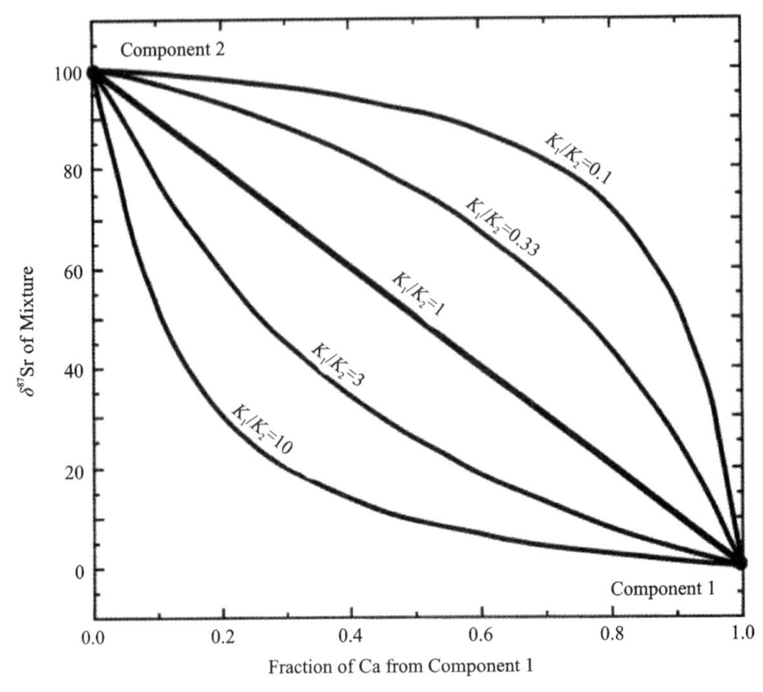

Figure 9.7 Mixing curves for different Sr/Ca ratios of a hypothetical two-component system. K_1 and K_2 represent the Sr/Ca ratios of components 1 and 2, respectively. The source of calcium in a two-component mixture can be determined from strontium isotopic data, provided the Sr/Ca of each component is known. In this way, strontium isotopes can be used for the determination of the atmospheric contribution to pedogenic calcium.

The two primary sources of cation nutrients such as Ca or Mg to the soil-biomass-water ecosystem is weathering of soil minerals and atmospheric deposition. Thus, mass balance studies of soils depend on an understanding of the rate and processes of weathering and of dust incorporation into surficial deposits. The $^{87}Sr/^{86}Sr$ ratios of weathering products reflect those of the parent material, so the Sr isotopic composition of secondary clays and soil minerals can be used as a tracer of parent material provenance (Brass, 1976). The strontium isotopic composition of pedogenic carbonate will reflect the sources of strontium available in the soil environment, both from atmospheric and *in situ* weathering sources (Graustein, 1989; Marshall et al., 1993; Capo and Chadwick, 1993; Marshall and Mahan, 1994; Capo et al., 1995). In some cases, the $^{87}Sr/^{86}Sr$ ratio can be used to discriminate between terrestrial lacustrine and marine paleoenvironments (McCulloch et al., 1989; Banner et al.,

1994; Johnson and DePaolo, 1994). If the external and in situ sources of strontium to the soil have distinct Sr isotopic signatures, the $^{87}Sr/^{86}Sr$ ratio can be used to track the movement and provenance of Sr through the soil-atmosphere system (Graustein, 1989). As a first approximation, knowledge of local bedrock geology can be used to estimate the $^{87}Sr/^{86}Sr$ range of parent materials in a particular area. Because Sr behaves like Ca, Sr can be used as a geochemical proxy for Ca, provided the relative concentrations of Sr and Ca in the system are known. With the appropriate assumptions, this can be extended to other major and trace elements in the soil-atmosphere system (Miller et al., 1993; Blum et al., 1993).

Strontium isotopic analysis in concert with mass balance studies can be used to quantify the amount, rate and sources of subsurface carbonate accumulation in desert ecosystems. The sources of solutes (e. g. Sr) for pedogenic carbonate include bedrock, surface and groundwater, airborne dust, precipitation, and seawater (Graustein, 1989). Sr isotopes can be used to characterize the secondary carbonate formed in arid soils and distinguish between eolian vs. *in situ* weathering components. Capo and Chadwick (1993) determined the Sr isotopic composition for carbonate and silicate fractions from A, upper B and petrocalcic Bkm (*K* horizon of Gile et al., 1966) of a soil developed on the Pleistocene Upper La Mesa surface in New Mexico Figure 9. 8. The $^{87}Sr/^{86}Sr$ ratios of local rain and the acetic acid-soluble fraction of dust (representing carbonate and labile Ca) range from 0. 708 8 to 0. 709 0 ($\delta^{87}Sr \approx -0.5$ to -0.2). The silicate fraction of the dust ranges from 0. 710 9 to 0. 711 2 ($\delta^{87}Sr \approx +2.4$ to $+2.9$), while the $^{87}Sr/^{86}Sr$ of the parent alluvial sediment is relatively high (0. 717; $\delta^{87}Sr \approx +11.0$). The acetic acid-soluble fractions of the A, B, and Bkm soil horizons range from 0. 708 7 to 0. 709 3 ($\delta^{87}Sr \approx -0.7$ to $+0.2$), and indicates that the Ca in the pedogenic carbonate is derived primarily from atmospheric sources, with an upper limit of ~5% on the silicate weathering contribution to soil carbonate.

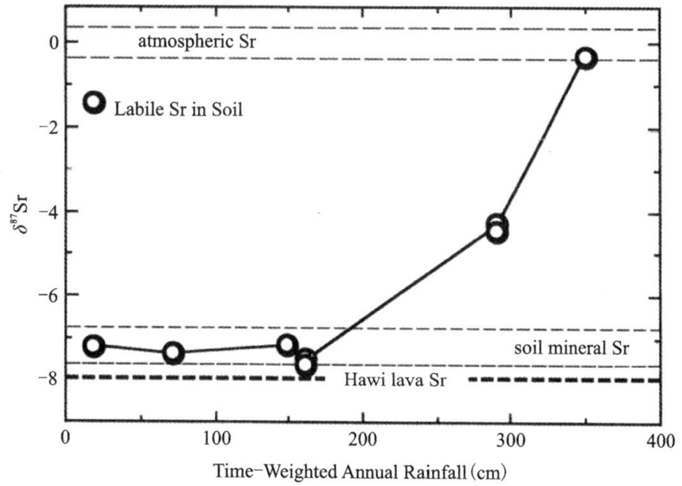

Figure 9. 8 Isotopic data for labile strontium at ~50cm depth as a function of time-weighted annual rainfall on the island of Hawaii. The soils are developed on the 170 000-year-old Hawi basalt. Soil mineral Sr isotope data are based on analysis of the silicate residue of the soil.

9.3 Variations of Calcium and Strontium isotopes in the environment

9.3.1 Variations of calcium isotopes

9.3.1.1 Silicate minerals

The total range in the $\delta^{44}Ca$ of terrestrial silicate rocks and minerals is significantly larger (>8‰) than the restricted, 2‰ range shown in Figure 9.9, likely due to significant radiogenic ^{40}Ca enrichments in potassic silicate minerals over long time scales. However, because not all published $^{44}Ca/^{40}Ca$ data have corresponding $^{44}Ca/^{42}Ca$ data, it is not completely clear how much of this range is attributable to radiogenic enrichments and how much of it reflects mass dependent fractionation. The silicate mean and median are 0.94‰, consistent with a small radiogenic enrichment compared to the bulk Earth value of 1‰ (Simon and DePaolo, 2010) and with Ca in the exogenic Ca cycle being derived from potassic rocks. The 2‰ range is relatively large for mass dependent variations during high temperature processes (i. e. melting and crystallization); Thus it has been proposed that some of this variability could be due to the recycling of carbonate (Huang et al., 2011). Given a favorable mass balance, and the data compiled herein for carbonates, such a mechanism is most likely to shift silicates to lighter isotopic compositions, but by less than a few tenths of a permil (i. e. the difference between the carbonate and silicate dataset means).

Although the large range in silicate mineral $\delta^{44}Ca$ suggests that Ca isotopes might be useful for tracing the weathering contribution of mineral phases with strong radiogenic compositions (especially potassic minerals such as biotite and potassium feldspar), few weathering studies to date have successfully utilized this approach (Farkaš et al., 2011). This is primarily because the proportion of Ca from weathering potassic minerals is low compared to the total dissolution flux. Because the halflife of K is long ($\tau_{1/2}=1.277\times10^{9}a$), significant accumulations of radiogenic ^{40}Ca require hundreds of millions of years, and can be reset at a mineralogical scale during metamorphism (Hindshaw et al., 2011). Because old rocks make up an increasingly smaller proportion of weathered substrates as time advances, there are likely few scenarios in which highly radiogenic substrates make up a significant proportion of the weathering flux to the ocean.

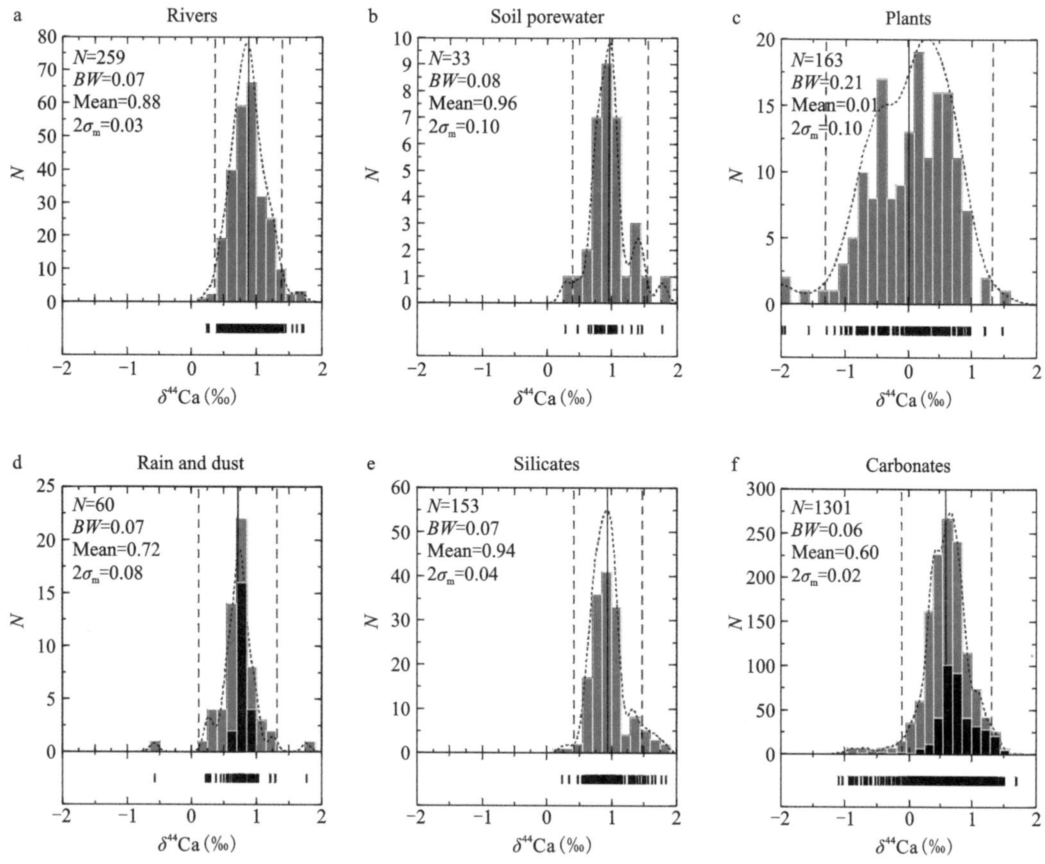

Figure 9. 9 Histograms of compiled Ca isotope data related to the modern Ca cycle, including (a) rivers (including groundwater and springs), (b) soil pore fluids, (c) plants, (d) rain and dust (gray and black respectively), (e) silicates, and (f) carbonates. Note that the carbonate dataset distinguishes Holocene (black) marine carbonates from carbonates sampled throughout all of geologic time (gray). The dashed lines in each panel indicate the δ^{44}Ca range over which 95% of the data fall. "N" is the number of samples reported, "BW" is the bandwidth used to generate the probability density function of the histogram (dotted lines), and $2\sigma_m$ is the standard error of the mean.

9.3.1.2 Carbonate minerals

The total range in the δ^{44}Ca of carbonate rocks (calcitic and aragonitic) measured to date is in excess of 3‰ (Figure 9. 9), about threequarters of the range observed in the entire compiled δ^{44}Ca dataset. The isotopic composition of carbonates is not constant in time (Fantle and DePaolo, 2005, 2007; Farkaš et al., 2007a,b; Blättler et al., 2012) and the distribution of carbonate δ^{44}Ca over hundreds of millions of years is quite broad. It has been proposed that the 1.8‰ range that encompasses 95% of the measured carbonate δ^{44}Ca values over time can be reconciled by different fractionation factors between different carbonate mineralogies and coexisting fluids (i. e., aragonite, calcite, vaterite, ikaite; Figure 9. 10;

Blättler et al. , 2012).

The mean carbonate $\delta^{44}Ca$ over geologic time is 0. 60‰ while Holocene carbonates average 0. 77‰, a significant 0. 2‰ difference. Interestingly, the mean of the carbonate dataset is about 0. 3‰~0. 4‰ lighter than the mean modern riverine and silicate $\delta^{44}Ca$. The cause of these differences is not yet clear. Assuming both isotopic steady state on million-year time scales and that the compiled carbonate data are representative of carbonates weathering in the terrestrial environment, the carbonate data suggest there is a greater range in riverine $\delta^{44}Ca$ than is presently observed. Alternatively, this difference could imply that hydrothermal and/or groundwater fluxes of Ca to the ocean (which are poorly constrained at present) might play a significant role in the evolution of seawater $\delta^{44}Ca$ over long time scales. Interestingly, Neoproterozoic carbonates have generally similar $\delta^{44}Ca$ values relative to the Holocene, which can be interpreted to mean that inputs to the Neoproterozoic oceans were not significantly different compared to inputs in the modern Ca cycle.

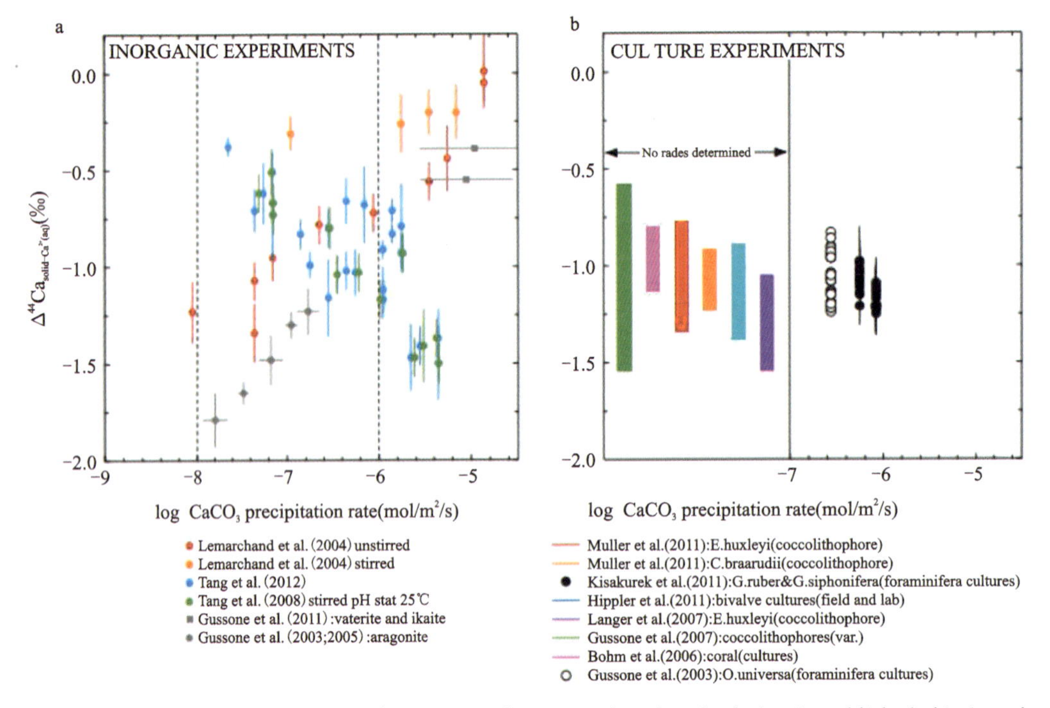

Figure 9. 10 Effective Ca isotopic fractionation between mineral and solution in published abiotic and culture experiments, as a function of mineral ($CaCO_3$, various polymorphs) precipitation rate (when reported). Dashed lines in (a) indicate an estimated range of foraminiferal biocalcification rates (e. g. , Carpenter and Lohmann, 1992).

9.3.1.3 Rivers

The range in riverine $\delta^{44}Ca$, which is representative of dissolved Ca^{2+} (aq), is rather restricted. In the riverine dataset, 95% of all published data fall within a 1. 1‰ range

(Figure 9.9), similar to that of silicate rocks but ~60% more restricted compared to the range of carbonate rocks. The mean riverine $\delta^{44}Ca$ is 0.86‰, which is similar to the flux-weighted mean of large rivers (0.80‰; Tipper et al., 2010). The detailed control on the Ca isotopic composition of rivers remains to be elucidated. While some authors have advocated the importance of conservative mixing to explain riverine $\delta^{44}Ca$ (Moore et al., 2013), such a control is not supported at the global scale. Because of mass dependent fractionation during terrestrial Ca cycling, interpreting the proportion of riverine Ca derived from carbonate, silicate, and evaporite lithologies is more complex than interpretations of radiogenic isotopes (e.g. $^{87}Sr/^{86}Sr$), where all process-related mass dependent fractionation is removed by normalization during data reduction (Tipper et al., 2008, 2010). While there are no coherent spatial trends in riverine $\delta^{44}Ca$, it is clear that the global average is strongly weighted towards the composition of Asian rivers because, after the Amazon, these are the major sources of Ca to the modern ocean (Figure 9.11; Gaillardet et al., 1999).

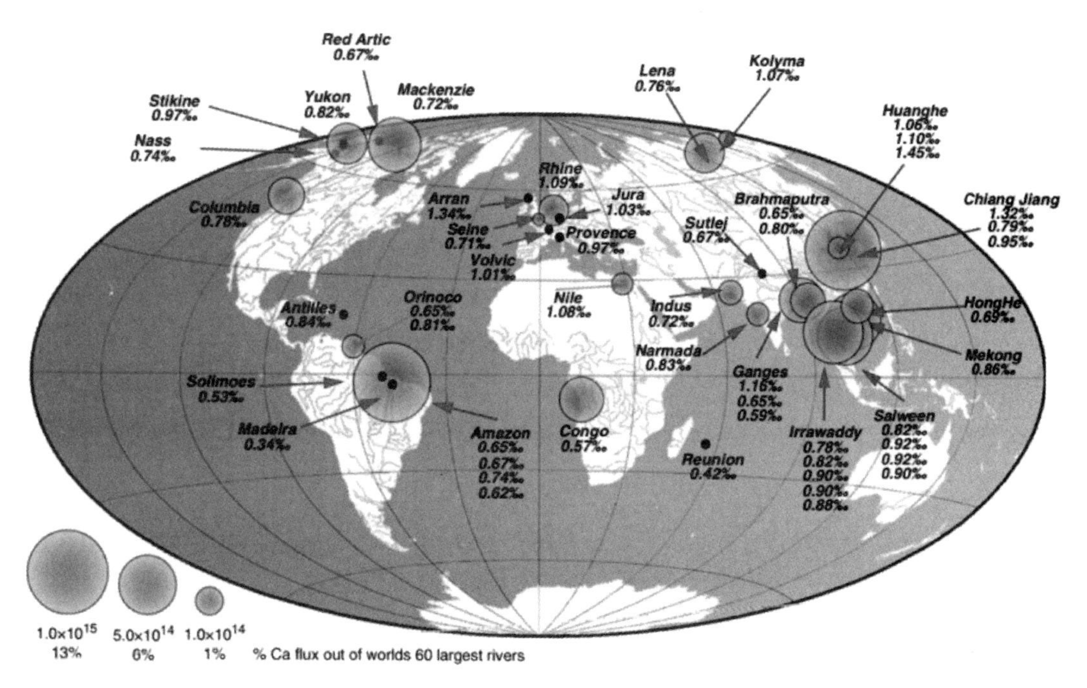

Figure 9.11 Summary of the global distribution of riverine Ca isotope data ($\delta^{44}Ca$, ‰); circles indicate the relative flux of Ca (in mol/a) supplied by each river system. Multiple values indicate cases in which multiple river samples that have been analyzed.

Not surprisingly, the mean riverine $\delta^{44}Ca$ value is intermediate between carbonate and silicate rocks. Unexpectedly, the mean riverine $\delta^{44}Ca$ is substantially closer to the silicate, rather than the carbonate, mean. Estimates of the global silicate weathering contribution to the global riverine Ca flux are between 10% and 26%; based on this, a simple mass balance suggests that global riverine $\delta^{44}Ca$ value should be 0.63‰ to 0.69‰ (assuming that the compiled dataset carbonate $\delta^{44}Ca$ is an appropriate endmember). Thus, the Ca isotope data indicate that either the estimates of global carbonate-silicate partitioning are incorrect, or there is considerable isotopic fractionation during weathering (including in limestone-dominated settings). Evidence to date, though scarce, has not revealed a sizeable fractionation associated with pedogenic processes such as mineral dissolution and the formation of secondary silicate phases (Hindshaw et al., 2011).

9.3.1.4 Rain and dust

To date, there are few measurements of sediments from dustproducing regions and no published measurements of mineral dust sampled in transit or during deposition. Fantle et al. (2012) measured the Ca isotopic composition of surface sediments from an active dust producing playa in Nevada, assuming this as a reasonable estimate of the initial $\delta^{44}Ca$ of playa-sourced dust. The primary soluble component in dust is calcite, and the average isotopic composition of this soluble component (0.78‰±0.08‰) is similar to that of rain and rivers (0.72‰ and 0.88‰, respectively; Figure 9.9), though the range is more restricted likely due to the small size of the dataset. This suggests that rain $\delta^{44}Ca$ is indeed controlled by calcite derived from semi-arid systems. Further, Ewing et al. (2008) measured the Ca isotopic composition of soils in the Atacama Desert and found systematically varying soil $\delta^{44}Ca$ values between −0.9‰ to 1.6‰, the lightest of which were near the surface. While not an active dust source, this dataset can provide insight into processes affecting Ca isotopes in arid and hyperarid settings. If such settings evolve into dust sources as climate and/or tectonic setting change, then mineral dust $\delta^{44}Ca$ may prove more variable than suggested by the playa data.

9.3.2 Variations of strontium isotopes

9.3.2.1 Terrestrial rocks and minerals

Weathering of bedrock or sediments can be a significant source of strontium to soil. Soil parent materials often have distinct Sr isotopic signatures. Precambrian granitic bedrock and alluvial sands derived from similar felsic rocks generally have high $^{87}Sr/^{86}Sr$ values ($\geqslant$0.71; $\delta^{87}Sr > 1.1$) that reflect their high Rb/Sr ratios and the age of the continental crust from which these rocks and sediments were derived. The upper continental crust exposed to weathering has an average Sr concentration of $\sim 340 \times 10^{-6}$ and $^{87}Sr/^{86}Sr$ ratio of 0.716 ($\delta^{87}Sr \approx +9.6$; Goldstein and Jacobsen, 1988). As discussed below, Sr derived from the weathering of Phanerozoic limestone and dolomite has a relatively low $^{87}Sr/^{86}Sr$ (0.707 to 0.709, $\delta^{87}Sr \approx -3$ to 0; Burke et al., 1982). Young oceanic basalts and sediments derived from them will generally have even lower $^{87}Sr/^{86}Sr$ values (typically 0.702 to 0.705; $\delta^{87}Sr \approx$

−10 to −6). Continental basalts have variable $^{87}Sr/^{86}Sr$ ratios that depend on the age and the extent of their pre-eruption interaction with continental crust. Table 9.3 lists typical values for common geologic materials.

Table 9.3 Ranges and average $^{87}Sr/^{86}Sr$ ratios of some geologic materials.

	$^{87}Sr/^{86}Sr$		$\delta^{87}Sr$		Reference[a]
	average	range	average	range	
Continental crust	0.716		9.7		1
Continental volcanics		0.702~0.714		−10.1 to +6.8	2
Oceanic island basalts	0.704	0.702~0.707	−7.3	−10.1 to −3.1	2
River water	0.712	0.704~0.922	4.0	−7.3 to 300	3~8
Modern seawater	0.709 2		0		9~10
Phanerozoic seawater		0.707~0.709		−3.1 to +0.2	9, 11~13
Proterozoic seawater		0.702~0.709		−10.1 to +0.2	14, 15

[a] 1. Goldstein and Jacobsen (1988); 2. Faure (1986); 3. Brass (1976); 4. Wadleigh et al. (1985); 5. Goldstein and Jacobsen (1987); 6. Palmer and Edmond (1989); 7. Palmer and Edmond (1992); 8. Krishnaswami et al. (1992); 9. Burke et al. (1982); 10. Capo and DePaolo (1992); 11. Hess et al. (1986); 12. DePaolo (1986); 13. Elderfield (1986); 14. Veizer et al. (1983); 15. Derry et al. (1992).

Minerals within a crystalline rock or sediment generally have variable Rb/Sr ratios and therefore a range of $^{87}Sr/^{86}Sr$ ratios. Rb/Sr ratios in basalt, granite, and lithic sediments can increase greatly during weathering due to preferential alteration of Sr-rich, Rb-poor minerals like plagioclase. Chemical weathering generally lowers the calculated Rb-Sr ages of both mineral separates and whole rocks, although the $^{87}Sr/^{86}Sr$ ratio of the bulk weathered rock is not always significantly altered (Kulp and Engels, 1963; Goldich and Gast, 1966; Bottino and Fullagar, 1968; Fullagar and Ragland, 1975). Decomposition of a high Rb/Sr mineral such as biotite during early stages of granite weathering can result in pore water enriched in ^{87}Sr and a weathered residuum depleted in ^{87}Sr relative to the whole-rock (Fritz et al., 1992; Blum et al., 1993), whereas early chemical weathering of Ca-plagioclase could result in the opposite effect. Thus, differential weathering rates for primary minerals can affect the isotopic composition of Sr released to the soil; realistic models must take parent rock mineralogy into account (Dasch, 1969; Brass, 1975; Graustein, 1989).

9.3.2.2 Rivers

The concentration of Sr in river water varies from 6×10^{-9} to 800×10^{-9}, and averages $\sim60\times10^{-9}$; Rb is even less abundant $(1\sim2)\times10^{-9}$ (Stallard, 1985). Surface and groundwater isotopic composition is a function of bedrock weathering and atmospheric inputs. Strontium from rivers in general has a high $^{87}Sr/^{86}Sr$ (~0.711; Wadleigh et al.,

1985), and reflects weathering and erosion rates of continents, which in turn are affected by sea level and climatic change (Dasch, 1969; Brass, 1976; Holland, 1984; Palmer and Edmond, 1989). The rapid rise in the seawater $^{87}Sr/^{86}Sr$ ratio over the last 2 million years is primarily due to a change in the Sr entering the oceans from rivers. This could result from an increase in the overall amount of chemical weathering on the continents, which would release more Ca and Sr into rivers, or by an increase in chemical weathering of old, highly radiogenic portions of the continental crust, which would increase the $^{87}Sr/^{86}Sr$ in the rivers entering the oceans (DePaolo, 1986; Capo and DePaolo, 1990; Hodell et al., 1990; Blum et al., 1993). Fluvial input is influenced by changes in rainfall, chemical weathering and the lithology of major drainage basins (e. g., Goldstein and Jacobsen, 1988; Palmer and Edmond, 1992; Krishnaswami et al., 1992).

9.3.2.3 Seawater and marine carbonates

Strontium in the oceans has a residence time of several million years and oceanic mixing time is on the order of a thousand years, so the Sr isotopic composition of today's oceans should be very homogeneous (Holland, 1984). The uniformity of the $^{87}Sr/^{86}Sr$ ratio of modern seawater, at least within the limits of present analytical uncertainty, is confirmed by measurements of Holocene shells and water from various depths from oceans around the world (Burke et al., 1982; Elderfield, 1986; Capo and DePaolo, 1992). Because of the lack of significant isotopic fractionation of Sr in nature, the $^{87}Sr/^{86}Sr$ ratios of the hard parts of calcareous marine organisms (such as coccolithophores, foraminifera and mollusks), and phosphatic fish debris, brachiopods, and conodonts record the Sr isotopic composition of all the world's oceans at the time the material formed. Rubidium is a very low-abundance element relative to strontium in carbonate, phosphate and seawater ($\sim 0.1 \times 10^{-6}$). Thus, addition of radiogenic ^{87}Sr from decay of ^{87}Rb after fossil formation is generally negligible (Faure, 1986). Ancient seawater Sr ratios can be determined by measurements of unaltered marine fossils, regardless of the environment, water depth, latitudinal range or species of the organism. The strontium isotopic composition of seawater has changed significantly over Earth's history and definition of these time-dependent variations provides a method for global correlation of marine sediments, as well as a tracer that reflects the cycling of material between the continental crust and the oceans (Armstrong, 1971; Brass, 1976; Palmer and Edmond, 1989; Capo and DePaolo, 1990; Hodell et al., 1990). The curve shown in Figure 9.12 (modified after Burke et al., 1982), shows that Phanerozoic seawater $^{87}Sr/^{86}Sr$ has fluctuated midway between the average ratio of rivers (~ 0.711; $\delta^{87}Sr \approx +2.6$) and that of hydrothermally altered oceanic basalt (~ 0.704; $\delta^{87}Sr \approx -7.3$). Changes in the Sr isotopic composition of the oceans over long-time scales (on the order of $10^7 \sim 10^9$ years) are associated with major tectonic events, such as uplift of the Himalayas (Palmer and Edmond,

1992; Krishnaswami et al. , 1992; Richter et al. , 1992). Shorter time-scale fluctuations on the order of a million years or less probably reflect climatic change, such as glaciation.

Global correlation of marine sediments and limestones of marine origin can be made by comparison of strontium isotopic compositions with an independently established record of the seawater $^{87}Sr/^{86}Sr$ ratio with time (DePaolo, 1987). This method has been used to correlate deep-sea sediments and subaerially exposed rocks of marine origin not amenable to dating by biostrati-graphic methods. Age resolution ranges from a few million years to less than a hundred thousand years, depending on the slope of the seawater curve and the precision of the reference curve data. High-resolution seawater Sr curves have been determined for various periods of the Phanerozoic; see Table 9. 3 for references.

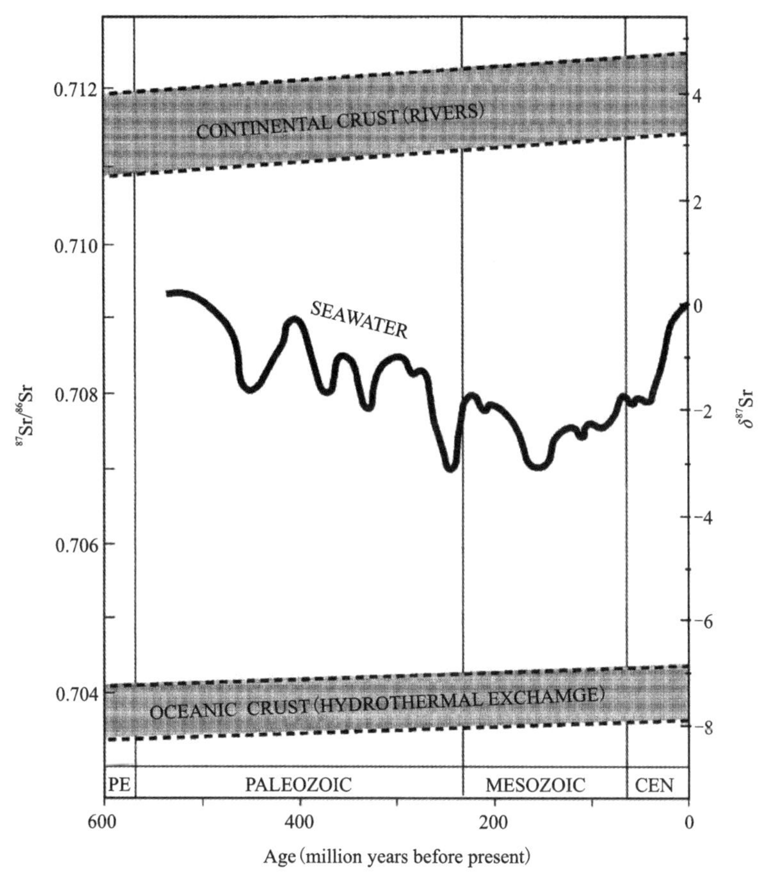

Figure 9. 12 Variation of the strontium isotopic composition of Phanerozoic oceans (modified from Burke et al. , 1982). The $^{87}Sr/^{86}Sr$ of seawater has fluctuated between the average values of terrestrial weathering and hydrothermal exchange with mid-ocean ridge basalts. The isotopic composition at any time in the past reflects that of unaltered marine carbonate of that age. Higher-precision reference data exist for specific periods of the geologic record with a corresponding improvement in age resolution (e. g. references 10～12 in Table 9. 3).

9.3.2.4 Precipitation and dryfall

Strontium isotopic composition can be used as a chemical tracer to determine the sources of atmospheric contributions to soil cation reservoirs over time (Graustein and Armstrong, 1983; Åberg et al., 1989; Graustein, 1989; Capo and Chadwick, 1993; Miller et al., 1993). In any given area, locally derived dust will have an $^{87}Sr/^{86}Sr$ ratio similar to the subaerially exposed soil, sediments, and bedrock. However, eolian processes can introduce isotopically distinct minerals to soils, such as zircons (Brimhall et al., 1994), or micas (Dymond et al., 1974). Dust from exposed Phanerozoic marine carbonate will have $^{87}Sr/^{86}Sr$ ratios between 0.707 and 0.709 ($\delta^{87}Sr \approx -3$ to 0). Terrigenous dust can have $^{87}Sr/^{86}Sr$ ratios much higher than seawater; dust from the Sahara, carried across on the northeast trade winds over the northern Atlantic Ocean, has values between 0.715 and 0.747 ($\delta^{87}Sr \approx 8$ to 53; Biscaye et al., 1974; Muhs et al., 1987; Muhs et al., 1990). Fly ash from coal combustion or woodburning can also have distinct $^{87}Sr/^{86}Sr$ ratios (Straughan et al., 1981; Andersson et al., 1990). Precipitation near ocean basins generally has a strontium isotopic composition similar to that of seawater (0.709; $\delta^{87}Sr = 0$). However, the concentration of Sr in rainfall is generally lower than that in seawater by several orders of magnitude (usually $< 1 \times 10^{-9}$; Graustein and Armstrong, 1983; Gosz and Moore, 1989; Åberg et al., 1990; Miller et al., 1993). Terrigenous components can also be entrained in precipitation; fine-grained mica and clay can contribute Sr to the dissolved portion of precipitation as well as to the particulate load (Andersson et al., 1990).

9.4 Application

9.4.1 Calcium isotope fractionation in a silicate dominated Cenozoic aquifer system (Li et al., 2018)

The hydrologic and geologic background of Datong Basin was well described in the section of 4.4.1. A total of 26 water samples (24 groundwater and 2 upstream reservoir water samples) and 5 rock samples were collected from the Datong Basin (Figure 9.13). All the sampling wells can be mainly divided into two types, including multiple screening wells (MSW) and confined screening wells (CSW). The groundwater from MSW is the mixture of water from all aquifers screening the well depth, while the groundwater from CSW is mainly from the lower confined aquifers. 16 of 24 groundwater samples were collected from MSW type, 6 samples were from CSW type, and the pump information of two groundwater sample was missing.

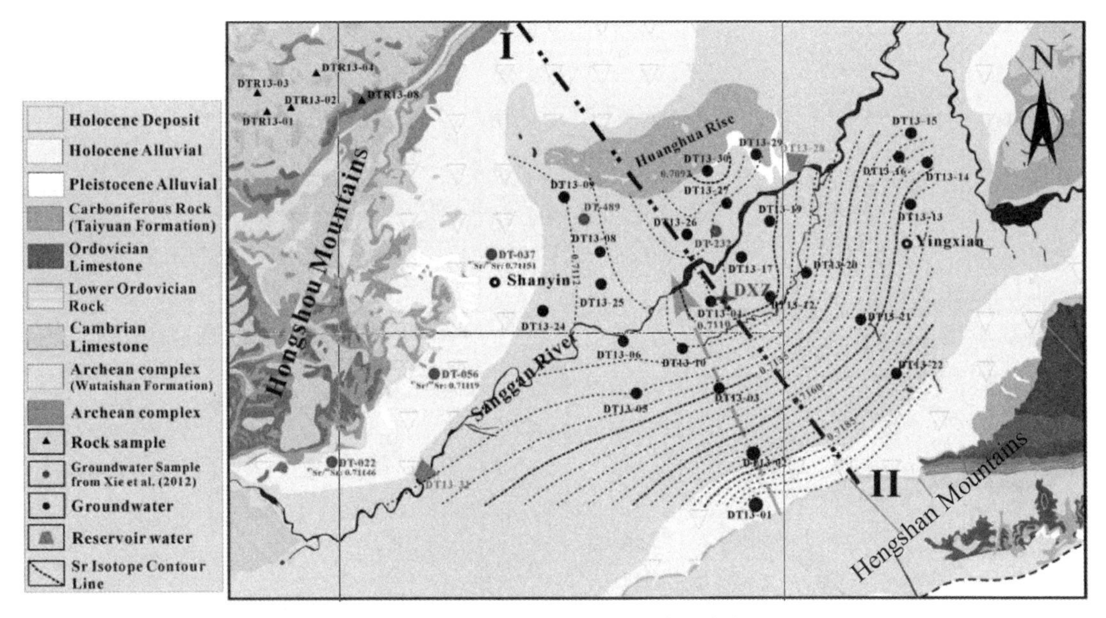

Figure 9.13 Locations of groundwater samples of the Datong Basin.

The groundwater $\delta^{44/40}Ca$ values in the Datong Basin had a range from $-0.11‰$ to $0.49‰$, lower than that of seawater ($0.9‰$) relative to the same standard values (Nielsen et al., 2012). The two reservoir samples had relatively higher $\delta^{44/40}Ca$ values of $0.55‰$ and $0.64‰$, respectively. Surface water samples from the reservoir had a temperature of 30℃, higher than those of groundwater samples (10.35~16.3℃). Temperature can influence the saturation state of carbonate minerals, and consequently, the saturation index of calcite in the two surface water samples were 0.998 6 and 1.042 1, which are higher than those of groundwater samples (0.311 7~1.013 8, except for one outlier: 1.269 9).

Along the groundwater flowpath, groundwater Ca decreases from 68.2mg/L to 3.3mg/L (Figure 9.14). As discussed above, several hydrochemical processes affect groundwater Ca cycling, including hydrolysis of metamorphic minerals, mixing of different groundwater sources, cation exchange on the surface of clay minerals, and dissolution/precipitation of carbonate minerals.

In the east margin, the groundwater is mainly recharged by the lateral infiltration of water in bedrock fractures of Hengshan Mountains, and therefore the groundwater chemistry would be comparable with that in the metamorphic complex, which is supported by the similarity in hydrochemistry between sample DT13-01 from the east margin and the spring sample DT-182 from Hengshan Mountains. The Ca-HCO_3 type water rich in potassium of groundwater DT13-01 mainly originates from the hydrolysis of Ca/K-bearing metamorphic minerals, such as anorthite and orthoclase:

$$CaAl_2Si_2O_8+3H_2O+2CO_2=Al_2Si_2O_5(OH)_4+Ca^{2+}+2HCO_3^- \tag{9.7}$$

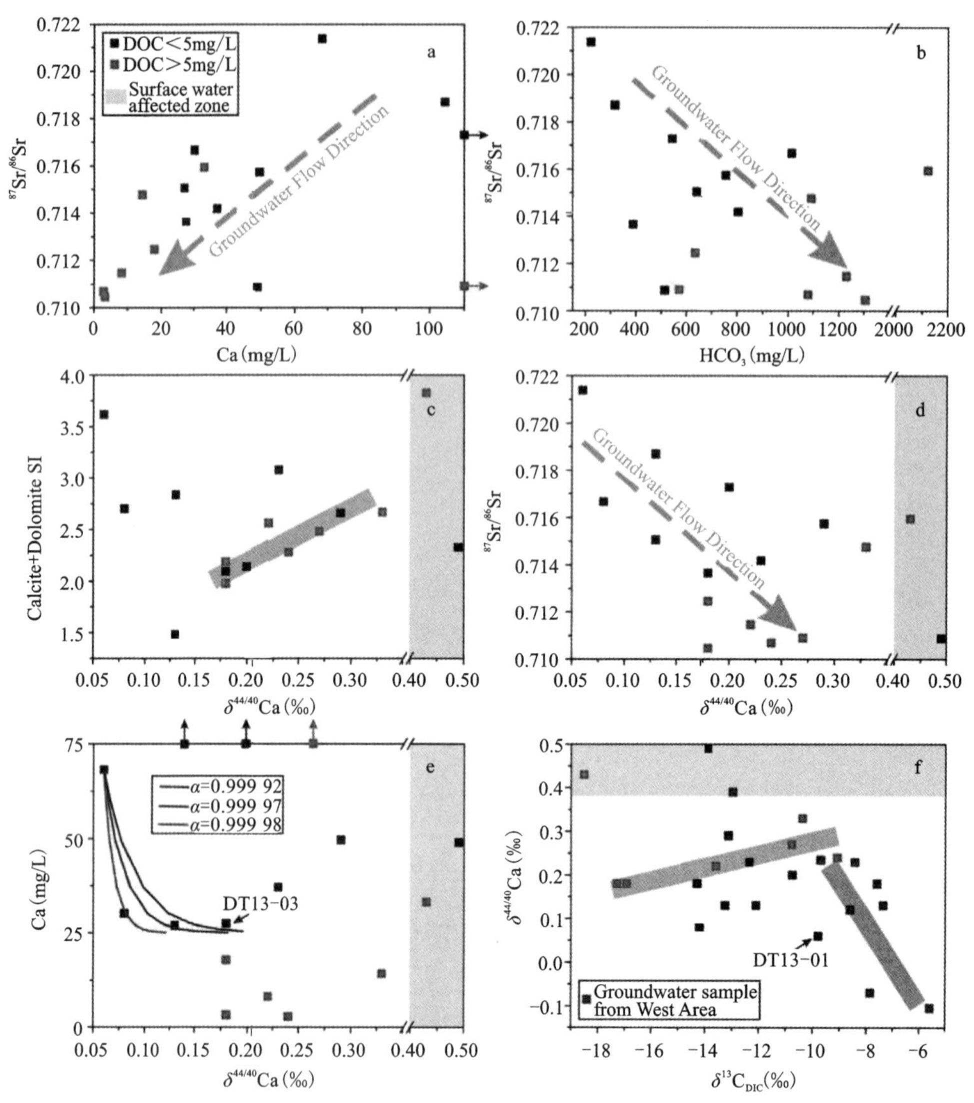

Figure 9.14 Plots of groundwater Ca concentration vs. $^{87}Sr/^{86}Sr$ (a), HCO_3 concentration vs. $^{87}Sr/^{86}Sr$ (b), groundwater $\delta^{44/40}Ca$ vs. calcite and dolomite SIs (c), groundwater $\delta^{44/40}Ca$ vs. $^{87}Sr/^{86}Sr$ (d), groundwater $\delta^{44/40}Ca$ vs. Ca concentration (e) for groundwater samples from east area of Datong Basin, and groundwater $\delta^{13}C_{DIC}$ vs. $\delta^{44/40}Ca$ (f) for all groundwater samples.

$$2KAlSi_3O_8 + 2CO_2 + 7H_2O = Al_2Si_2O_5(OH)_4 + 2K^+ + 2HCO_3^- + 4H_2SiO_3 \quad (9.8)$$

As a result, the clay mineral kaolinite is formed, which is the dominant type of clay mineral in bedrock, and the concentration of groundwater HCO_3 increase along groundwater flowpath (Figure 9.14b). The chemistry of water infiltrating from the silicate bedrock with little carbonate fraction generally has higher $^{87}Sr/^{86}Sr$ ratios (Jacobson and Blum, 2000), which is consistent with the groundwater sample DT13-01 with the highest $^{87}Sr/^{86}Sr$ ratio. The congruent dissolution of Ca-bearing plagioclase is unlikely to induce Ca isotope fractionation (Hindshaw et al., 2013; Ryu et al., 2011). It explains the relatively depleted

$\delta^{44/40}$Ca values of groundwater sample DT13-01 from the east margin.

Due to the prevalence of irrigate practice using reservoir water in the central area, the vertical mixing process can introduce surface water into shallow groundwater system with the depth<30m, and the contribution of reservoir water to shallow groundwater approximately varies from 29% to 93%, depending on well depth (Li et al., 2016). The introduction of ^{44}Ca enriched surface water leads to the elevation of $\delta^{44/40}$Ca value of shallow groundwater. As a result, the higher values of $\delta^{44/40}$Ca are mainly observed in the shallow groundwater (Figure 9.15). Assuming that 30% ~ 90% of shallow groundwater was recharged by reservoir water which had the $\delta^{44/40}$Ca value of 0.595‰ (the mean value of two reservoir water) and the groundwater sample DT13-04 from central area with well depth of 75m acts as the groundwater end-member, the calculated result of two end-member model shows that the mixed water sample between groundwater and surface water will have the $\delta^{44/40}$Ca range from 0.305‰ to 0.554‰. The $\delta^{44/40}$Ca ratios of four shallow groundwater samples fall into the mixing range, indicating that the vertical introduction of surface water leads to the elevation of their $\delta^{44/40}$Ca ratios (Figure 9.15).

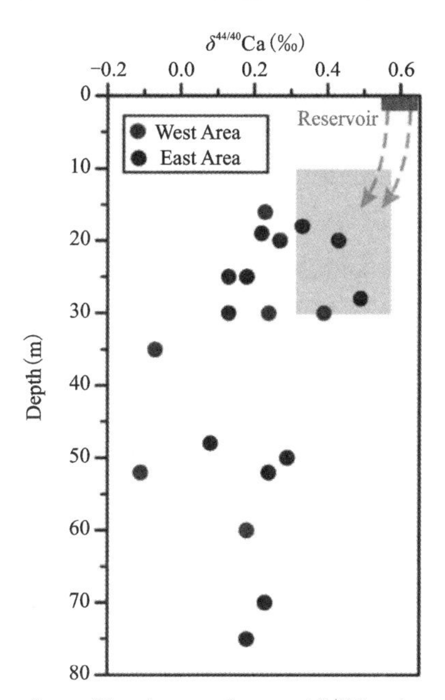

Figure 9.15 The depth profile of groundwater $\delta^{44/40}$Ca signatures and the effects of vertical irrigation process on $\delta^{44/40}$Ca signatures of shallow groundwater.

Compared to studies on Ca isotope fractionation in carbonate-dominated settings, little work has been conducted on Ca cycling and isotope fractionation in silicate systems (Jacobson et al., 2015; Moore et al., 2013). Silicate rocks and minerals have a somewhat larger range of $\delta^{44/40}$Ca than carbonate rocks (Fantle and Tipper, 2014). Therefore, it is not surprising that water in silicate systems would exhibit different Ca isotope signatures from those in carbonate systems. Up to now, the Ca isotope signatures of groundwater in silicate aquifers have not been systematically evaluated. Therefore, theoretical and case studies on silicate-dominated Cenozoic aquifer systems under different hydrogeological conditions may provide new insights about Ca isotope signature and fractionation.

9.4.2 Using strontium isotopes to evaluate the spatial variation of groundwater recharge(Christensen et al., 2018)

The Rifle Site is situated along the north bank of the Colorado River near the city of Rifle in western Colorado. The following description comes in part from the final site report (DOE, 1999). The site is located within the Colorado River floodplain where a former uranium vanadium ore mill and its associated tailings were situated (Figure 9.16). The tailings pile was a source of U contamination of the local groundwater, and was removed and the site remediated from 1992 to 1996 through the Uranium Mill Tailings Remedial Action (UMTRA) program under the U. S. Department of Energy. As part of the remediation process, along with the removal of tailing piles, contaminated surface soils across the site were excavated to an average depth of ~2m and replaced with material derived from excavation of the near-by Estes Gulch Disposal Site. The ~2m of fill is a key feature of the site in that it provides a relatively uniform vadose source with a potentially contrasting Sr isotopic signature compared to the aquifer sediments. The fill material consisting of cobbly loam overlies ~4m of Quaternary alluvial material consisting of quartz/feldspar sands, with silty and coarser pebble/cobble interbeds. The Quaternary alluvium overlies the Tertiary age Wasatch formation which forms the bottom boundary at 6~7m depth of the upper unconfined aquifer. The water table depth within the floodplain is seasonally variable, ranging from a low of about 4m below the soil surface to a maximum of about 1.5m depth below the soil surface during the late spring producing a typical swing of ~2.5m (Williams et al., 2011; Yabusaki et al., 2017). However, during 2013, the year of this study, the spring water table rise was greatly muted to about a 1m rise due to a relatively dry year resulting in low peak-flow of the Colorado River. The groundwater flow direction is generally from 220° to 180° from N across the site towards the Colorado River, with groundwater velocities in the range of 0.1m/d to 0.6m/d with slower velocities associated with high river stands (Williams et al., 2011).

Samples collected and analyzed for this study include surface water, groundwater seeps, groundwater from monitoring wells, and samples from multilevel pore-water samplers installed (TT1) within the vadose zone near Well SY08 (Figure 9.16). All samples were filtered to 0.22μm and subsequently acidified with high-purity concentrated nitric acid to a pH of ~2. Filtering was done to avoid contributions of Sr from suspended particulate matter, while acidification was done to avoid absorption of Sr to the HDPE of the sample bottles. Groundwater monitoring wells were sampled in May and again in July 2013. This period brackets the peak in water table elevation. In situ, direct pore water samples at TT1 were taken in April, May and June 2013. As separate samples of pore water composition, sediment samples collected during the boring of TT1 (Figure 9.16) were also analyzed. These were treated with DI H_2O, and the leachates separated by centrifugation to produce representations of the porewater present in the sediment samples at the time of collection.

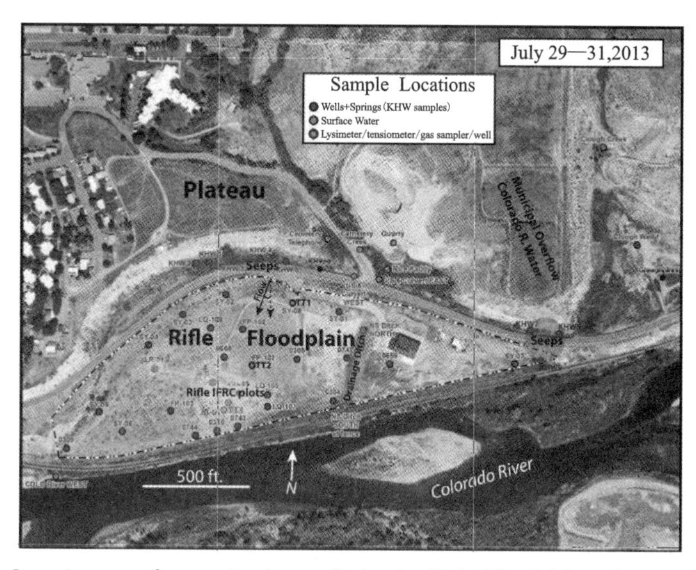

Figure 9.16 Location map for monitoring wells in the Rifle Floodplain, along with locations of sampled surface water features. Black arrows indicate the typical range in flow direction (220° to 180° from N) over the course of a year (Williams et al., 2011).

The Sr isotopic compositions and Sr concentration of the groundwater, seep, and surface water samples are provided in Table 9.4. In Figure 9.17, contours of average (May & July) groundwater and surface water $^{87}Sr/^{86}S$ ratios have been overlaid onto the location map. Seeps (samples KHW 1 through KHW 5) along the base of the bluffs, representing upgradient groundwater from beneath the plateau, have relatively low $^{87}Sr/^{86}Sr$ ratios ranging from 0.710 75 to 0.710 92 (ave. 0.710 84±0.000 06) for samples taken in May and July 2013. Samples from groundwater monitoring Wells SY02, SY03, SY04, SY05 and 309 along the western edge of the site also have relatively low $^{87}Sr/^{86}Sr$ (0.710 63~0.710 92, ave. =0.710 79±0.000 14) and also represent upgradient groundwater. Higher $^{87}Sr/^{86}Sr$ ratios are found in the central portion of the Rifle Floodplain down towards the river, ranging up to 0.711 14. The highest $^{87}Sr/^{86}Sr$ measured (~0.711 30) are for groundwater samples from Well SY08 adjacent to TT1, and create a steep, local high in the contour map (Figure 9.17). Relatively low $^{87}Sr/^{86}Sr$ at Wells 304 and 742 coincide with low $^{87}Sr/^{86}Sr$ measured in surface water from the North-South Ditch. Surface water from nearby locations (US6 Culvert East & West, Quarry, Rice Paddy, Quarry, Cemetery Ck., and Cemetery

Telephone) that potentially contribute water to the NS Ditch range in $^{87}Sr/^{86}Sr$ ratios from 0.710 95～0.711 01, falling along a rough mixing line with the NS Ditch samples. Samples from wells 304 and 742 fall near this mixing line, close to the NS Ditch samples.

Table 9.4 Sr isotopic compositions of groundwater and surface water samples taken in July 2013 and May 2013. Included are Sr concentrations for the July samples.

	Location/name	7/29/2013—7/31/2013			5/28/2013—5/30/2013	
		$^{87}Sr/^{86}Sr$	±2s	[Sr], $\times10^{-9}$	$^{87}Sr/^{86}Sr$	±2s
Monitoring wells	305	0.710 951	6×10^{-6}	4439	0.710 922	6×10^{-6}
	743	0.711 098	6×10^{-6}	6005	0.711 082	7×10^{-6}
	FP-101	0.711 102	6×10^{-6}	7536	0.711 083	6×10^{-6}
	LQ-105	0.711 124	6×10^{-6}	4527	0.711 125	7×10^{-6}
	LQ-107	0.711 136	7×10^{-6}	4412	0.711 119	6×10^{-6}
	309	0.710 631	6×10^{-6}	4563	0.710 629	7×10^{-6}
	655	0.711 055	5×10^{-6}	4749	0.711 000	6×10^{-6}
	744	0.710 990	6×10^{-6}	3728	0.710 968	6×10^{-6}
	FP-103	0.710 893	5×10^{-6}	3381	0.710 879	8×10^{-6}
	LQ-109	0.710 861	6×10^{-6}	4893	0.710 853	5×10^{-6}
	LR-01	0.710 898	5×10^{-6}	4033	0.710 881	5×10^{-6}
	304	0.710 843	7×10^{-6}	3191	0.710 873	6×10^{-6}
	656	0.711 204	6×10^{-6}	4562	0.711 190	6×10^{-6}
	742	0.710 838	5×10^{-6}	3452	0.710 816	6×10^{-6}
	B01	0.711 053	6×10^{-6}	5680	0.711 036	7×10^{-6}
	FP-102	0.710 975	5×10^{-6}	5381	0.710 944	7×10^{-6}
	SY-01	0.711 025	6×10^{-6}	4354	0.711 041	7×10^{-6}
	SY-02	0.710 919	7×10^{-6}	4633	0.710 891	7×10^{-6}
	SY-03	0.710 881	6×10^{-6}	3573	0.710 981	5×10^{-6}
	SY-04	0.710 883	6×10^{-6}	3525	0.710 881	8×10^{-6}
	SY-05	0.710 649	5×10^{-6}	5620	0.710 678	6×10^{-6}
	SY-06	0.710 862	5×10^{-6}	5011	0.710 872	8×10^{-6}
	SY-08	0.711 298	7×10^{-6}	8539	0.711 300	7×10^{-6}
	310	0.711 048	5×10^{-6}	3878	0.711 027	8×10^{-6}
	JB01	0.711 057	6×10^{-6}	5203		
	SY-07	0.711 566	5×10^{-6}	3749		
	U01	0.711 084	6×10^{-6}	5650	0.711 081	9×10^{-6}
	Clough Well 1	0.711 239	6×10^{-6}	5334		

Continue

	Location/name	7/29/2013—7/31/2013			5/28/2013—5/30/2013	
		$^{87}Sr/^{86}Sr$	±2s	[Sr], $\times 10^{-9}$	$^{87}Sr/^{86}Sr$	±2s
Seeps	KHW 1	0.710 801	6×10^{-6}	4272	0.710 803	8×10^{-6}
	KHW 2	0.710 769	6×10^{-6}	2725	0.710 745	6×10^{-6}
	KHW 3	0.710 858	6×10^{-6}	4034	0.710 835	6×10^{-6}
	KHW 4	0.710 895	6×10^{-6}	4386	0.710 879	6×10^{-6}
	KHW 5	0.710 916	7×10^{-6}	4901	0.710 899	6×10^{-6}
	KHW 7	0.711 153	5×10^{-6}	1159		
	KHW 8	0.711 221	6×10^{-6}	1141		
	KHW 9	0.711 308	6×10^{-6}	2130		
Surface water	Cemetery Ck.	0.710 987	9×10^{-6}	4550		
	Cemetery telephone	0.710 953	5×10^{-6}	4376		
	Ditch north	0.710 874	7×10^{-6}	3420	0.710 860	6×10^{-6}
	Quarry	0.710 983	5×10^{-6}	5185		
	Rice paddy	0.710 895	5×10^{-6}	3416		
	U6 east	0.711 009	6×10^{-6}	2411		
	Cem ditch			1301		
	Ditch south	0.710 856	6×10^{-6}	2697	0.710 861	7×10^{-6}
	Upper Cem Ck	0.710 452	6×10^{-6}	3361		
	US6 west	0.710 961	5×10^{-6}	4872		
Colorado River	Co. R. East 7/30/2013	0.711 318	5×10^{-6}	658		
	Co. R. West 7/29/2013	0.711 287	5×10^{-6}	740		

The Sr isotopic and concentration data can be used as a basis for a mixing model between Rifle groundwater and vadose zone porewater to quantify the recharge of the aquifer from the vaodse zone and its spatial variation. Eqs. 9.9 and 9.10 can be used to model the Sr concentrations and the $^{87}Sr/^{86}Sr$ of mixtures of various proportions of the model end-members.

The Sr concentration of the mixture of two water, $[Sr]_M$, is:

$$[Sr]_M = f \cdot ([Sr]_A - [Sr]_B) + [Sr]_B \tag{9.9}$$

where f is the fraction of water A, and $[Sr]_A$ and $[Sr]_B$ are the concentrations of Sr in water A and water B respectively (Faure, 1986).

The $^{87}Sr/^{86}Sr$ ratio of that mixture, $(^{87}Sr/^{86}Sr)_M$, is given by (Faure, 1986):

$$(^{87}Sr/^{86}Sr)_M = (a/[Sr]_M) + b \tag{9.10}$$

Where $a = \{([Sr]_A \cdot [Sr]_B) \cdot ((^{87}Sr/^{86}Sr)_B - (^{87}Sr/^{86}Sr)_A)\}/([Sr]_A - [Sr]_B)$, and $b =$

Figure 9. 17 Contour map of $^{87}Sr/^{86}Sr$ in groundwater samples from monitoring wells and seeps (Red contours), along with $^{87}Sr/^{86}Sr$ in surface water samples (Blue contours). Data from Table 9. 4.

$\{[Sr]_A \cdot [(^{87}Sr/^{86}Sr)_A - [Sr]_B \cdot (^{87}Sr/^{86}Sr)_B]\}/([Sr]_A - [Sr]_B)$.

Eqs. 9. 9 and 9. 10 are used to calculate model mixing-lines for mixtures between vadose porewater and upgradient groundwater (called the Plateau Groundwater endmember).

In the Sr isotopic mixing models, one two end-member pair represents mixing between Plateau groundwater, and Rifle Floodplain vadose zone porewater as represented by May and June samples from the 2. 5m deep and 3m deep lysimeters situated at that time above the water table. A second Sr mixing model involves mixing between Rifle Floodplain groundwater and upstream Colorado River. For the U mixing model, again two pairs of end-members are used: one pair representing mixing between Plateau groundwater and the Rifle U groundwater, and the second pair representing mixing between water from the NS Ditch and Rifle U groundwater.

The Sr isotopic model for mixing between Rifle Floodplain groundwater and Colorado River water is shown in Figure 9. 18 as a green line. The model suggests that the Sr isotopic difference seen between Colorado River samples taken upstream (East) and downstream (West) of the Rifle Site can be explained by 2% Rifle Groundwater in the downstream sample. Since the river samples were taken close to shore, this percentage cannot be extrapolated to the entire flow of the Colorado and that with dilution by the entire flow this percentage is likely much lower. However, it does suggest that with sampling profiles across the Colorado situated upstream and downstream of the Rifle Site that it may be possible to constrain the flux of groundwater to the river. For seep samples KHW 7 and KHW 8, which appear to have been affected by seepage of Colorado River water from an overflow pond as described above, the model suggests the addition of about 10% Colorado River water.

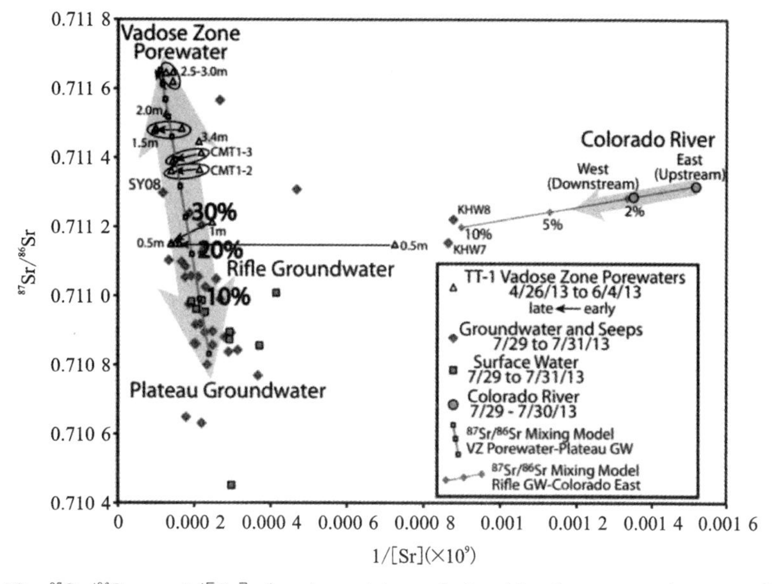

Figure 9.18 $^{87}Sr/^{86}Sr$ vs. 1/[Sr] showing mixing relationships between vadose samples at TT1 and Rifle Floodplain groundwater. Blue line shows the mixing model of vadose zone water with upgradient (Plateau) groundwater with bold numbers showing the percentage of vadose zone water in Rifle groundwater. Green line shows the mixing model between Rifle Floodplain groundwater and near-shore Colorado River samples with numbers showing the percentage of Rifle Floodplain groundwater. Narrow black arrows point from early to late samples from the same depth in TT1. Samples of the Colorado taken upstream and downstream of the Rifle site show the net effect of exported Floodplain groundwater on the local Colorado $^{87}Sr/^{86}Sr$, suggesting ~2% Rifle Groundwater in the near-shore downstream sample. However, the percentage contribution to the whole river flow would be orders of magnitude less.

Our results demonstrate a strategy for assessing groundwater recharge, using $^{87}Sr/^{86}Sr$ ratio and Sr concentration data for groundwater and vadose zone porewater. It is likely other kinds of isotopic or geochemical data can be used in such a fashion as long as there is a contrast between vadose zone porewater and groundwater that can be sufficiently resolved by measurement techniques and the chosen tracer has effectively conservative behavior. Other former U-mill tailings sites across the Western U. S., many of which are situated along river floodplains, may benefit from similar evaluation of recharge spatial variation given the common practice of multiple-well groundwater monitoring programs. This would allow for more responsible management of those sites in a way that is protective of water quality. In addition, with the availability of the appropriate samples, this strategy can also be applied to other semi-arid aquifers where vadose zone recharge can be an important component of understanding the water balance. In this way, our method could be a valuable tool for water managers in water scarce regions to develop effective water resource management plans.

References

ÅBERG G, JACKS G, HAMILTON P J, 1989. Weathering rates and $^{87}Sr/^{86}Sr$ ratios: an isotopic approach[J]. J. Hydrol., 109:65-78.

ÅBERG G, JACKS G, WICKMAN T, et al., 1990. Strontium isotopes in trees as an indicator for calcium availability[J]. Catena., 17:1-11.

ANDERSSON P, LOFVENDAHL R, ÅBERG G, 1990. Major element chemistry, δ^2H, $\delta^{18}O$ and $^{87}Sr/^{86}Sr$ in a snow profile across central Scandinavia[J]. Atmos. Environ., 24: 2601-2608.

ARMSTRONG R L, 1971. Glacial erosion and the variable isotopic composition of strontium in seawater[J]. Nature, 230:132-133.

BAISDEN W T, BLUM J D, MILLER E K, et al., 1995. Elemental concentrations in fresh snowfall across a regional transect in the northeastern U. S.: Apparent sources and contribution to acidity[J]. Water Air Soil Pollut., 84:269-286.

BALCAEN L, DE SCHRIJVER I, MOENS L, et al., 2005. Determination of the $^{87}Sr/^{86}Sr$ isotope ratio in USGS silicate reference materials by multi-collector ICP-mass spectrometry[J]. International Journal of Mass Spectrometry, 242:251-255.

BANNER J L, MUSGROVE M, CAPO R C, 1994. Tracing groundwater evolution in a Pleistocene limestone aquifer using Sr isotopes: Effects of multiple sources of dissolved ions and mineral-solution reactions[J]. Geology, 22: 687-690.

BECKER J S, 2005. Recent developments in isotope analysis by advanced mass spectrometric techniques-Plenary lecture[J]. J. Anal. At. Spectrom., 20:1173-1184.

BISCAYE P E, CHESSELET R, PROSPERO J M, 1974. Rb/Sr, $^{87}Sr/^{86}Sr$ isotope systems as an index of provenance of continental dusts in open Atlantic Ocean[J]. J. Rech. Atmos., 8:819-829.

BLAND J M, ALTMAN D G, 1986. Statistical methods for assessing agreement between two methods of clinical measurement[J]. Lancet, 1:307-310.

BLÄTTLER C L, HENDERSON G M, JENKYNS H C, 2012. Explaining the Phanerozoic Ca isotope history of seawater[J]. Geology, 40:843-846.

BLUM J D, EREL Y, BROWN K, 1993. $^{87}Sr/^{86}Sr$ ratios of Sierra Nevada stream water: Implications for relative mineral weathering rates[J]. Geochim. Cosmochim. Acta, 57: 5019-5025.

BOHN H L, MCNEAL B L, O'CONNOR G A, 1979. Soil Chemistry[M]. New York: Wiley.

BOTTINO M L, FULLAGAR P D, 1968. Effects of weathering on whole rock Rb-Sr ages of granitic rock[J]. Am. J. Sci. , 266:661-670.

BRASS G W,1976. The variation of the marine $^{87}Sr/^{86}Sr$ ratio during Phanerozoic time: Interpretation using a flux model[J]. Geochim. Cosmochim. Acta, 40:721-730.

BRIMHALL G H, CHADWICK O A, LEWIS C J,et al. ,1991. Deformational mass transport and invasive processes in soil evolution[J]. Science,255:695-702.

BRIMHALL G H, DONOVAN J, SINGH B, 1994. Provenance and discordance mechanism of zircons in the Australian regolith using SHRIMP and electron microprobe mapping[R]. 8th Int. Conf. Geochronol. Cosmochronol. Isot. Geol. U. S. Geol. Surv. Circ. , 1107:39.

BRUBAKER T M, STEWART B W, CAPO R C,et al. ,2013. Coal fly ash interaction with environmental fluids: Geochemical and strontium isotope results from combined column and batch leaching experiments[J]. Applied Geochemistry, 32:184-194.

BULLEN T, WHITE A, BLUM A, et al. , 1997. Chemical weathering of a soil chronosequence on granitoid alluvium, II. Mineralogic and isotopic constraints on the behavior of strontium[J]. Geochim. Cosmochim. Acta,61:291-306.

BURKE W H, DENISON R E, HETHERINGTON E A,et al. ,1982. Variation of seawater $^{87}Sr/^{86}Sr$ throughout Phanerozoic time[J]. Geology, 10:516-519.

CAPO R C, CHADWICK O A, 1993. Partitioning of atmospheric and silicate weathering sources in the formation of desert soil carbonate using strontium isotopes[J]. EOS, Trans. Am. Geophys. Union, 74: 263.

CAPO R C, DEPAOLO D J,1992. Homogeneity of Sr isotopes in the oceans[J]. EOS, Trans. Am. Geophys. Union, 73: 272.

CAPO R C, HSIEH J C C, CHADWICK O A,1995. Pedogenic origin of dolomite and calcite in a basaltic weathering profile,Kohala Peninsula, Hawaii[J]. V. M. Goldschmidt Conf. Progr. Abstr. , 34.

CARPENTER S J, LOHMANN K C,1992. Sr/Mg ratios of modern marine calcite-empirical indicators of ocean chemistry and precipitation rate[J]. Geochim. Cosmochim. Acta, 56:1837-1849.

CHADWICK O A, BRIMHALL G H, HENDRICKS D M,1990. From a black box to a gray box-a mass balance interpretation of pedogenesis[J]. Geomorphology, 3:369-390.

CHAPMAN E C, CAPO R C, STEWART B W, et al. , 2013. Strontium isotope quantification of siderite, brine and acid mine drainage contributions to abandoned gas well discharges in the Appalachian Plateau[J]. Applied Geochemistry, 31:109-118.

CHAPMAN E C, CAPO R C, STEWART B W, et al. , 2012. Geochemical and Strontium Isotope Characterization of Produced water from Marcellus Shale Natural Gas Extraction[J]. Environ. Sci. Technol. , 46:3545-3553.

Chen C H, Nakada S, Shieh Y N,et al. ,1999. The Sr, Nd, and O isotopic studies of the 1991—1995 eruption at Unzen, Japan[J]. J. Volc. Geothermal. Res., 89:243-253.

CLEMENTZ M T, HOLDEN P, KOCH P L, 2003. Are calcium isotopes a reliable monitor of trophic level in marine settings? [J]. Int. J. Osteoarchaeology, 13:29-36.

DASCH E J, 1969. Strontium isotopes in weathering profiles, deep-sea sediments, and sedimentary rocks[J]. Geochim. Cosmochim. Acta, 33:1521-1552.

DePaolo D J, 1986. Detailed record of the Neogene Sr isotopic evolution of seawater from DSDP Site 590B[J]. Geology, 14:103-106.

DERRY L A, KAUFMAN A J, JACOBSEN S B, 1992. Sedimentary cycling and environmental change in the Late Proterozoic: Evidence from stable and radiogenic isotopes [J]. Geochim. Cosmochim. Acta, 56:1317-1329.

DYMOND J, BISCAYE P E, REX R W, 1974. Eolian origin of mica in Hawaiian soils [J]. Geol. Soc. Am. Bull., 85: 37-40.

EILER J M, FARLEY K A, VALLEY J W, et al., 1997. Oxygen isotope variations in ocean island basalt phenocrysts[J]. Geochim. Cosmochim. Acta, 61:2281-2293.

EILER J M, VALLEY J W, STOLPER E M, 1996. Oxygen isotope ratios in olivine from the Hawaii Scientific Drilling Project[J]. J. Geophys. Res-Solid Earth, 101: 11 807-11 813.

ELDERFIELD H, 1986. Strontium isotope stratigraphy[J]. Palaeogeogr. Palaeoclimatol. Palaeoecol., 57:71-90.

EPA, 2010. Rapid Radiochemical Methods for selected radionuclides in water for environmental restoration following homeland security events, U. S. Environmental Protection Agency[R]. U. S. EPA, Ed.

EWING S A, YANG W, DEPAOLO D J, et al., 2008. Non-biological fractionation of stable Ca isotopes in soils of the Atacama Desert, Chile[J]. Geochim. Cosmochim. Acta, 72:1096-1110.

FANTLE M S, DEPAOLO D J, 2005. Variations in the marine Ca cycle over the past 20 million years[J]. Earth Planet. Sci. Lett., 237:102-117.

FANTLE M S, DEPAOLO D J, 2007. Ca isotopes in carbonate sediment and pore fluid from ODP site 807A: the Ca^{2+} (aq)-calcite equilibrium fractionation factor and calcite recrystallization rates in Pleistocene sediments[J]. Geochim. Cosmochim. Acta, 71:2524-2546.

FANTLE M S, TIPPER E T, 2014. Calcium isotopes in the global biogeochemical Ca cycle: implications for development of a Ca isotope proxy[J]. Earth Sci. Rev., 129:148-177.

FANTLE M S, TOLLERUD H, EISENHAUER A, et al., 2012. The Ca isotopic composition of dust-producing regions: measurements of surface sediments in the Black Rock Desert, Nevada[J]. Geochim. Cosmochim. Acta, 87:178-193.

FARKAŠ J, BÖHM F, WALLMANN K, et al., 2007a. Calcium isotope record of Phanerozoic oceans: implications for chemical evolution of seawater and its causative mechanisms[J]. Geochim. Cosmochim. Acta, 71:5117-5134.

FARKAŠ J, BUHL D, BLENKINSOP J, et al., 2007b. Evolution of the oceanic calcium cycle during the late Mesozoic: evidence from $\delta^{44/40}$Ca of marine skeletal carbonates [J]. Earth Planet. Sci. Lett., 253:96-111.

FARKAŠ J, DÉJEANT A, NOVÁK M, et al., 2011. Calcium isotope constraints on the uptake and sources of Ca^{2+} in a base-poor forest: a new concept of combining stable ($\delta^{44/42}$Ca) and radiogenic signals[J]. Geochim. Cosmochim. Acta, 75:7031-7046.

FAURE G, 1986. Principles of Isotope Geology[M]. New York: Wiley.

FAURE G, POWELL J L, 1972. Strontium Isotope Geology[M]. New York: Springer-Verlag.

FORTUNATO G, MUMIC K, WUNDERLI S, et al., 2004. Application of strontium isotope abundance ratios measured by MC-ICP-MS for food authentication[J]. J. Anal. At. Spectrom., 19:227-234.

FRITZ B, RICHARD L, MCNUTT R H, 1992. Geochemical modelling of Sr isotopic signaturesin the interaction between granitic rocks and natural solutions[R]//Kharaka Y K, Maest A S (Eds.). Proc. 7th Internat. Symp. Water-Rock Interaction[M]. Balkema: 927-930.

FULLAGAR P D, RAGLAND P C, 1975. Chemical weathering and Rb-Sr whole rock ages[J]. Geochim. Cosmochim. Acta, 39:1245-1252.

GAILLARDET J, DUPRE B, LOUVAT P, et al., 1999. Global silicate weathering and CO_2 consumption rates deduced from the chemistry of large rivers[J] Chem. Geol., 159: 3-30.

GALLER P, LIMBECK A, UVEGES M, et al., 2008. Automation and miniaturization of an on-line flow injection Sr/matrix separation method for accurate, high throughput determination of Sr isotope ratios by MC-ICP-MS [J]. J. Anal. At. Spectrom., 23: 1388-1391.

GARCIA-RUIZ S, MOLDOVAN M, ALONSO J I G, 2008. Measurement of strontium isotope ratios by MC-ICP-MS after on-line Rb-Sr ion chromatography separation[J]. J. Anal. At. Spectrom., 23:84-93.

GETTY S J, DEPAOLO D J, 1995. Quaternary geochronology by the U-Th-Pb method [J]. Geochim. Cosmochim. Acta, 59:3267-3272.

GILE L H, 1979. The Desert Project Soil Monograph[M]. Washington D C: SCS-USDA.

GILE L H, Peterson F F, Grossman R B, 1966. Morphological and genetic sequences of carbonate accumulation in desert soils[J]. Soil Sci., 101:347-360.

GOLDICH S S, GAST P W, 1966. Effects of weathering on the Rb-Sr and K-Ar ages of biotite from the Morton Gneiss, Minnesota[J]. Earth Planet. Sci. Lett., 1:372-375.

GOLDSTEIN S L, JACOBSEN S B, 1987. The Nd and Sr isotope systematics of river water dissolved material: Implications for the source of Nd and Sr in sea water[J]. Chem. Geol. Isot. Geosci. Sect., 66:245-272.

GOLDSTEIN S L, JACOBSEN S B, 1988. Nd and Sr isotopic systematics of river water suspended material: implications for crustal evolution[J]. Earth Planet. Sci. Lett., 87: 249-265.

GOSZ J R, BROOKINS D G, MOORE D I, 1983. Using strontium isotope ratios to estimate inputs to ecosystems[J]. Bio. Science, 33:23-30.

GOSZ J R, MOORE D I, 1989. Strontium isotope studies of atmospheric inputs to forested waterheds in New Mexico[J]. Biogeochemistry, 8:115-134.

GRAUSTEIN W C, 1989. $^{87}Sr/^{86}Sr$ ratios measure the sources and flow of strontium in terrestrial ecosystems[J]//Rundel P W, Ehleringer J R, Nagy K A. Stable Isotopes in Ecological Research[M]. New York: Springer-Verlag: 491-512.

GRAUSTEIN W C, ARMSTRONG R L, 1983. The use of strontium-87/strontium-86 ratios to measure transport into forested waterheds[J]. Science, 219:289-292.

HERUT B, STARINSKY A, KATZ A, 1993. Strontium in rainwater from Israel: Sources, isotopes and chemistry[J]. Earth Planet. Sci. Lett., 120:77-84.

HESS J, BENDER M, SCHILLING J G, 1986. Seawater $^{87}Sr/^{86}Sr$ evolution from Cretaceous to present[J]. Science, 231:979-984.

HINDSHAW R S, BOURDON B, POGGE VON STRANDMANN P A E, et al., 2013. The stable calcium isotopic composition of rivers draining basaltic catchments in Iceland[J]. Earth Planet. Sci. Lett., 374:173-184.

HINDSHAW R S, REYNOLDS B C, WIEDERHOLD J G, et al., 2011. Calcium isotopes in a proglacial weathering environment: Damma Glacier, Switzerland[J]. Geochim. Cosmochim. Acta, 75:106-118.

HODELL D A, MEAD G A, MUELLER P A, 1990. Variation in the strontium isotopic composition of seawater (8Ma to present): Implications for chemical weathering rates and dissolved fluxes to the oceans[J]. Chem. Geol. Isot. Geosci. Sect., 80:291-307.

HOLLAND H D, 1984. The Chemical Evolution of the Atmospheres and Ocean[D]. Princeton, N J: Princeton University.

HORWITZ E P, CHIARIZIA R, DIETZ M L, 1992. A Novel Strontium-Selective Extraction Chromatographic Resin[J]. Solvent Extraction and Ion Exchange, 10:313-336.

HORWITZ E P, DIETZ M L, FISHER D E, 1991. Separation and Preconcentration of Strontium from Biological, Environmental, and Nuclear Waste Samples by Extraction Chromatography Using a Crown-Ether[J]. Analytical Chemistry, 63:522-525.

HUANG S, FARKAŠ J, JACOBSEN S B, 2011. Stable calcium isotopic compositions of Hawaiian shield lavas: evidence for recycling of ancient marine carbonates into the mantle [J]. Geochim. Cosmochim. Acta, 75:4987-4997.

JACOBSON A D, GRACE ANDREWS M, LEHN G O, et al., 2015. Silicate versus carbonate weathering in Iceland: New insights from Ca isotopes[J]. Earth Planet. Sci. Lett., 416:132-142.

JOHNSON T M, DEPAOLO D J, 1994. Interpretation of isotopic data in groundwater-rock systems: Model development and application to Sr isotope data from Yucca Mountain [J]. Water Resour. Res., 30:1571-1587.

KRISHNASWAMI S, TRIVEDI J R, SARIN M M, et al., 1992. Strontium isotopes and rubidium in the Ganga-Brahmaputra river system: Weathering in the Himalaya, fluxes to the Bay of Bengal and contributions to the evolution of oceanic $^{87}Sr/^{86}Sr$[J]. Earth Planet. Sci. Lett., 109:243-253.

KULP J L, ENGELS J, 1963. Discordance in K-Ar and Rb-Sr isotopic ages[R]. In: Radioactive Dating. Int. Atomic Energy Agency, Vienna, 219-238.

LI J X, DEPAOLO D J, WANG Y X, et al., 2018. Calcium isotope fractionation in a silicate dominated Cenozoic aquifer system[J]. Journal of Hydrology, 559:523-533.

LI J X, WANG Y X, XIE X J, 2016. Cl/Br ratios and chlorine isotope evidences for groundwater salinization and its impact on groundwater arsenic, fluoride and iodine enrichment in the Datong Basin[J]. China Sci. Total Environ., 544:158-167.

MARSHALL B D, DEPAOLO D J, 1982. Precise age determinations and petrogenetic studies using the K-Ca method[J]. Geochim. Cosmochim. Acta, 46:2537-2545.

MARSHALL B D, DEPAOLO D J, 1989. Calcium isotopes in igneous rocks and the origin of granite[J]. Geochim. Cosmochim. Acta, 53:917-922.

MAXWELL S L, 2006. Rapid column extraction method for actinides and Sr-89/90 in water samples[J]. Journal of Radioanalytical and Nuclear Chemistry, 267:537-543.

MAXWELL S L, CULLIGAN B K, NOYES G W, 2010. Rapid separation of actinides and radio strontium in vegetation samples [J]. Journal of Radioanalytical and Nuclear Chemistry, 286:273-282.

MCCULLOCH M T, DE DECKKER P, CHIVAS A R, 1989. Strontium isotope variations in single ostracod valves from the Gulf of Carpentaria, Australia: A palaeo environmental indicator[J]. Geochim. Cosmochim. Acta, 53:1703-1710.

MILLER E K, BLUM J D, FRIEDLAND A J, 1993. Determination of soil exchangeable-cation loss and weathering rates using Sr isotopes[J]. Nature, 362:438-441.

MOORE J, JACOBSON A D, HOLMDEN C, et al., 2013. Tracking the relationship between mountain uplift, silicate weathering, and long-term CO_2 consumption with Ca isotopes: Southern Alps, New Zealand[J]. Chem. Geol., 341:110-127.

MUHS D R, BUSH C A, TRACY R R, et al., 1990. Geochemical evidence of Saharan dust parent material for soils developed on Quaternary limestones of Caribbean and western Atlantic islands[J]. Quat. Res., 33:157-177.

MUHS D R, CRITTENDEN R C, ROSHOLT J N, et al., 1987. Genesis of marine terrace soils, Barbados, West Indies: Evidence from mineralogy and geochemistry, Earth Planet[J]. Sci. Lett., 12:605-618.

NIELSEN L C, DEPAOLO D J, DE YOREO J J, 2012. Self-consistention-by-ion growth model for kinetic isotopic fractionation during calcite precipitation[J]. Geochim. Cosmochim. Acta, 86:166-181.

Palmer M R, Edmond J M, 1989. The strontium isotope budget of the modern ocean [J]. Earth Planet. Sci. Lett., 92:11-26.

PALMER M R, EDMOND J M, 1992. Controls over the strontium isotope composition of river water[J]. Geochim. Cosmochim. Acta, 56:2099-2111.

Pin C, Joannon S, Bosq C, et al., 2003. Precise determination of Rb, Sr, Ba, and Pb in geological materials by isotope dilution and ICP-quadrupole mass spectrometry following selective separation of the analytes[J]. J. Anal. At. Spectrom., 18:135-141.

RABENHORST M C, WILDING L P, GIRDNER C L, 1984. Airborne dusts in the Edwards Plateau region of Texas[J]. Soil Sci. Soc. Am. J., 48:621-627.

REHEIS M C, GOODMACHER J C, HARDEN J W, et al., 1995. Quaternary soils and dust deposition in southern Nevada and California[J]. Geol. Soc. Am. Bull., 107: 1003-1022.

REHEIS M C, KIHL R, 1995. Dust deposition in southern Nevada and California, 1984-1989: Relations to climate, source area, and source lithology[J]. J. Geophys. Res., 100:8893-8918.

RICHTER F M, DAVIS A M, DEPAOLO D J, et al., 2003. Isotope fractionation by chemical diffusion between molten basalt and rhyolite[J]. Geochim. Cosmochim. Acta, 67: 3905-3923.

RUSSELL W A, PAPANASTASSIOU D A, 1978. Calcium isotope fractionation in ion-exchange chromatography[J]. Anal. Chem., 50:1151-1153.

RYU J S, JACOBSON A D, HOLMDEN C, et al., 2011. The major ion, $\delta^{44/40}$Ca, $\delta^{44/42}$Ca, and $\delta^{26/24}$Mg geochemistry of granite weathering at pH=1 and T=25℃: power-law processes and the relative reactivity of minerals[J]. Geochim. Cosmochim. Acta, 75: 6004-6026.

SHARMA S, SACK A, ADAMS J P, et al., 2013. Isotopic evidence of enhanced carbonate dissolution at a coal mine drainage site in Allegheny County, Pennsylvania, USA. [J]. Applied Geochemistry, 29:32-42.

SIMON J I, DEPAOLO D J, 2010. Stable calcium isotopic composition of meteorites and rocky planets[J]. Earth Planet. Sci. Lett., 289:457-466.

SIMS K W W, DEPAOLO D J, MURRELL M T, et al., 1999. Porosity of the melting zone and variations in solid mantle upwelling rate beneath Hawaii: inferences from ^{238}U-

^{230}Th-^{226}Ra and ^{235}U-^{231}Pa[J]. Geochim. Cosmochim. Acta, 63:4119-4138.

SKULAN J, DEPAOLO D J, 1999. Calcium isotope fractionation between soft and mineralized tissues as a monitor of calcium use in vertebrates[J]. Proc. Nat. Acad. Sci., 96:13,709-13,713.

SKULAN J, DEPAOLO D J, OWENS T L, 1997. Biological control of calcium isotopic abundances in the global calcium cycle[J]. Geochim. Cosmochim. Acta, 61:2505-2510.

SKULAN J L, 1999. Calcium isotopes and the evolution of terrestrial reproduction in vertebrates[D]. Berkeley: University of California.

SPOSITO G, 1989. The Chemistry of Soils[M]. New York: Oxford Univ. Press: 277.

STALLARD R F, 1985. River chemistry, geology, geomorphology and soils in the Amazon and Orinoco basins[J]//Drever J I, The Chemistry of Weathering[M]. New York: D Reidel: 293-316.

STEWART B W, CAPO R C, CHADWICK O A, 1998. Quantitative strontium isotope models for weathering, pedogenesis and biogeochemical cycling[J]. Geoderma, 82:173-195.

STRAUGHAN I R, ELSEEWI A A, PAGE A L, et al., 1981. Fly ash-derived strontium as an index to monitor deposition of coal-fired power plants[J]. Science, 212: 1267-1269.

TIPPER E T, GAILLARDET J, GALY A, et al., 2010. Calcium isotope ratios in the world's largest rivers: a constraint on the maximum imbalance of oceanic calcium fluxes[J]. Glob. Biogeochem. Cycles, 24, GB3019.

TIPPER E T, GALY A, BICKLE M J, 2008. Calcium and Magnesium isotope systematics in rivers draining the Himalaya-Tibetan-Plateau region: lithological or fractionation control? [J]. Geochim. Cosmochim. Acta, 72:1057-1075.

VANHAECKE F, BALCAEN L, MALINOVSKY D, 2009. Use of single-collector and multi-collector ICP mass spectrometry for isotopic analysis[J]. J. Anal. At. Spectrom., 24: 863-886.

VEIZER J, COMPSTON W, CLAUER N, et al., 1983. $^{87}Sr/^{86}Sr$ in Late Proterozoic carbonates: Evidence for a "mantle event" at 900Ma ago[J]. Geochim. Cosmochim. Acta, 47:295-302.

WADLEIGH M A, VEIZER J, BROOKS C, 1985. Strontium and its isotopes in Canadian rivers: Fluxes and global implications [J]. Geochim. Cosmochim. Acta, 49: 1727-1736.

WEST L T, DREES L R, WILDING L P, et al., 1988. Differentiation of pedogenic and lithogenic carbonate forms in Texas[J]. Geoderma., 43:271-287.

WICKMAN T, JACKS G, 1993. Base cation nutrition for pine stands on lithic soils near Stockholm, Sweden[J]. Appl. Geochem. Suppl., 2:199-202.

WILLIAMS K H, LONG P E, DAVIS J A, et al., 2011. Acetate availability and its influence on sustainable bioremediation of uranium-contaminated groundwater[J]. Geomicrobiol J. , 28:519-539.

YABUSAKI S B, WILKINS M J, FANG Y, et al. , 2017. Water table dynamics and biogeochemical cycling in a shallow, variably-saturated floodplain[J]. Enviro. Sci. Technol. , 51:3307-3317.

YANG C, TELMER K, VEIZER J, 1996. Chemical dynamics of the "St. Lawrence" riverine system: δD, $\delta^{18}O$, $\delta^{13}C$, $\delta^{34}S$, and dissolved $^{87}Sr/^{86}Sr$[J]. Geochim. Cosmochim. Acta, 60:851-866.

Chapter 10 Iron Isotope

Iron (Fe) is the fourth most abundant element in the Earth's crust. It occurs mainly in rocks, minerals, fluids and organisms, being involved in a wide range of geochemical and biochemical processes. It consists of four stable isotopes (^{54}Fe, ^{56}Fe, ^{57}Fe and ^{58}Fe) with the following abundances (Beard and Johnson 1999):

$^{54}Fe = 5.84\%$

$^{56}Fe = 91.76\%$

$^{57}Fe = 2.12\%$

$^{58}Fe = 0.28\%$

Iron is a redox-sensitive element. Consequently, the cycling of Fe is largely related to reduction and oxidation processes, and frequently coupled to the cycles of other elements such as carbon (C), nitrogen (N), phosphorus (P), sulfur (S), and manganese (Mn) (Straub et al., 1996; Weber et al., 2006). Iron has a variety of important bonding partners and ligands, forming sulfide, oxide and silicate minerals as well as complexes with water. As is well known, bacteria can use Fe during both dissimilatory and assimilatory redox processes. Because of its high abundance and its important role in high and low temperature processes, isotope studies of iron have received the most attention of the transition elements. Fe isotope analysis is highly challenging, because of interferences from $^{40}Ar^{14}N^{+}$, $^{40}Ar^{16}O^{+}$ and $^{40}Ar^{16}OH^{+}$ at masses 54, 56 and 57 respectively. Nevertheless δ-values can be measured routinely with a precision of $\pm 0.1‰$ or better (Craddock and Dauphas, 2010).

10.1 Iron species in groundwater

Most of the iron in the earth's crust is dispersed in various magmatic and sedimentary rocks and the quaternary strata, they get into the groundwater in a number of ways: ①The groundwater containing carbonic acid can gradually dissolve the oxides containing iron divalent in the strata and produce soluble ferrous bicarbonate during the filtration process. ②In the Organic-rich formations, the organic matter is decomposed by anaerobic oxidation, producing a considerable amount of reducing gas, which result in the iron trivalent to be reduced to iron divalent and dissolved in water. ③Organic matter plays an important role in iron entering groundwater. Some organic acids can dissolve iron divalent in rock and soil layers. Part of organic matters can reduce iron trivalent to iron divalent and dissolve it in water. And some organic matters could combine with iron to form complex organic iron, which helps it dissolve in water.

Due to the filtering effect of strata, generally there are only soluble iron compounds in groundwater. Iron trivalent has low solubility, so it is hard to be found in groundwater especially the pH is greater than 5. In general, iron-bearing groundwater mainly contains bivalent iron bicarbonate. Ferrous bicarbonate is a kind of strong electrolyte and can be fully dissociated in water, therefore iron divalent exists mainly in the form of ion in groundwater. The so-called underground water contains ferrous bicarbonate, which refers to its imaginary combination form. Ferrous sulfate in groundwater is relatively rare. In the presence of dissolved oxygen, iron divalent ions are readily oxidized to iron trivalent. The generated trivalent iron will be precipitated in the form of $Fe(OH)_3$ because of its very small solubility in water. And when there is a lot of organic matter in shallow groundwater, iron will combine with it to form organic iron.

10.2 Analytical techniques for the Iron isotope

Iron isotopic composition has been of interest since the invention of isotope ratio mass spectrometry. The biggest analytical challenges for analysis of Fe that preserves naturally occurring mass-dependent isotopic fractionation include correcting for instrumental mass bias, and resolving Fe isotopes from isobaric interferences that can include elemental isobars such as ^{58}Ni on ^{58}Fe, or molecular isobars such as $^{40}Ar^{16}O$ on ^{56}Fe.

10.2.1 Mass Spectrometry

Prior to the wide adoption of multi-collector inductively-coupled-plasma mass-spectrometry (MC-ICP-MS) in the early 2000s, Fe isotope ratios were analyzed using numerous techniques, including TIMS, SIMS and a variety of mass spectrometry techniques.

Today, the most common method for Fe isotope analysis is by Multi Collector Inductively Coupled Plasma Mass Spectrometry (MC-ICP-MS). This instrumentation benefits from a highly efficient ionization source coupled with a stable instrumental mass bias during the course of a sample analysis, as compared to TIMS, where the mass bias of the sample continuously changes as the sample is evaporated from the filament. The first MC-ICP-MS study (Belshaw et al., 2000) only reported $^{57}Fe/^{54}Fe$ measurements because it was not able to limit ArO from the mass spectrum. The formation of ArN in their study was minimized by using HCl as a diluting acid instead of HNO_3, and oxides (ArO and ArOH) were suppressed by using a desolvating nebulizer. The next generation of MC-ICP-MS was developed (Nu Instruments Sapphire, and Thermo Fisher Proteus) including a collision cell, which uses either H_2 or D_2 as a collision gas with N_2 addition to the injector. Use of H_2 and D_2 allows evaluation of the formation of argon hydroxide (e. g. ArOH or ArOD) caused by reactions with the collision gas as compared to formation of argon hydroxide not associated with the collision gas.

The techniques that minimized the effects of argide isobars for the first-generation MC-ICP-MS instruments, although successful, were not as robust as the method developed for the 2nd generation of MC-ICP-MS, such as the Thermo Neptune/Neptune Plus and the Nu Instruments Plasma500/Plasma2/Plasma3. These newer instruments used a technique frequently described as "pseudo high mass resolution". For this technique, the beam width is reduced by using a narrow defining slit. A disadvantage of this technique is that, in multi-collection mode, parallel resolution of all interfering species is only possible if they appear at the same side of the ions of interest. This is true for all significant mass interferences on Fe isotopes, as well as for all molecular interferences on other first-row transition-or lighter metals (double-charged ions are exceptions that may cause problems in some cases). Another disadvantage of running at pseudo-high mass resolution with a narrow defining slit is a significant decrease in ion transmission, typically only ~10% of the ion is transmitted through the narrow slit.

In all ICP-MS, the measured analyte always has a "heavier" isotopic composition compared to its true isotopic composition for ratios where the numerator is the heavier mass. Workers have developed methods to improve instrumental mass-bias corrections. The double spike method is the most robust, but requires accurate measurement of the abundance of ^{58}Fe, which is affected by an isobaric interference from ^{58}Ni (Finlayson et al., 2015). The ^{58}Ni isobar can be from Ni not removed from the sample during chemical purification or from sampler and skimmer cones if these are made of Ni. A second method to correct for mass bias is to add an element with two or more isotopes of similar atomic Z, and use the measured isotope composition of the added element to infer the instrumental mass bias for the analyte (Devulder et al., 2013; Marechal et al., 1999).

10.2.2 In Situ Techniques

A number of studies have made use of in situ methods to analyze Fe-rich minerals (including oxides, sulfides, carbonates, and silicates), metals, and glasses. These studies used either secondary ionization mass spectrometry (SIMS) (Galic et al., 2017; Virtasalo et al., 2013; Whitehouse and Fedo, 2007) or laser ablation (LA) coupled to MC-ICP-MS to analyze Fe isotope compositions of minerals (Collinet et al., 2017; Czaja et al., 2013; Sio et al., 2013; Steinhoefel et al., 2009a, 2009b, 2010; Yoshiya et al., 2015).

Laser ablation techniques for Fe isotope analysis have focused on using two broad categories of lasers: those with nanosecond (ns, 10 ~ 9s) pulse widths and those with femtosecond(fs, 10 ~ 15s) pulse widths. During laser ablation with a ns-laser, energy delivery by the pulsed phonons is long enough to allow for thermal diffusion within the target mineral, while the fs-laser pulse is too short for significant thermal diffusion, although considerable time still exists for laser-matter interaction. Contrary to ns-LA, fs-LA does not produce significant amounts of these large particles during ablation of conductors such as Fe metal (Gonzalez et al. ,2007b). The differences in particle sizes and heat-affected zones is likely one of the reasons that Fe isotope analysis of Fe metal by ns-LA is inaccurate compared to fs-LA (Kosler et al. ,2006).

10.3 Fractionation processes

Various studies have found that the largest Fe isotope fractionations are associated with redox processes. For example, the fractionation between aqueous Fe^{3+} and aqueous Fe^{2+} is 3‰ in $^{56}Fe/^{54}Fe$ ratios at 20℃. These studies have also revealed that the magnitude of Fe isotope redox fractionations is a function of the Fe mineral produced and the aqueous Fe speciation, and that Fe redox couples such as Fe oxides and aqueous Fe^{2+} have rapid Fe isotope exchange rates that result in near-complete Fe isotope exchange even at low temperature.

The cycling of iron (Fe) is often closely linked with that of carbon, nitrogen, phosphorus and manganese. Therefore, alterations in the Fe cycle may be indicative of concurrent overall changes in the biogeochemistry of terrestrial and aquatic ecosystems. Biogeochemical processes taking part in the Fe cycle frequently fractionate stable Fe isotopes, leaving soil, plant and other compartments of the ecosystems with varied Fe isotopic signatures. Biogeochemical processes often lead to Fe isotope fractionation, leaving behind an Fe isotopic fingerprint on the compartments of the systems (e. g. , vegetation, topsoil, subsoil, parent rock, or different water reservoirs, Figure 10. 1).

Dissolution of primary and secondary Fe minerals, transformation between Fe^{2+} and Fe^{3+} , adsorption and precipitation of Fe species, changes of Fe-binding ligands, uptake by and translocation within plants, and microbial activities, are often accompanied by Fe isotope fractionation. Iron isotope fractionation can be due to either kinetic or equilibrium fractionation effects, or a combination of both.

10.3.1 Redox transformations

An apparent pathway in which Fe isotope fractionation is expected to occur is redox transformation. In the presence of oxygen, the oxidation process leaves a solution enriched in

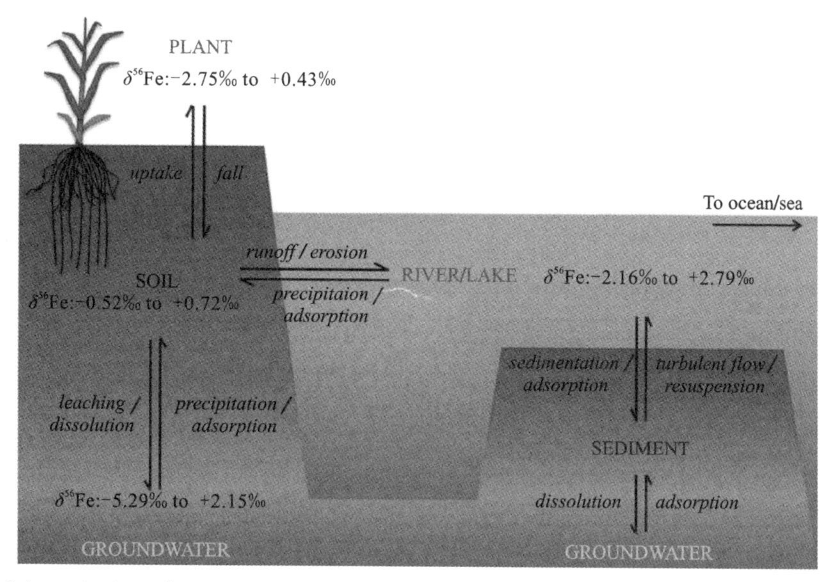

Figure 10.1 Schematic view of main processes of Fe cycles in soil-plant-freshwater ecosystems(Wu B et al. ,2019).

light Fe isotopes, while heavier Fe^{3+} aq is rapidly precipitated as ferrihydrite. The overall fractionation at equilibrium of this process shows that the isotopic signature of the ferrihydrite is about 1‰ heavier than the Fe^{2+} in solution, as a result of the combination of Fe^{3+} aq-Fe^{2+} aq equilibrium fractionation (~3‰) and a kinetic isotope effect favoring light Fe isotopes to precipitate as ferrihydrite from Fe^{3+} aq (Anbar et al. , 2005; Bullen et al. , 2001; Dauphas and Rouxel, 2006; Johnson et al. , 2002; Welch et al. , 2003). In addition, Fe^{2+} aq can be adsorbed on the surface of the newly formed Fe^{3+} (hydr.)oxides, which also favors heavy Fe isotopes, thus producing an isotopically very light Fe^{2+} in solution (Icopini et al. , 2004; Teutsch et al. , 2005). When oxygen is absent, Fe^{2+} can be transformed via anaerobic photoautotrophic Fe^{2+} oxidation. This oxidation process also favors heavy Fe isotopes leaving relatively light Fe^{2+} aq compared with the precipitated poorly crystalline ferrihydrite (Croal et al. , 2004).

Microorganisms play an important role in Fe redox transformation either utilizing the energy produced during oxidation by Fe-oxidizing bacteria (known as FeOB, Emerson et al. , 2010) or using Fe^{3+} as terminal electron acceptor during Fe^{3+} reduction by DIR bacteria (Lovley et al. , 2004). The processes mediated by these two types of bacteria including those of nitrate-reducing Fe^{2+} oxidizing bacteria have been shown to induce Fe isotope fractionation. Both inorganic and microbially mediated redox transformations between Fe^{2+} and Fe^{3+} show a preferential removal of heavy Fe isotope from the solution.

10.3.2 Dissolution of iron minerals

Four major dissolution mechanisms of Fe minerals (proton-promoted, ligand-controlled, reductive and oxidative dissolution) have been demonstrated in laboratory experiments showing distinct isotope fractionation effects.

Reductive dissolution of Fe minerals is largely studied with the presence of DIR bacteria. The Fe^{2+} aq is commonly depleted in heavy Fe isotopes compared to the substrate. This fractionation can remain constant over a long period and is independent of the Fe mineral substrate (e. g. hematite or goethite), indicating a common mechanism for Fe isotope fractionation during DIR of different substrates (Crosby et al. , 2007). This leads to changing proportions of Fe species within the system, which in turn results in variations of the absolute δ values for Fe^{2+} aq, especially at the early stage of the reduction process. Reductive dissolution of ferrihydrite by DIR bacteria results not only in Fe^{2+} aq but also in biogenic Fe oxides and carbonates, such as magnetite and siderite, through reprecipitation or reaction with anions. The isotope fractionation between Fe^{2+} aq and ferrihydrite and related biogenic products is a function of Fe^{3+} reduction rates and pathways by which the biogenic minerals are formed (Johnson et al. , 2005).

Proton-promoted dissolution is due to the reaction of H with O and hydroxyl groups (OH) at the mineral surface, which promotes Fe release from the mineral surface into solution (Cornell and Schwertmann, 2003). Laboratory data indicate distinct isotope effects during proton-promoted dissolution that are related to the Fe oxide mineral form. Proton-promoted dissolution of goethite (Wiederhold et al. , 2006) and hematite (Skulan et al. , 2002; Johnson et al. , 2004) do not fractionate Fe isotopes, even when the dissolution is incomplete. However, proton-promoted dissolution of biotite and chlorite shows a preferential release of light Fe isotopes into solution by up to $-1.4‰$ in $\delta^{56}Fe$ compared with the bulk phyllosilicates, although the fractionation is less pronounced at the later stage of the dissolution (Kiczka et al. , 2010a).

Organic ligands in soils such as siderophores and low molecular weight organic acids are reported to participate in mineral dissolution (Brantley et al. , 2001, 2004; Wiederhold et al. , 2006) and in Fe uptake by plant roots (Takagi et al. , 1984; Curie et al. , 2001; Schaaf et al. , 2004). Experimentally, the dissolution of Fe-bearing minerals (e. g. hornblende, goethite) by organic ligands (e. g. , siderophores, oxalic acid) results in significant Fe isotope fractionation (Brantley et al. , 2001, 2004; Wiederhold et al. , 2006; Chapman et al. , 2009; Kiczka et al. , 2010). The extent of this isotope fractionation is a function of the binding strength of the ligands (Brantley et al. , 2004). Stronger Fe-binding chelates preferentially extract lighter Fe isotopes (Brantley et al. , 2001, 2004). Abiotic and biotic ligand-controlled dissolution of Fe minerals can lead to considerably different isotope fractionation effects. Abiotic and biotic ligand-controlled dissolution of Fe minerals can lead to considerably different isotope fractionation effects. Brantley et al. (2004) found that when goethite was dissolved by siderophores in the presence of bacteria, lighter Fe isotopes were preferentially dissolved, while no significant fractionation was observed without the mediation of bacteria.

During abiotic oxidative leaching of iron-sulfide-bearing rocks (rich in pyrite, chalcopyrite, and sphalerite) under strongly acidic conditions (pH = 2), the isotope fractionation at the initial stage was found to be towards the accumulation of heavy Fe isotopes in the leachates, leaving behind a relatively light rock surface (Fernandez and Borrok, 2009).

10.3.3 Plant uptake and redistribution

The growth and metabolism of plants involve the transformation and redistribution of the nutrients they take in. These processes consist of various linked series of chemical reactions catalyzed by enzymes, including reduction and oxidation, complexation and complex break-down, which have the potential to fractionate an element's isotopes. Iron isotope fractionation in plants depends not only on the metal uptake strategy of the plant but also on Fe availability in the growth substrate (Kiczka et al., 2010b). Within the plant, Fe can be further fractionated showing more negative $\delta^{56}Fe$ values in leaves and flowers than in the roots (Kiczka et al., 2010b).

10.4 Variations of Iron isotopes

Since the first investigations on Fe isotope variations by Beard and Johnson (1999), the number of studies on Fe isotope variations has increased exponentially. Reviews on Fe-isotope geochemistry have been given by Anbar (2004a, b), Beard and Johnson (2004), Dauphas and Rouxel (2006), Anbar and Rouxel (2007). Figure 10.2 summarizes Fe-isotope variations in important geological reservoirs.

Bullen et al. (2001) showed that abiotic oxidation of ferrous iron into ferric iron in nature can cause isotopic fractionation of Fe. They analyzed precipitated ferrihydrite and found it to be enriched in the heavy isotopes of Fe by approximately 0.45%/amu relative to the source material. The authors argued that this probably reflected equilibrium fractionation between Fe^{2+} aq and Fe^{3+} aq.

Mineral dissolution tends to enrich the fluid in the light isotopes (Brantley et al., 2001, 2004). The magnitude of this effect, documented for hornblende crystals, depends on the affinity of Fe with the ligands that are present in the fluid. Many factors come into play and it is very difficult to disentangle the various mechanisms that are responsible for the net isotopic fractionation that is observed.

It is difficult to probe equilibrium fractionation between aqueous liquids and minerals because the exchange kinetics are very slow at low temperature. For this reason, one has to

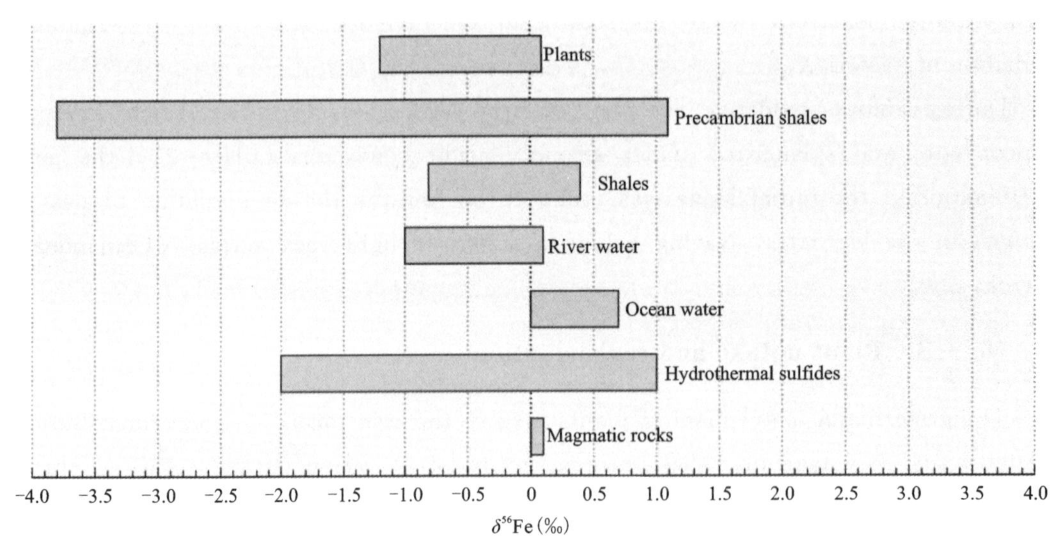

Figure 10.2 δ^{56}Fe-values of important geological reservoirs.

rely on precipitation experiments done at low precipitation rates to infer the equilibrium fractionation factor. Experiments of hematite (Fe_2O_3) precipitation from Fe^{3+} aq show that there is a correlation between the isotopic fractionation and the rate of precipitation (Skulan et al., 2002).

Strong concentration contrasts may develop in natural systems, notably when precipitation occurs. As discussed previously, during diffusion, the light atoms tend to move faster than the heavy ones and the region from which the atoms diffuse out gets enriched in the heavy isotopes compared to the region where the atoms diffuse in. One recent study documented the isotopic fractionation of transition metals during diffusion in water (Rodushkin et al., 2004). The diffusion profile in the diffusion cell is not perfect for Fe but it seems to be similar to that observed for Zn, which is better defined.

The importance of Fe^{3+} oxides as electron acceptors for anaerobic respiration in Fe-rich modern sediments is widely recognized (Lovley et al., 1987; Roden, 2004). Mineralogical products of dissimilatory Fe^{3+} reduction that may be preserved in the rock record include magnetite, Fe carbonates, and sulfides. In pioneering studies, Bullen and McMahon (1998) and Beard et al. (1999) reported that δ^{56}Fe-value of dissolved Fe^{2+} produced by dissimilatory Fereducing bacteria was fractionated by ~−1.3% relative to the ferrihydrite substrate. Significant Fe isotope fractionations have also been reported between dissolved Fe^{2+} and magnetite, siderite, and Ca-substitute siderite produced by Fe-reducing bacteria grown on ferrihydrite (Johnson et al., 2005).

Although chemical oxidation of ferrous iron is thermodynamically favored during the interaction of reduced fluids with oxygenated water, bacterial Fe^{2+} oxidation may prevail in microaerobic or anoxic environments. Croal et al. (2004) investigated Fe isotope fractionation produced by Fe^{2+}-oxidizing phototrophs under anaerobic conditions. Among key results, the ferrihydrite precipitate has δ^{56}Fe-value that is ~1.5‰ higher than the aqueous Fe^{2+} source. The enrichment of heavy Fe isotope in the oxidation product is globally consistent with equilibrium effects between Fe^{2+} and Fe^{3+} (Welch et al., 2003). The fractionation factor, however, is higher than for abiotic Fe^{2+} oxidation (Bullen et al., 2001) and lower than for equilibrium fractionation between aqueous Fe^{2+} and Fe^{3+} (Welch et al., 2003; Anbar et al., 2005), possibly reflecting kinetic effects during Fe-oxyhydroxide precipitation.

Magnetotactic bacteria (MB) are prokaryotes participating in the chemical transformation of Fe and S species via both redox and mineral precipitation processes. Initial Fe isotope studies of magnetite produced by magnetotactic bacteria (Mandernack et al., 1999) have shown no detectable fractionation when either Fe^{2+} or Fe^{3+} source were used in the growth media. These results contrast strongly with recent experimental work of Fe isotope fractionation during magnetite formation coupled to dissimilatory hydrous ferric oxide reduction, which shows large isotopic fractionation between Fe in magnetite and Fe in the fluid (Johnson et al., 2005). However, this does not preclude that there is an isotopic effect produced by MB because only two strains were investigated over a restricted range of laboratory conditions. In particular, Fe isotope fractionation might be dependent on the kinetics of Fe uptake by MB, which may vary with Fe concentration, Fe redox state, and the presence of Fe chelators (Schuler and Baeuerlein, 1996). It is presently unknown if any Fe isotope variability occurs in naturally occurring magnetite produced by MB.

Iron is an essential element for the biogeochemical and physiological functioning of terrestrial and oceanic organisms, and in particular phytoplankton, which is responsible for the primary productivity in the world's ocean Significant. Fe isotopic fractionation may be produced during mineral dissolution in the presence of organic ligands such as siderophore produced by soil bacteria (Brantley et al., 2001; Brantley et al., 2004). The magnitude of the Fe isotope fractionation during hornblende dissolution correlates with the binding constant of the chelating agent. Higher binding strength ligands such as siderophores produce larger δ^{56}Fe differences (up to 0.8‰) between the aqueous Fe and Fe remaining in the mineral. The Fe isotope fractionation observed during mineral dissolution with organic ligands is attributed predominantly to the preferential retention of the heavy isotopes in an altered surface layer. Mineral precipitation of isotopically heavy alteration phases may also contribute to the observed fractionation (Brantley et al., 2004). Experimental data of Fe isotope fractionation during Fe complexation by organic ligands confirm that the coordination chemistry of Fe exhibits a profound control on its isotopic behavior and may be of major importance in natural systems.

Iron is essential to the human body for oxygen transport in blood, for oxygen storage in muscle tissue, and as an enzyme co-factor. Recent studies have indicated natural variations of Fe isotope abundances in whole blood, hemoglobin, and tissues by as much as ~3% in the $^{56}Fe/^{54}Fe$ ratios (Walczyk and von Blanckenburg, 2002, 2005; Zhu et al., 2002; Stenberg et al., 2003, 2005; Ohno et al., 2004). Blood and, tissue have $\delta^{56}Fe$ values lower by up to 2.6% relative to dietary Fe, and the results have been interpreted to reflect Fe isotope fractionation during intestinal Fe absorption (Walczyk and von Blanckenburg, 2002; Figure 10.3). Interestingly, animals in general are enriched in lighter Fe isotopes relative to dietary Fe suggesting that Fe isotopes can be used to study trophic level along a food chain (Walczyk and von Blanckenburg, 2002).

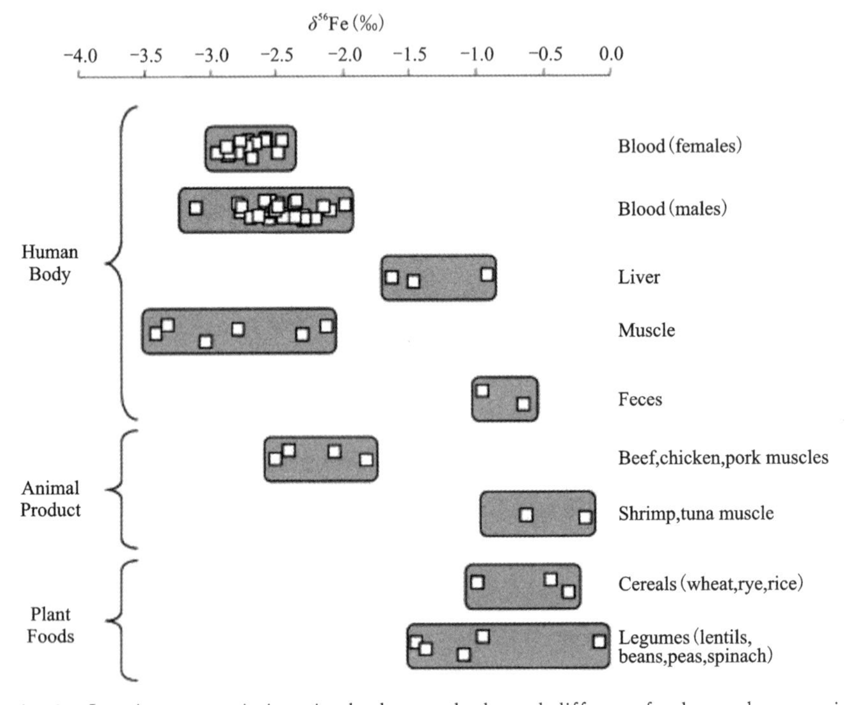

Figure 10.3 Iron isotope variations in the human body and different food samples covering the most relevant dietary Fe sources. Data are from Walczyk and von Blanckenburg (2002).

In theory, continental run-off may lead to Fe isotope variability in rivers that is dependend on the weathering regime (chemical vs. physical erosion). Despite its importance, the isotopic composition of the riverine-dissolved Fe flux remains largely unexplored. Dissolved Fe concentrations in estuaries are strongly controlled by non-conservative (removal) behavior during mixing of river water and seawater. Because the removal of Fe involves complex transformations between dissolved, colloidal, and particulate phases, it is possible that the Fe isotope composition of dissolved Fe is modified in estuaries. Consequently, the Fe isotope composition of the dissolved riverine Fe flux to the ocean remains poorly constrained but is likely characterized by lower $d^{56}Fe$ values relative to the bulk continental crust.

10.5 Application

Pathways of arsenic from sediments to groundwater in the hyporheic zone: Evidence from an iron isotope study by Xie et al. (2014). In the subsurface environment, the biogeochemical cycling of Fe and As are closely coupled (O'Day, 2002; Saalfield and Bostick, 2009; Wang et al., 2012; Xie et al., 2009). Due to the strong association between Fe and As, Fe biogeochemical cycling can provide significant clues to geochemical pathways of As. The prevailing consensus is that microbial reductive dissolution of As-associated Fe^{3+} oxides, hydroxides and oxyhydroxides is the main process that controls the level of As in groundwater (Nickson et al., 2000). Thus, As mobilization associated with Fe redox cycling in the hyporheic zone can be demonstrated by Fe isotope data. In recent years, the use of iron isotopes to study arsenic has become a hot topic. Arsenic concentrations as high as 1820μg/L have been detected in groundwater of the Datong Basin (Xie et al., 2008), which greatly exceeds the maximum limited value (10μg/L) for drinking water recommended by the WHO. Here, we take the south bank of the Sanggan River, in Shanxi Provence China, as an example to understand the complex (bio.) geochemical processes that cause As mobilization.

10.5.1 Methods

In this study, four 20m sediment cores were collected using rotary techniques at distances of 0m (Core A), 10m (Core B), 40m (Core C) and 80m (Core D) from the river (Figure 10.4). Water is pumped to the surface through plastic tubing. Slow pumping and immediate filtration using syringes keeps the ambient air out of the samples and minimizes the oxidation of dissolved Fe^{2+} to particulate Fe^{3+} oxides prior to acidification and storage. A total of six hyporheic water samples were collected from the study site. Chemical and physical parameters were measured on the site using portable meters made by Hach Instruments. Two filtered ($<$0.45μm) acidified samples (acidified to pH$<$ 2 using ultra-pure HNO_3) were collected in 50mL HDPE bottles for the laboratory analysis of their chemical composition, including As and Fe concentrations and Fe isotopic ratios.

The Fe isotope composition was determined by multicollector inductively coupled plasma mass spectrometry (MC-ICP-MS)(Nu Plasma Nu Instrument) equipped with a Cetac ASX-110 automatic sampler and a DSN-100 Desolvating Nebulizer System at the Department of Geology at the University of Illinois at Urbana Champaign (UIUC). Fe isotopic ratios were reported in delta notation as defined by the following relationship:

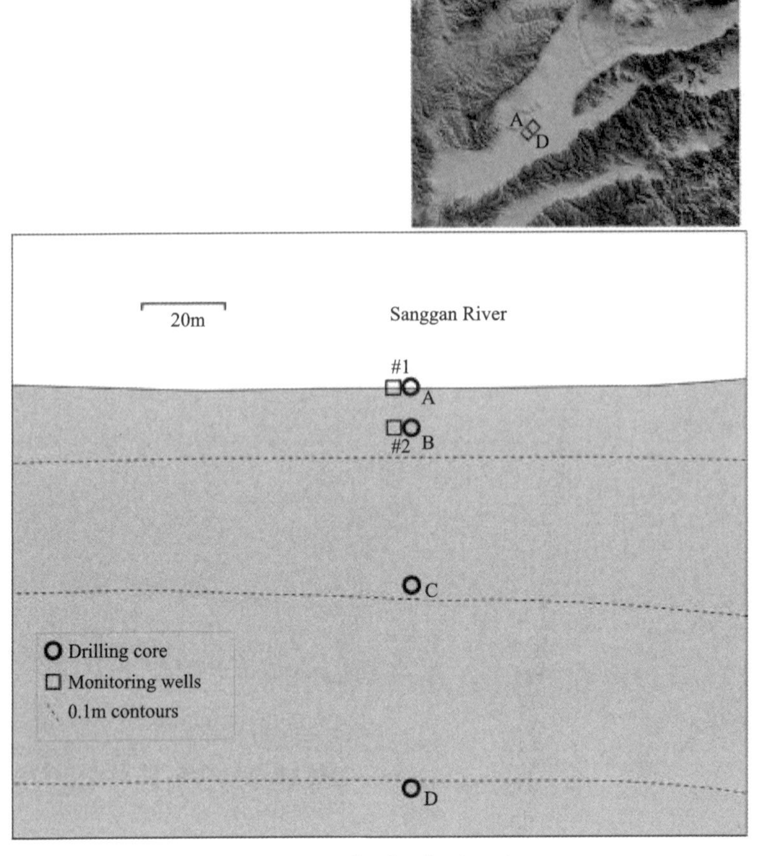

Figure 10.4 Study site location.

$$\delta^{56}Fe=\left[\frac{(^{56}Fe/^{54}Fe)_{sample}}{(^{56}Fe/^{54}Fe)_{standard}}-1\right]\times 1000$$

10.5.2 Isotope characteristics in hyporheic water

The water samples had high As concentrations, varying between 12μg/L and 132.3μg/L. The hyporheic water samples from Well 1 had relatively higher As concentrations than those from Well 2. In Well 1, high As concentrations were detected in the samples from the shallow (10m depth) and deep (20m depth) water samples. However, high As concentrations were only observed in the deep sample (20m) in Well 2. Fe concentrations in water samples ranged from 206μg/L to 863μg/L. Like HS^-, Fe concentrations were generally low in the high As samples. The $\delta^{56}Fe$ values in the hyporheic water samples ranged widely (from 0.04‰ to 0.71‰). Unlike HS^- and Fe, the $\delta^{56}Fe$ values were relatively higher in the samples with high As.

Table 10.1 Hydrochemical Fe-isotope composition of hyporheic water.

Depth (m)	pH	ORP (mV)	HS^- (μg/L)	Fe (μg/L)	Mn (mg/L)	As (μg/L)	$\delta^{56}Fe$	SD
Well 1								
10	7.66	−174.1	3	206	0.22	114.2	0.67	0.02
14	7.27	− 211.4	9	863	1.68	14.54	0.21	0.03
20	7.70	−171.8	2	236	0.22	132.3	0.71	0.07
Well 2								
10	7.20	−167.2	3	588	0.94	14.48	0.04	0.02
14	7.20	−199.6	8	620	0.85	12	0.37	0.02
20	7.68	−196.0	3	251	0.20	81.6	0.56	0.04

Table 10.2 Chemical and Fe-isotope composition of core sediments.

Sample ID	Depth (m)	As (mg/kg)	Fe (mg/g)	Mn (mg/g)	TOC (%)	$S_{sulfide}$ (%)	$\delta^{56}Fe$	SD
A1	5.8	25.79	26.88	0.580	1.36	0.34	0.35	0.072
A2	8	28.93	39.14	0.859	2.56	0.43	0.30	0.077
A3	10	14.69	24.92	0.630	1.67	0.35	0.27	0.037
A4	12.8	21.70	25.43	0.974	2.17	0.38	0.19	0.058
A5	13.8	14.09	22.47	0.667	1.96	0.34	0.29	0.021
A6	19.1	18.84	24.98	0.560	1.45	0.37	0.41	0.067
B1	8.5	31.11	37.99	0.936	2.24	0.35	0.04	0.063
B2	10.2	17.56	24.79	0.580	1.69	0.38	0.42	0.036
B3	13.4	18.27	17.27	0.540	2.32	0.41	0.36	0.057
B4	18.9	27.29	35.29	0.748	1.74	0.63	0.30	0.079
B5	19.8	21.23	29.76.	0.655	1.41	0.52	0.36	0.039
C1	7.1	—	—	—	1.71	0.36	0.37	0.027
C2	8.9	26.45	21.71	0.627	1.33	0.26	0.36	0.043
C3	13.6	13.45	26.50	0.740	2.28	0.32	0.26	0.017
C4	15.6	—	—	—	2.99	0.32	0.18	0.097
C5	20.4	18.92	25.44	0.540	1.06	0.38	0.23	0.032
D1	8.1	—	—	—	1.41	0.30	0.48	0.048
D2	10.5	—	—	—	3.19	0.32	0.22	0.049
D3	12.9	46.92	30.50	0.681	1.41	0.35	0.19	0.030
D4	15.9	37.59	33.80	0.790	0.98	0.37	0.14	0.042
D5	19.1	12.09	20.32	0.390	0.67	0.40	0.32	0.051
D6	19.3	15.71	23.35	0.505	0.65	0.32	0.29	0.024
D7	20.9	10.29	17.91	0.429	0.75	0.25	0.26	0.044

— Not determined.

The $\delta^{56}Fe$ values of water at various depths in Well 1 mirrored the bulk $\delta^{56}Fe$ values in Core A (Table 10.1 and Table 10.2). Although the Fe-reducing bacteria preferentially release light Fe from silicates and Fe-oxides (Brantley et al., 2001, 2004; Emmanuel et al., 2005), enriched-$\delta^{56}Fe$ values bulk sediments can produce relatively enriched-$\delta^{56}Fe$ values of Fe^{2+} aq due to Rayleigh distillation. Accordingly, the bulk sediment $\delta^{56}Fe$ values in Core A can account for the observed vertical variation of $\delta^{56}Fe$ values of water in Well 1. The high $\delta^{56}Fe$ value and low Fe concentration in the water sample from Well 1 at 10m depth could be due to a moderate reduction of enriched-$\delta^{56}Fe$ Fe^{3+} minerals in hyporheic sediments at the same depth. Taken together, the low $\delta^{56}Fe$ value and high Fe and HS^- concentrations in the water sample from Well 1 at 14m depth and the Fe concentrations and $\delta^{56}Fe$ values in core sediments suggest that there was intensive reduction of Fe^{3+} and SO_4^{2-}. High concentration of Fe and HS^- can result in the formation of Fe sulfide precipitate and depletion of $\delta^{56}Fe$ in water samples according to the study conducted by Butler et al. (2005). It is well documented that FeS precipitate can retain a large amount of As (Smedley and Kinniburgh, 2002; Han et al., 2011; Wolthers et al., 2005). Therefore, this group of reactions may account for the high HS^- and Fe and the low As and $\delta^{56}Fe$ values in water at 14m in Well 1 (Table 10. 1). The high $\delta^{56}Fe$ value in the water sample at 20m can be attributed to reduction of the lithogenic Fe oxides/hydroxides with high $\delta^{56}Fe$ values in the core sediments at the corresponding depth. However, this appears inconsistent with the low Fe concentrations measured in water from the bottom of the Well 1. The microbial reduction of Fe^{3+} does not necessarily release all of the Fe^{2+} into groundwater because some is retained in the solid phase (Fredrickson et al., 1998). Therefore, high Fe concentrations would not be expected along with the microbial reduction of Fe^{3+} minerals at the bottom of Well 1. The vertical distribution of As concentrations and $\delta^{56}Fe$ values in water samples from Well 1 were strikingly similar (Figure 10. 5a). It is well known that Fe^{3+} oxides and hydroxides are the critical sequesters in sediments in the study area (Xie et al., 2008, 2013), and the close relationship between Fe and As contents in the bulk core sediments offers further evidence for this (Figure 10. 6a). Therefore, the high As concentrations at 10m and 20m in Well 1 can be attributed to the microbial reduction and dissolution of As-bearing Fe^{3+} minerals, such as Fe^{3+} oxides/hydroxides.

The $\delta^{56}Fe$ values in Well 2 differed from those in Well 1 because they increased with depth from 0.04‰ to 0.56‰ (Figure 10. 6b). The $\delta^{56}Fe$ value in water at 10m (0.04‰) are consistent with the bulk $\delta^{56}Fe$ value of sediments from Core B at 8.5m (0.04‰) (Table 10.1 and Table 10.2), indicating that Fe does not experience significant redox cycling. It is worthy of noting that water sample from Well 1 at 14m had similar $\delta^{56}Fe$ value with sediment samples from 13.4m depth. Importantly, the sample also contained high

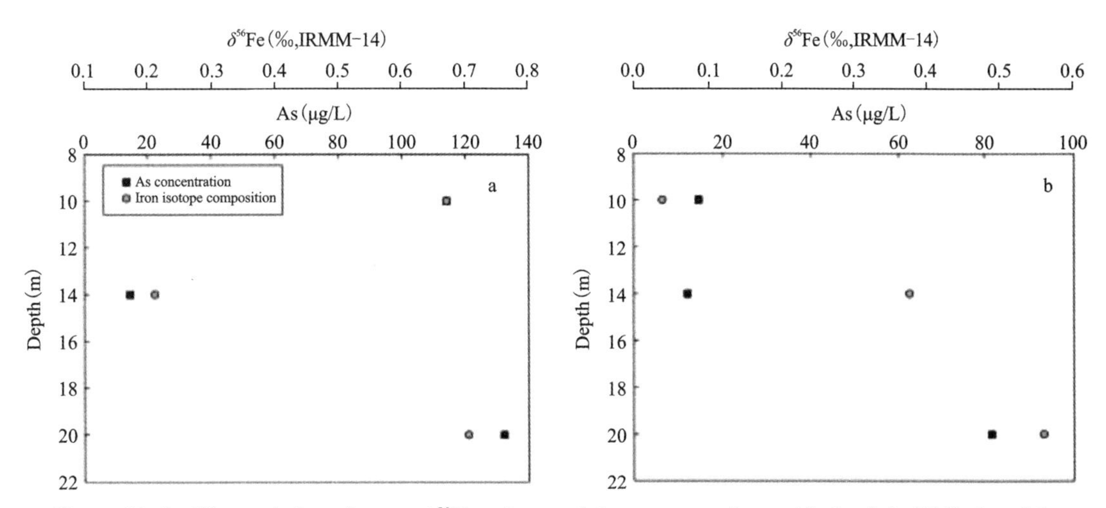

Figure 10.5 The variation of water δ^{56}Fe values and As concentrations with depth in Wells 1 and 2.

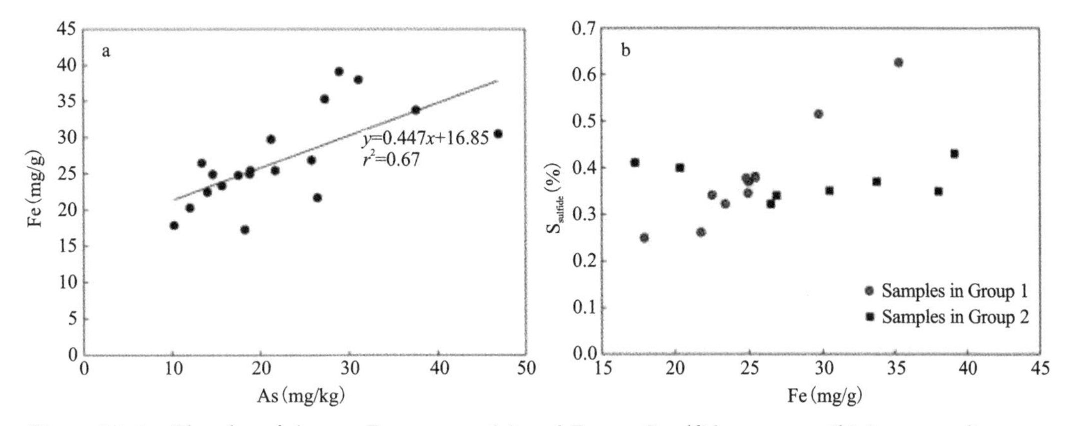

Figure 10.6 The plot of As vs. Fe contents (a) and Fe vs. S sulfide contents (b) in core sediments.

concentration of Fe(620μg/L) and HS^-(8μg/L). Sulfide can directly react with Fe^{3+} oxides, hydroxides and oxyhydroxides to abiotically reduce Fe^{3+} to Fe^{2+} under high concentrations of HS^- conditions (Yao and Millero, 1996). No significant isotopic fractionation can be expected between Fe^{2+} aq and the residual Fe^{3+} minerals in this process. At the same time, high concentration of Fe and HS^- resulted in the formation of Fe sulfide precipitate and sequester As from water (Smedley and Kinniburgh, 2002; Han et al., 2011; Wolthers et al., 2005). Thus, the observed iron isotopic and hydrochemical characteristics can be attributed to HS^- abiotical reduction of Fe^{3+} minerals. In contrast, the relative enriched-δ^{56}Fe value of water sample in Well 2 at 20m (0.56‰) indicates Fe^{3+} minerals have experienced an intensive microbial redox cycling. Therefore, the vertical distribution of As concentrations and δ^{56}Fe values apparently supports that As levels in water from Well 2 governed by abiotical and microbial Fe^{3+} reduction.

10.5.3 Implications for arsenic mobilization in the hyporheic zone

We constructed a conceptual model of Fe geochemical pathways and As mobilization that incorporates data on the sulfide and Fe contents and $\delta^{56}Fe$ value of the bulk core sediments in hyporheic water (Figure 10.7). In the hyporheic zone, the $\delta^{56}Fe$ of the bulk core sediments and hyporheic water at each depth is governed by the intensity of microbial Fe^{3+} and SO_4^{2-} reduction and the formation of secondary sulfidic and non-sulfidic Fe^{2+} minerals. In the upper sections (with depth less than 10m) of the hyporheic zone, Fe^{3+} reduction catalyzed by microbes is the dominant biogeochemical reaction. The reduction of As-bearing Fe^{3+} minerals can release As into the hyperheic water and formation of nonsulfidic Fe^{2+} minerals. This explains the high As and low Fe concentration and high $\delta^{56}Fe$ value in water at 10m in Well 1. Under strongly reducing conditions, intense SO_4^{2-} reduction and Fe-sulfide precipitation will prevail in the midsections (≈13～19m) of the hyporheic zone. Because it is sequestered by the Fe^{2+}-sulfides, As concentration is limited in water at corresponding depths. This may explain the low As concentrations in water at 14m in both cores. However, Fe isotope compositions in water at this depth may be controlled by different processes in Core A and Core B. In core A, interaction between Fe^{2+} aq and produced Fe sulfide can be used to account for the low $\delta^{56}Fe$ value in water sample. However, relative enriched-$\delta^{56}Fe$ value in water sample from Core B may be due to HS^- abiotical reduction of Fe^{3+} minerals under high HS^- concentration environment. According

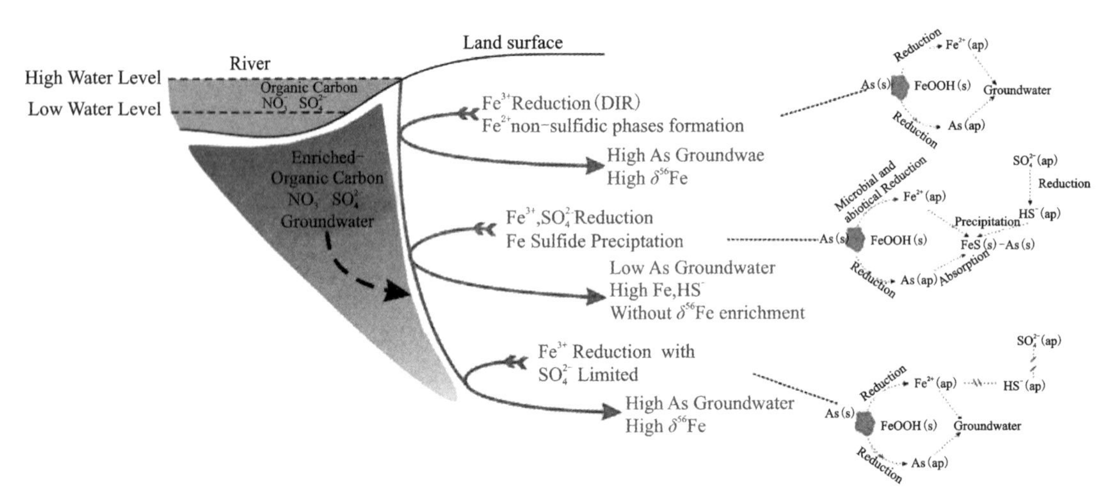

Figure 10.7 Conceptual hydrogeochemical model for As pathways in the hyporheic zone at the study site. This model integrates porewater Fe isotope compositions and As profile presented in Figure 10.6. The gray area represents high concentration of organic matter, NO_3^- and SO_4^{2-} groundwater domain. The gray gradient indicates the change from high (dark gray) to low (light gray) concentration of chemicals.

to the sequence of redox reactions (Borch et al., 2010), reduction of SO_4^{2-} to HS^- is followed by the reduction of crystalline Fe oxide/hydroxides. At the SO_4^{2-} reduction stage, most of SO_4^{2-} is likely used up, and then the formation of Fe^{2+}-sulfides will be limited. Subsequently, the reduction of crystalline Fe oxide/hydroxides can promote the mobilization of As. As a result, high As concentrations can be detected in water samples at ~20m depth in both Core A and Core B.

References

ANBAR A D, 2004a. Iron stable isotopes: beyond biosignatures[J]. Earth Planet Sci. Lett., 217:223-236.

ANBAR A D, 2004b. Molybdenum stable isotopes: observations, interpretations and directions[J]. Rev. Mineral Geochem., 55:429-454.

ANBAR A D, ROUXEL O, 2007. Metal stable isotopes in paleoceanography[J]. Ann. Rev. Earth Planet Sci., 35:717-746.

ANBAR A D, JARZECKI A A, SPIRO T G, 2005. Theoretical investigation of iron isotope fractionation between $Fe(H_2O)_6^{3+}$ and $Fe(H_2O)_6^{2+}$: Implications for iron stable isotope geochemistry[J]. Geochim. Cosmochim. Acta, 69: 825-837.

BEARD B L, JOHNSON C M, 1999. High-precision iron isotope measurements of terrestrial and lunar materials[J]. Geochim. Cosmochim. Acta, 63:1653-1660.

BEARD B L, JOHNSON C M, 2004. Fe isotope variations in the modern and ancient Earth and other planetary bodies[J]. Rev. Mineral Geoch., 55:319-357.

BEARD B L, JOHNSON C M, COX L, et al., 1999. Iron isotope biosignatures[J]. Science, 285(5435):1889-1892.

BELSHAW N, ZHU X, GUO Y O, et al., 2000. High precision measurement of iron isotopes by plasma source mass spectrometry[J]. Inter. J. Mass Spectrom., 197:191-195.

BRANTLEY S L, LIERMANN L, BULLEN T D, 2001. Fractionation of Fe isotopes by soil microbes and organic acids[J]. Geology, 29:535-538.

BRANTLEY S L, LIERMANN L J, GUYNN R L, et al., 2004. Fe isotopic fractionation during mineral dissolution with and without bacteria[J]. Geochim. Cosmochim. Acta, 68:3189-3204.

BULLEN T D, MCMAHON P M, 1998. Using stable Fe isotopes to assess microbially mediated Fe^{3+} reduction in a jet-fuel contaminated aquifer[J]. Mineral Mag., 62A:255-256.

BULLEN T D, WHITE A F, CHILDS C W, et al., 2001. Demonstration of significant abiotic iron isotope fractionation in nature[J]. Geology, 29(8):699-702.

BUTLER I B, ARCHER C, VANCE D, et al., 2005. Fe isotope fractionation on FeS formation in ambient aqueous solution[J]. Earth Planet. Sci. Lett., 236 (1-2): 430-442.

CHAPMAN J B, WEISS D J, SHAN Y, et al., 2009. Iron isotope fractionation during leaching of granite and basalt by hydrochloric and oxalic acids[J]. Geochim. Cosmochim. Acta, 73:1312-1324.

COLLINET M, CHARLIER B, NAMUR O, et al., 2017. Crystallization history of enriched shergottites from Fe and Mg isotope fractionation in olivine megacrysts [J]. Geochim. Cosmochim. Acta, 207:277-297.

CORNELL R M, SCHWERTMANN U, 2003. The Iron Oxides — Structure, Properties, Reactions, Occurrence and Uses[M]. Weinheim: Wiley-VGH Verlag GmbH & Co. KGaA.

CRADDOCK P R, DAUPHAS N, 2010. Iron isotopic compositions of geological reference materials and chondrites[J]. Geostand. Geoanal. Res., 35:101-123.

CROAL L R, JOHNSON C M, BEARD B L, et al., 2004. Iron isotope fractionation by Fe^{2+}-oxidizing photoautotrophic bacteria[J]. Geochim. Cosmochim. Acta, 68:1227-1242.

CROSBY H A, RODEN E E, JOHNSON C M, et al., 2007. The mechanisms of iron isotope fractionation produced during dissimilatory Fe^{3+} reduction by Shewanella putrefaciens and Geobacter sulfurreducens[J]. Geobiology, 5:169-189.

CURIE C, PANAVIENE Z, LOULERGUE C, et al., 2001. Maize yellow stripel encodes a membrane protein directly involved in Fe^{3+} uptake[J]. Nature, 409: 346-349.

CZAJA A D, JOHNSON C M, BEARD B L, et al., 2013. Biological Fe oxidation controlled deposition of banded iron formation in the ca. 3770Ma Isua Supracrustal Belt (West Greenland)[J]. Earth Planet. Sci. Lett., 363:192-203.

DAUPHAS N, ROUXEL O, 2006. Mass spectrometry and natural variations in iron isotopes[J]. Mass Spectrom. Rev. 25:515-550.

Dauphas N, John S G, Rouxel O, 2017. Iron isotope systematics[J]. Rev. Mineral Geochem., 82:415-510.

DEVULDER V, LOBO L, VAN HOECKE K, et al., 2013. Common analyte internal standardization as a tool for correction for mass discrimination in multi-collector inductively coupled plasma-mass spectrometry[J]. Spectrochim. Acta Part B At Spectrosc, 89:2029.

EMERSON D, FLEMING E J, MCBETH J M, 2010. Iron-oxidizing bacteria: An environmental and genomic perspective[J]. Annu. Rev. Microbiol., 64:561-583.

EMMANUEL S, EREL Y, MATTHEWS A, et al., 2005. A preliminary mixing model for Fe isotopes in soils[J]. Chem. Geol., 222: 23-34.

FERNANDEZ A, BORROK D M, 2009. Fractionation of Cu, Fe, and Zn isotopes during the oxidative weathering of sulfide-rich rocks[J]. Chem. Geol., 264: 1-12.

FINLAYSON V A, KONTER J G, MA L, 2015. The importance of a Ni correction with ion counter in the double spike analysis of Fe isotope compositions using a $^{57}Fe/^{58}Fe$ double spike[J]. Geochem. Geophys. Geosyst., 16(12):4209-4222.

FREDRICKSON J K, ZACHARA J M, KENNEDY D W, et al. , 1998. Biogenic iron mineralization accompanying the dissimilatory reduction of hydrous ferric oxide by a groundwater bacterium[J]. Geochim. Cosmochim. Acta, 62: 3239-3257.

GALIC A, MASON P R D, MOGOLLON J M, et al. , 2017. Pyrite in a sulfate-poor Paleoarchean basin was derived predominantly from elemental sulfur: evidence from 3. 2Ga sediments in the Barberton Greenstone Belt, Kaapvaal Craton[J]. Chem. Geol. , 449: 135-146.

GONZALEZ J J, LIU C Y, WEN S B, et al. , 2007b. Metal particles produced by laser ablation for ICP-MS measurements[J]. Talanta, 73(3):567-576.

HAN Y-S, GALLEGOS T J, DEMOND A H, et al. , 2011. FeS-coated sand for removal of arsenic(Ⅲ) under anaerobic conditions in permeable reactive barriers[J]. Water Res. , 45: 593-604.

ICOPINI G A, ANBAR A D, RUEBUSH S S, et al. , 2004. Iron isotope fractionation during microbial reduction of iron: The importance of adsorption[J]. Geology, 32: 205-208.

JOHNSON C M, RODEN E E, WELCH S A, et al. , 2005. Experimental constraints on Fe isotope fractionation during magnetite and Fe carbonate formation coupled to dissimilatory hydrous ferric oxide reduction[J]. Geochim. Cosmochim. Acta, 69:963-993.

JOHNSON C M, SKULAN J L, BEARD B L, et al. , 2002. Isotopic fractionation between Fe^{3+} and Fe^{2+} in aqueous solutions[J]. Earth Planet Sci. Lett. , 195(1-2): 141-153.

JOHNSON C M, BEARD B L, RODEN E E, et al. , 2004. Isotopic constraints on biogeochemical cycling of Fe Rev. Mineral[J]. Geochem. , 55:359-408.

KICZKA M, WIEDERHOLD J G, FROMMER J, et al. , 2010a. Iron isotope fractionation during proton-and ligand-promoted dissolution of primary phyllosilicates[J]. Geochim. Cosmochim. Acta, 74: 3112-3128.

KICZKA M, WIEDERHOLD J G, KRAEMER S M, et al. , 2010b. Iron isotope fractionation during plant uptake and translocation in alpine plants[J]. Environ. Sci. Technol. , 44: 6144-6150.

KOSLER J, PEDERSEN R B, KRUBER C, et al. , 2005 Analysis of Fe isotopes in sulfides and iron meteorites by laser ablation high-mass resolution multi-collector ICP mass spectrometry[J]. J. Anal. At. Spectrom. , 20 (3):192-199.

LOVLEY D R, STOLZ J F, NORD G L, et al. , 1987. Anaerobic production of magnetite by a dissimilatory iron-reducing microorganism[J]. Nature, 330:252-254.

LOVLEY D R, HOLMES D E, NEVIN K P, 2004. Dissimilatory Fe^{3+} and Mn^{4+} reduction[J]. Adv. Microb. Physiol. , 49: 219-286.

MANDERNACK K W, BAZYLINSKI D A, SHANKS W C, et al. , 1999. Oxygen and iron isotope studies of magnetite produced by magnetotactic bacteria[J]. Science, 285:1892-1896.

MARECHAL C N, TELOUK P, ALBARÈDE F,1999. Precise analysis of copper and zinc isotopic compositions by plasma-source mass spectrometry[J]. Chem. Geol., 156(14): 251-273.

NICKSON R T, MCARTHUR J M, RAVENSCROFT P,et al., 2000. Mechanism of arsenic release to groundwater, Bangladesh and West Bengal[J]. Appl. Geochem., 15: 403-413.

O'Day P A, 2002. Spatial distribution and energetics of microbially mediated arsenic, iron, and sulfur redox reactions in shallow subsurface sediments[J]. Abstr. Pap. Am. Chem. Soc., 223: U596-U597.

Ohno T, Shinohara A, Kohge I,et al., 2004. Isotopic analysis of Fe in human red blood cells by multiple collector-ICP-mass spectrometry[J]. Anal. Sci., 20:617-621.

RAISWELL R, CANFIELD D E, 2012. The iron biogeochemical cycle past and present [J]. Geochem. Perspect., 1: 1-232.

RODEN E, 2004. Analysis of long-term bacterial vs. chemical Fe^{3+} oxide reduction kinetics[J]. Geochim. Cosmochim. Acta, 68:3205-3216.

RODUSHKIN I, STENBERG A, ANDRÉN H,et al., 2004. Isotopic fractionation during diffusion of transition metal ions in solution[J]. Anal. Chem., 76:2148-2151.

SAALFIELD S L, BOSTICK B C, 2009. Changes in iron, sulfur, and arsenic speciation associated with bacterial sulfate reduction in ferrihydrite-rich systems[J]. Environ. Sci. Technol., 43: 8787-8793.

SCHAAF G, LUDEWIG U, ERENOGLU B E,et al., 2004. ZmYS1 functions as a proton-coupled symporter for phytosiderophore-and nicotianamine-chelated metals[J]. J. Biol. Chem., 279:9091-9096.

SCHULER D, BAEUERLEIN E,1996. Iron-limited growth and kinetics of iron uptake in Magnetospirilum gryphiswaldense[J]. Arch. Microbiol., 166: 301-307.

SIO C K, DAUPHAS N, TENG N Z,et al.,2013. Discerning crystal growth from diffusion profiles in zoned olivine by in-situ Mg-Fe isotopic analyses[J]. Geochim. Cosmochim. Acta,123:302-321.

SKULAN J L, BEARD B L, JOHNSON C M, 2002. Kinetic and equilibrium Fe isotope fractionation between aqueous Fe^{3+} and hematite[J]. Geochim. Cosmochim. Acta, 66:2995-3015.

SMEDLEY P L, KINNIBURGH D G, 2002. A review of the source,behaviour and distribution of arsenic in natural water[J]. Appl. Geochem., 17 (5): 517-568.

STEINHOEFEL G, HORN I, VON BLANCKENBURG F,2009a. Matrix-independent Fe isotope ratio determination in silicates using UV femtosecond laser ablation[J]. Chem.

Geol., 268(1-2):67-73.

STEINHOEFEL G, HORN I, VON BLANCKENBURG F, 2009b. Micro-scale tracing of Fe and Si isotope signatures in banded iron formation using femtosecond laser ablation [J]. Geochim. Cosmochim. Acta, 73(18):5343-5360.

STENBERG A, MALINOVSKY D, O'HLANDER B, et al., 2005. Measurement of iron and zinc isotopes in human whole blood: Preliminary application to the study of HFE genotypes[J]. J. Trace Elem. Med. Bio., 19:55-60.

STENBERG A, MALINOVSKY D, RODUSHKIN I, et al., 2003. Separation of Fe from whole blood matrix for precise isotopic ratio measurements by MC-ICP-MS: A comparison of different approaches[J]. J. Anal. At. Spectrom., 18:23-28.

STRAUB K L, BENZ M, SCHINK B, et al., 1996. Anaerobic, nitrate-dependent microbial oxidation of ferrous iron[J]. Appl. Environ. Microbiol., 62:1458-1460.

TAKAGI S, NOMOTO K, TAKEMOTO T, 1984. Physiological aspect of mugineic acid, a possible phytosiderophore of graminaceous plants[J]. J. Plant Nutr., 7:469-477.

TEUTSCH N, VON GUNTEN U, PORCELLI D, et al., 2005. Adsorption as a cause for iron isotope fractionation in reduced groundwater[J]. Geochim. Cosmochim. Acta, 69: 4175-4185.

VIRTASALO J J, WHITEHOUSE M J, KOTILAINEN A T, 2013. Iron isotope heterogeneity in pyrite fillings of Holocene worm burrows[J]. Geology, 41(1):39-42.

WALCZYK T, VON BLANCKENBURG F, 2002. Natural iron isotope variations in human blood[J]. Science, 295:2065-2066.

WALCZYK T, VON BLANCKENBURG F, 2005. Deciphering the iron isotope message of the human body[J]. Int. J. Mass Spectrom., 242:117-134.

WANG S F, XU L Y, ZHAO Z X, et al., 2012. Arsenic retention and remobilization in muddy sediments with high iron and sulfur contents from a heavily contaminated estuary in China[J]. Chem. Geol., 314:57-65.

WEBER K A, ACHENBACH L A, COATES J D, 2006. Microorganisms pumping iron: anaerobic microbial iron oxidation and reduction[J]. Nature, 4: 752-764.

WELCH S A, BEARD B L, JOHNSON C M, et al., 2003. Kinetic and equilibrium Fe isotope fractionation between aqueous Fe^{2+} and Fe^{3+} [J]. Geochim. Cosmochim. Acta, 67 (22):4231-4250.

WHITEHOUSE M J, FEDO C M, 2007. Microscale heterogeneity of Fe isotopes in>3.71Ga banded iron formation from the Isua Greenstone Belt, Southwest Greenland[J]. Geology, 35(8):719-722.

WIEDERHOLD J G, KRAEMER S M, TEUTSCH N, et al., 2006. Iron isotope fractionation during proton-promoted, ligand-controlled, and reductive dissolution of goethite[J]. Environ. Sci. Technol., 40:3787-3793.

WOLTHERS M, CHARLET L, VAN DER WEIJDEN C H, et al., 2005. Arsenic mobility in the ambient sulfidic environment: Sorption of arsenic(V) and arsenic(Ⅲ) onto disordered mackinawite[J]. Geochim. Cosmochim. Acta, 69: 3483-3492.

XIE X, JOHNSON T M, WANG Y, et al., 2014. Pathways of arsenic from sediments to groundwater in the hyporheic zone: Evidence from an iron isotope study[J]. J. Hydrol., 511: 509-517.

XIE X, JOHNSON T M, WANG Y X, et al., 2013. Mobilization of arsenic in aquifers from the Datong Basin, China: evidence from geochemical and iron isotopic data[J]. Chemosphere, 90 (6): 1878-1884.

XIE X, WANG Y, SU C, et al., 2008. Arsenic mobilization in shallow aquifers of Datong Basin: hydrochemical and mineralogical evidences[J]. J. Geochem. Explor., 98: 107-115.

XIE X J, ELLIS A, WANGY X, et al., 2009. Geochemistry of redox-sensitive elements and sulfur isotopes in the high arsenic groundwater system of Datong Basin, China [J]. Sci. Total Environ., 407: 3823-3835.

YOSHIYA K, SAWAKI Y, HIRATA T, et al., 2015a. In-situ iron isotope analysis of pyrites in similar to 3.7Ga sedimentary protoliths from the Isua supracrustal belt, southern West Greenland[J]. Chem. Geol., 401: 126-139.

ZHU X K, GUO Y, WILLIAMS R J P, et al., 2002. Mass fractionation processes of transition metal isotopes[J]. Earth Planet Sci. Lett., 200: 47-62.

Chapter 11 Compound-Specific Isotope

Compound-specific isotope ratio mass spectrometry enables the molecular specificity and isotopic signature of compounds to be exploited concomitantly, to provide a powerful tool for tracing the origin and fate of organic matter or organic contaminants in the groundwater system. Such approaches can be used to investigate chemical and biological processes and metabolism in both whole ecosystems (marine, terrestrial, freshwater) and individual organisms of all classes, i. e. microbes to mammals.

Even before the advent of new on-line technologies, the stable isotopic compositions of individual compounds had begun to be rigorously investigated. The introduction of high-performance liquid chromatography provided an effective means of isolating milligram quantities of highly purified compounds of widely varying polarity for off-line combustion and stable isotope analysis. The major breakthrough in compound specific stable isotope analysis came with the commercial production of the gas chromatograph/combustion/isotope ratio mass spectrometer (GC/C/IRMS) in the early 1990s.

Compound-specific stable isotope analysis (CSIA) quantifies this isotopic composition and hence provides additional and often unique means to ① allocate and distinguish sources of organic compounds, and ② identify and quantify transformation reactions, sometimes even on a mechanistic level. This has been found for $\delta^{13}C$ in BTEX (Dempster et al., 1997), $\delta^{13}C$ in MTBE (Smallwood et al., 2001), $\delta^{13}C$ in PCBs (Drenzek et al., 2002; Yanik et al., 2003), $\delta^{13}C$, $\delta^{2}H$, and $\delta^{37}Cl$ in chlorinated solvents (Shouakar-Stash et al., 2003), and $\delta^{13}C$ and $\delta^{15}N$ in trinitrotoluene (Coffin et al., 2001).

11.1 Definitions

Compound-specific isotope analysis (CSIA) yields data of the isotopic composition of a single compound x relative to an international standard that are usually expressed as δ values in per mil (‰) according to the equation:

$$\delta_x = \left(\frac{R_x - R_{\text{reference}}}{R_{\text{reference}}}\right) \times 1000 \tag{11.1}$$

where R_x and $R_{\text{reference}}$ are the ratios of the heavy isotope to the light isotope (e. g. $^{13}C/^{12}C$ or

D/H) in compound x and an international standard, respectively. Thus, rather than absolute values, the differences in relative ratios are reported to allow a correction for mass-discriminating effects in a single instrument and to facilitate the comparison of published GC/IRMS data. Only such relative isotope ratios can be determined with the required precision. A $\delta^{13}C$ value of $+10‰$ then corresponds to a sample with an isotope ratio one percent higher than that of the international standard (usually Vienna Peedee Belemnite, VPDB). For VPDB a ratio of ^{13}C versus ^{12}C of 0.011 180 has been reported (Werner and Brand, 2001). The $\delta^{13}C$ value of $+10‰$ for the sample then corresponds to a ^{13}C to ^{12}C ratio of 0.011 292, which demonstrates the very subtle changes that need to be measured.

The isotope fractionation between two compounds (e.g. a substrate and its degradation product) can be expressed either with the fractionation factor α or the enrichment factor ε according to Eqs. 11.2 and 11.3:

$$\alpha_{pr} = \frac{R_{\text{product}}}{R_{\text{reactant}}} = \frac{10^{-3}\delta_p + 1}{10^{-3}\delta_r + 1} \tag{11.2}$$

$$\varepsilon_{pr} = \left(\frac{R_{\text{product}}}{R_{\text{reactant}}} - 1\right) \times 1000 = (\alpha - 1) \times 1000 \tag{11.3}$$

where subscripts r and p refer to reactant and product, respectively, and R_{reactant} and R_{product} are the ratios of the heavy isotope to the light isotope in the substrate and the degradation product, respectively. For small molecules in which all isotopes are located in the same reactive position, α can also be interpreted according to follow equation:

$$\alpha = \frac{^{\text{heavy}}k}{^{\text{light}}k} = \text{KIE}^{-1} \tag{11.4}$$

where $^{\text{heavy}}k$ and $^{\text{light}}k$ are the rate constants of compounds containing heavy and light isotopes at the reactive position and $\text{KIE} = {}^{\text{light}}k/{}^{\text{heavy}}k$, which is the kinetic isotope effect of the reaction.

The enrichment factor ε or the fractionation factor α is usually determined by using the relationship between substrate concentration change and isotope fractionation given in follow equation:

$$\frac{R_t}{R_0} = \left[f\frac{(1+R_0)}{(1+R_t)}\right]^{\alpha-1} = \left[f\frac{(1+R_0)}{(1+R_t)}\right]^{\varepsilon} \tag{11.5}$$

where R_t and R_0 are the ratios of the heavy isotope to the light isotope in the reactant at time $t=0$ and t, respectively, and f is the remaining fraction of the reactant at time t according to follow equation:

$$f = \frac{L_t + H_t}{L_0 + H_0} = \frac{L_t(1+R_t)}{L_0(1+R_0)} \tag{11.6}$$

where L_0 and H_0 are the concentrations of the light isotope and the heavy isotope at time $t=0$, respectively, and L_t and H_t are the concentrations of the light isotope and the heavy isotope at time t, respectively.

If studies at the low natural abundance level of the heavy isotopes are carried out (i. e. $H+L\approx L$) or the fractionation is very small (i. e. $1+R_t\approx 1+R_0$), Eq. 11.5 can be approximated by following the classical Rayleigh-type equation originally derived by Lord Rayleigh to describe fractional distillation of mixed liquids:

$$\frac{R_t}{R_0}=f^{\alpha-1} \tag{11.7}$$

After ln transformation and combination with Eq. 11.3, we obtain:

$$\ln\left(\frac{R_t}{R_0}\right)=(\alpha-1)\ln f=\frac{\varepsilon}{1000}\ln f \tag{11.8}$$

which yields:

$$1000\ln\left(\frac{10^{-3}\delta_{r,t}+1}{10^{-3}\delta_{r,0}+1}\right)=\varepsilon\ln f \tag{11.9}$$

where $\delta_{r,0}$ and $\delta_{r,t}$ are the ratios of the heavy isotope to the light isotope in the reactant at time $t=0$ and t, respectively, expressed in the δ notation. For enrichment factors typically obtained during transformations ($|\varepsilon|<20‰$), $\ln(1+10^{-3}\delta_r)\approx 10^{-3}\delta_r$, and Eq. 11.9 is often simplified to:

$$\delta_{r,t}-\delta_{r,0}\approx\varepsilon\ln f \tag{11.10}$$

This equation is the most frequently used in environmental sciences to derive isotope enrichment factors ε or fractionation factors α, which is presented in Chapter 2.

From Eqs. 11.2 and 11.3 it follows that the difference between the reactant and the instantaneous product formed always equals ε_{pr} normalized to δ_r according to:

$$\delta_p-\delta_r=\varepsilon_{pr}\left(1+\frac{\delta_t}{1000}\right) \tag{11.11}$$

It can be easily seen that the extent of fractionation (expressed in δ values) depends on the remaining reactant fraction. Note that this equation is only applicable if several conditions are fulfilled: we have a closed system (i. e. if the reactant pool is limited), only one product is formed in the transformation, and the product does not react further (which would change its isotopic composition). The last two conditions, however, are only relevant foı the isotope ratio and its change in the product which is rarely measured.

11.2 Analytical techniques-Gas chromatography coupled to isotope-ratio mass spectrometry

To date, compound-specific isotope-ratio measurements can, in principle, be carried out for most elements present in organic compounds at or near the natural isotope abundances (Table 11.1). Isotopic analyses of the elements C, H, and N are becoming routine for some typical pollutants (e. g. , fuel components, chlorinated solvents, and some agrochemicals and explosives).

Table 11. 1 Stable isotope systems and natural abundance isotope ratios for typical elements in organic contaminants.

Isotope system	Isotope ratio (%)
$^{2}H/^{1}H$	0. 015 58
$^{13}C/^{12}C$	1. 123 0
$^{15}N/^{14}N$	0. 366 3
$^{18}O/^{16}O$	0. 200 5
$^{34}S/^{32}S$	4. 416 0
$^{37}Cl/^{35}Cl$	31. 96
$^{81}Br/^{79}Br$	97. 27

GC-IRMS is currently the most widely used instrumental setup for CSIA of organic contaminants. These systems usually comprise units for sample pre-concentration and injection, pollutant separation, conversion to analyte gases and isotope-selective detection (Table 11. 2). Rather few compound classes have been made accessible for CSIA for date, due to the great structural diversity of micropollutants and thus the need for developing and calibrating analytical procedures on a compound by-compound basis (Schwarzenbach et al. , 2010). Compared to concentration measurements of pollutants by (high-resolution) mass spectrometry which is routinely done even in the ng/L range, CSIA requires efficient pre-concentration steps such as purge and trap, solid-phase (micro.) extraction, and vacuum extraction to enable isotope-ratio analysis of contaminated soils or groundwater at more than 1000-fold higher concentrations (i. e. 10 ~ 100lg/L) for routine operations (Berg et al. , 2007; Amaral et al. , 2010; Penning and Elsner, 2007; Meyer et al. , 2008). The challenges of separating components of contaminant mixtures are similar to those in standard GC; however, GC/IRMS needs to deal with larger amounts of analytes and matrix effects.

In the GC/IRMS-interface system, organic compounds are reacted to H_2, CO_2, N_2, or CO for measuring H, C, N, or O isotopes, respectively, through optimized combustion, pyrolysis, or combustion coupled to reduction processes, followed by the removal of reaction byproducts such as water or corrosive gases (Figure 11. 1). Quantitative chemical conversion of organic molecules is achieved in narrow bore reactor tubes, usually containing CuO, NiO, and/or Pt as catalysts, which are operated at high temperatures depending mostly on the isotope system investigated (Table 11. 2). Finally, the sample gas isotopologues, for example $^{14}N_2$, ^{14}N, ^{15}N, and $^{15}N_2$ of N_2 for $^{15}N/^{14}N$ ratios, are analyzed in magnetic sector field mass spectrometers, which are specialized for maximizing ionbeam

Table 11.2 Overview of instrumental setups for compound-specific analyses of stable isotope ratios in organic contaminants (GC. Gas chromatography; IRMS. Isorope ratio mass spectrometry; LC. Liquid chromatography; qMS. Quadrupol mass spectrometry; MC-ICP-MS. Multicollector inductively coupled plasma mass spectrometry; CRDS. Cavity ring-down spectroscopy.

Instrumentation	Separation	Interface system	Analyte	Lonization	Mass analysis/ion detection	Isotope ratio
GC/IRMS	GC	Comb[a]	CO_2	EI[b]	Magnetic sector/Faraday cups	$^{13}C/^{12}C$
	GC×GC[c]	Comb/Red[d]	N_2			$^{15}N/^{14}N$
		Pyr[e]	H_2			$^{2}H/^{1}H$
		Pyr	CO			$^{18}O/^{16}O$
LC/IRMS	LC	Wet oxidation	CO_2	EI	Magnetic sector/Faraday cups	$^{13}C/^{12}C$
GC/IRMS[f]	GC	none	fragment ions	EI	Magnetic sector/Faraday cups	$^{37}Cl/^{35}Cl$, $^{81}Br/^{79}Br$
GC/qMS	GC	none	molecular ion & fragment ions	EI	Quadrupole electron multiplier	$^{37}Cl/^{35}Cl$
GC/MC-ICPMS	GC	ICP[g]	Cl	ICP	Magnetic sector/Faraday cups	$^{37}Cl/^{35}Cl$
			Br			$^{81}Br/^{79}Br$
			S			$^{34}S/^{32}S$
GC/CRDS[h]	GC	Comb	CO_2	none	Infrared spectroscopy	$^{13}C/^{12}C$

Comb[a]. combustion at 900～950℃; EI[b]. electron ionization; GC×GC[c]. applications reported exclusively for analysis of C isotopes; Comb/Red[d]. combustions followed by reduction at 600～650℃; Pyr[e]. pyrolysis at 1200～1450℃; GC/IRMS[f]. direct injection GC/IRMS, see text for details; ICP[g]. inductively coupled plasma; GC/CRDS[h]. also denoted as GC/C/CRDS owing to the use of a combustion interface.

currents and stability. The high precision arises from the system of differential measurements of analyte and standard gases with known isotopic composition simultaneously for at least two masses using multiple detectors. The difference in ion-current ratios measured in the detectors is exactly proportional to the difference in isotope ratios, even though the absolute isotopic abundances are poorly constrained (Brand, 2004), thus requiring isotope ratios to be reported relative to reference materials (i. e. in the "delta notation", Equation 11. 1).

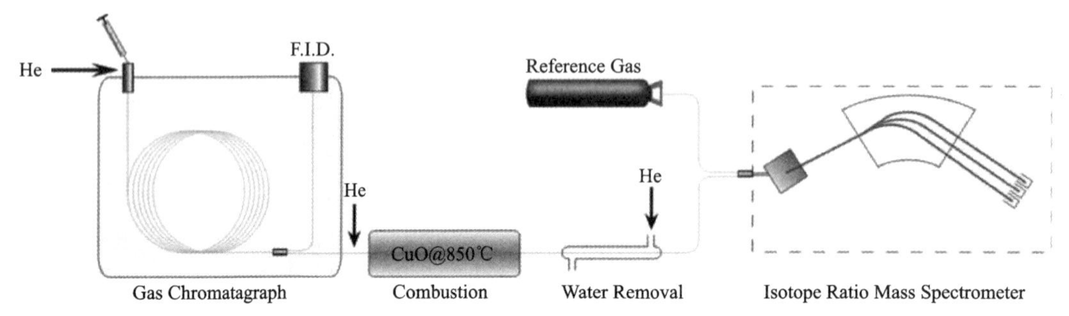

Figure 11. 1 A schematic gas chromatography ecombustione isotope ratio mass spectrometer (GCeCeIRMS) for carbon isotope ratio determination.

Establishing procedures for organic-contaminant CSIA by GC/IRMS comes with a series of tests to ensure accuracy and precision. Given that many steps of the analytical procedure such as sample preparation, analyte enrichment, and conversion can give rise to isotope fractionation, referencing strategies for comparing the isotope signatures of known and unknown compounds on the GC/IRMS are key for accurate isotope-ratio analysis (Werner and Brand, 2001). Such comparisons are based on standard compounds, whose isotope ratios have been measured independently by different techniques (EA-IRMS or DI-IRMS, see above), and, in most cases, they allow one to identify suitable operating conditions for GC/IRMS of organic contaminants. As the attainable precision of an isotope-ratio measurement increases with the amount of isotopologue ions in the mass spectrometer and with increasing abundance of the rare isotope, optimum concentration ranges for analysis need to be identified for every compound separately.

Instrument performance is, in principle, limited at low concentrations by intolerable loss of precision and by the amount of analyte that can be loaded onto the chromatographic column and converted adequately in the interface system. However, acceptable limits of measurement uncertainty lack clear-cut definitions, as they not only include the reproducibility of repeated measurements but also have to account for the rather narrow linear range of continuous-flow IRMS (typically one order of magnitude in contaminant concentration) (Jochmann et al., 2006; Hunkeler et al., 2008; Lollar et al., 2007). Deviations of isotope ratio measurements due to nonlinearity effects arise from the signal sizes of standard and sample being too different. The lack of accuracy from multiple sample measurements carried out over a range of signal sizes thus additionally reduces precision (Lollar et al., 2007). Current experience suggests that total instrumental uncertainties are ±0.5‰ for $\delta^{13}C$ and $\delta^{15}N$ while they are ±0.8‰ and ±5‰ for $\delta^{18}O$ and $\delta^{2}H$, respectively. Depending on the compound and the sample matrix, these limits can be surpassed by variations imposed by sample preparation. Method detection limits for GC/IRMS are linked to these definitions of uncertainty in that they reflect the lowest concentration in a(n) (environmental) sample, for which the measured isotope signature does not deviate by more than the total instrumental uncertainty from the accurate value (Jochmann et al., 2004).

Finally, chromatographic resolution to baseline separation of the analyte is essential for unambiguous quantification of isotope ratios. Even though algorithms are used in standard software solutions that can deconvolute isotopologue signals from partially coeluting peaks, this step can be a source of error. Substantially improved separation can be achieved through the on-line coupling of comprehensive two-dimensional GC (GC × GC) to IRMS. This approach, which is in an early stage of development, follows the general set-up of GC/IRMS systems (Table 11.2) but requires complex instrumental modifications such as microreactors for analyte conversion and accelerated IRMS-signal processing (Tobias et al., 2008, 2010; Tobias et al., 2007).

To date, Liquid Chromatography (LC) is coupled to IRMS exclusively for compound-specific analysis of $^{13}C/^{12}C$ ratios (Godin et al., 2007). In commercialized LC/IRMS interfaces, a wet oxidation of organic compounds to CO_2 is carried out in a heated reactor by peroxodisulfate followed by a quantitative, membrane-based extraction of CO_2 under acidic conditions into a counter flow of helium (Krummen et al., 2004). This approach has enabled CSIA of many additional compound classes, despite constraints regarding mobile-phase composition (buffers, organic modifiers), which compromise the use of reversed-phase LC. Alternative strategies for chromatographic separation include temperature programmed LC, as used in GC, coupled to wet-oxidation IRMS. In a less widespread approach, CSIA of liquid sample can be carried out with moving-wire devices after preparative separation of analytes (Sessions et al., 2005).

Connecting GC to multicollector inductively coupled plasma MS (GC/MC-ICP-MS, Table 11.2) provides more versatile avenue to measuring Cl-isotope ratios, even though it has been shown primarily for PCE and TCE. In this setup, the ICP functions as a conversion and ionization unit to ionize and filter off the carbon skeleton of organic compounds and ionize Cl isotopes for detection in multiple collectors, thus simplifying sample preparation procedures. The high mass resolution of the MC-ICP-MS is essential to separate Cl-isotopologue signals from mass-interferences of the Ar plasma. However, Cl ionization potentials are higher than those of the heavy elements, whose isotope ratios are typically analyzed by MC-ICP-MS and Cl ionization yields are therefore low (Vanhaecke et al., 2009; Weiss et al., 2008). The ensuing low degree of ionization reduces the signal intensity of Cl isotopes and can make the operation of GC/MC-ICP-MS more challenging to obtain accurate and precise results. Together with the need for independently calibrated standard materials and the high costs of instrumentation, these obstacles currently limit more widespread application of GC/MC-ICP-MS for polychlorinated organic contaminants.

Benchtop quadrupole mass spectrometers (qMS) can measure chlorine isotopologues is inferior to that of the multi-collector devices, the GC/qMS setup probably has the greatest potential to propel Cl-isotope analysis. This approach is favored by the large relative abundance of heavy halogen isotopes (Table 11.1), which enables quantification of both isotopes at lower concentrations without specialized mass spectrometers. Isotope ratios are obtained from the abundance of Cl isotopologues measured in the molecular ion and in (dechlorination) fragment ions after electron ionization (Aeppli et al., 2010; Sakaguchi-SÖder et al., 2007). However, to obtain accurate and precise $\delta^{37}Cl$ values, a series of procedural measures have been proposed. Currently, they include extensive bracketing of samples with standards containing the target analyte of known $^{37}Cl/^{35}Cl$ ratios in identical concentrations as well as optimization of peak integration parameters (Aeppli et al., 2010). Thus, while the GC/qMS approach can, in principle, be implemented with standard analytical equipment, it still requires rather large amounts of standard materials that need to be calibrated by conventional isotope ratio mass spectrometers.

11.3 Fractionation processes

In the groundwater system, the fractionation of compound specific isotope is commonly affected by physical processes, abiotic and biotic processes. In this section, taking the Cl isotope in the compound specific isotope as an example is to introduce the effects of several processes on the isotope fractionation.

11.3.1 Chlorine isotope fractionation by physical processes

Physical processes, such as evaporation (volatilization/liquidvapor partioning), diffusion, sorption and dissolution can lead to an isotopic fractionation due to slight differences in the thermodynamic properties of isotopologues. Table 11.3 summarizes the chlorine isotope enrichment factors caused by physical processes that have been obtained through laboratory studies.

Table 11.3 Bulk carbon and chlorine isotope enrichment factorsby physical processes based on laboratory experiments.

Mechanism	Compound	εC, bulk (‰)	εCl, bulk (‰)	Reference
Continuous evaporation of pure compound	DCM	+0.65±0.02	−0.48+0.06	Huang et al. (1999)
Continuous evaporation of pure compound	TCE	+0.31±0.04	−1.82±0.22	Huang et al. (1999)
Continuous evaporation of pure compound	TCE	+0.24±0.06 to +0.35±0.02	−1.64±0.13	Poulson and Drever (1999)
Continuous evaporation of pure compound	TCE	+0.28±0.03	−1.35±0.03	Jeannottat and Hunkeler (2012)
Stepwise evaporation of pure compound	TCE	+0.75±0.04	−0.39±0.03	Jeannottat and Hunkeler (2012)
Diffusion-controlled evaporation of pure compound	TCE	+0.10±0.05	−1.39±0.06	Jeannottat and Hunkeler (2012)
Stepwise evaporation of aqueous phase compound	TCE	+0.38±0.04	−0.06±0.05	Jeannttat and Hunkeler (2012)
Aqueous phase back-diffusion (enrichment in reservoir)	*cis*-1,2-DCE		−0.28 to −1.33	Vakili (2017)
Aqueous phase back-diffusion (enrichment in reservoir)	1,2-DCA	−0.23±0.04	−0.61±0.03	Wanner and Hunkeler (2015)
Aqueous phase sorption (enrichment in solution)	1,2-DCA	−0.40±0.06	−0.55±0.13	Wanner et al. (2017)
Aqueous phase sorption (enrichment in solution)	*cis*-1,2-DCE	<0.1	−0.2	Vakili (2017)
Aqueous phase sorption (enrichment in solution)	TCE	<0.1	−0.85	Vakili (2017)

Liquid-vapor chlorine isotope fractionation in chlorinated VOCs has been studied from an early date in the laboratory. Baertschi et al. (1953) and Bradley (1954) observed that for carbon tetrachloride (CT) and chloroform (CF), the liquid phase became enriched in ^{37}Cl, while $\delta^{13}C$ showed an inverse isotope trend, leading to an enrichment in the vapor phase. These opposite trends were confirmed by Huang et al. (1999) for TCE and dichloromethane (DCM) as well as by Poulson and Drever (1999) and Jeannottat and Hunkeler (2012) for TCE. They are believed to be caused by the differing effects of isotopic substitution on intermolecular forces in the condensed phase. In general, heavy isotope substitution leads to weaker intermolecular forces and thus a lower boiling point due to lower frequencies of internal vibration, however the effect of ^{13}C substitution is considerably larger than that of ^{37}Cl substitution. The weak inverse chlorine isotope effect is thus counteracted by other intermolecular forces in the condensed phase, while it is possible to observe the strong inverse carbon isotope effect. When the chlorinated VOC is dissolved in water, hydrogen bonding and dipole-dipole interactions, which are relevant between the more polar molecules, may become strong enough to completely cancel out the inverse carbon isotope effect (Horst et al., 2016).

Vapor-phase diffusion has been shown to cause a significant chlorine isotope fractionation in laboratory experiments (Jeannottat and Hunkeler, 2012). This arises due to heavier isotopocules generally being less mobile than their lighter counterparts. The difference of two atomic units between stable chlorine isotopes compared to only one atomic unit between stable carbon isotopes explains the pronounced chlorine isotope fractionation.

Isotope fractionation due to aqueous-phase diffusion also occurs in the subsurface. Several models for the diffusion coefficient have been proposed. Wanner and Hunkeler (2019) presented a comprehensive review of five relevant models and existing experimental data that have been obtained for several compounds. In general, models for the mass dependency of a compound's diffusion coefficient reduce to the form:

$$D \propto m^{-\beta} \tag{11.12}$$

where m is the isotopologue molecular mass and β is a number generally between 0 and $\sim 1/2$. This relation is sometimes written as a function of the reduced mass $[\mu_r = m \cdot m_{solvent}/(m + m_{solvent})]$. However, as many theories that assume that the water molecule network acts as a massive effective particle ($m_{solvent} \gg m$) (Bourg and Sposito, 2007; Wanner and Hunkeler, 2019), Eq. 11.12 is generally valid.

Various models such as the well-known Chapman-Enskog relation have been developed for diffusion. However, only some are relevant in the case of aqueous solutions in porous media. Extensions of classical Fickian kinetic theory imply that $\beta = 1/2$ in Eq. 11.12. The Einstein (1905) and Langevin (1908) models of diffusion are based on the random trajectory of a single solute particle (i. e. Brownian motion), but do not integrate mass effects. The

Maxwell Stefan diffusion model, like the Fick model, considers the ensemble behavior of solute particles. In this model, mass is explicitly taken into account with the resulting relationship of $\beta = 1/2$. This theory assumes that solute molecules move according to solute concentration gradients, but are subject to a friction-like force from the solvent molecules. A final theory of note is mode-coupling theory analysis (MCTA) where in $0 < \beta < 1/2$. In MCTA, the diffusion coefficient depends on the interplay between extremely frequent intermolecular collisions, which are strongly mass dependent, and longer-term hydrodynamic modes which exhibit little mass dependence (Wanner and Hunkeler, 2019).

Finally, sorption can also induce isotopic fractionation in the adsorbed and desorped substances. While Vakili (2017) detected only an insignificant carbon isotope effect, the chlorine isotope enrichment in the solution reached −0.2‰ for *cis*-1,2-DCE and −0.85‰ for TCE, which may also be explained by the difference of two atomic units between stable chlorine isotopes compared to only one atomic unit between stable carbon isotopes. The preferential sorption of the light isotopologues leads to an enrichment of ^{37}Cl in the solution. Similarly to evaporation, this may be explained by the weaker intermolecular forces between heavy isotopologues and the sorbent compared to light isotopologues. Sorption is of concern primarily in organic-rich sediments as well as low permeability sediments with small pore sizes, such as shales and clays, due to their high specific surface area. It is also relevant in zero-valent iron systems which are of increasing interest in remediation efforts (Crane and Scott, 2012).

Physical processes are often not well constrained to a single process. In laboratory experiments, Jeannottat and Hunkeler (2012) found a cumulative normal chlorine isotope effect by both liquid vapor partioning and vapor phase diffusion. Vakili (2017) investigated chlorine isotope effects caused by back-diffusion from a low permeability zone as well as sorption on the laboratory scale, although it is unclear how the set-up discriminated between diffusion and sorption processes. Wanner et al. (2017) attempted to isolate the isotopic effects of these processes by including field data into a numerical model. The theoretical sorption-induced isotope shift was calculated to reach up to 2‰.

Chlorine isotope effects caused by physical processes have not been widely studied on the field scale. While gas-phase diffusion may cause a significant chlorine isotope effect, its relevance in vadose zone transport has not been established (Jeannottat and Hunkeler, 2012; Hunkeler et al., 2011b). The isotope effects of physical processes are believed to have a minor influence compared to the isotope effects observed during (bio-) chemical degradation processes and are less pronounced once steady-state conditions are reached (Schmidt and Jochmann, 2012).

11.3.2 Chlorine isotope fractionation during biotic processes

Biotic degradation of chlorinated contaminants in the subsurface is a process that causes isotope fractionation due to bond cleavage. Chemical bonds involving the heavier isotope are slightly stronger than those involving the light isotope, thus the lighter isotopologue of a molecule degrades preferentially resulting in a heavy isotope enrichment in the residual compound. Figure 11.2 gives an overview of potential degradation pathways for the example of PCE. One advantage of compound-specific isotope analysis is that degradation can be proven even if there are no characteristic degradation products. If they however do occur, it is advisable to monitor them, also their isotope ratios, to provide confirmation about the degradation path.

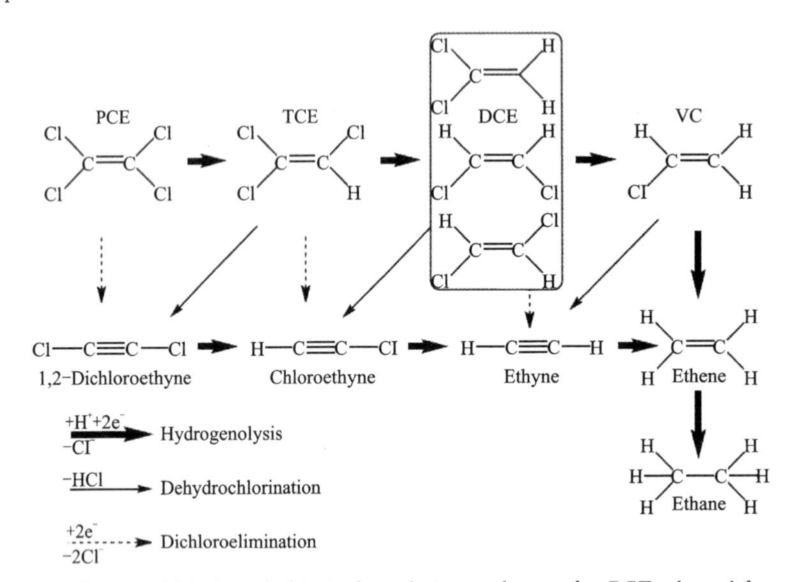

Figure 11.2 Potential biotic and abiotic degradation pathways for PCE adapted from Brown et al. (2006). Biotic degradation typically takes the hydrogenolysis pathway from PCE to VC.

The biotic degradation pathway is a function of the redox potential in the soil and the availability of electron donors and adapted bacteria. Chlorinated hydrocarbons either serve as electron acceptors during anaerobic organohalide respiration or else as electron donors during anaerobic as well as aerobic metabolic degradation. If degradation of xenobiotics is not the primary metabolic process, it is referred to as co-metabolism and attributed to the activity of non-specific enzymes. Anaerobic dechlorination is usually the predominant process in biotic degradation on the field scale, as aerobic dechlorination is limited by the lack of electron donor, specifically molecular oxygen (Numata et al., 2002a). The aerobic process does however become more relevant with decreasing chlorination, e.g. during the sequential degradation of PCE, where a purely anerobic process may stall at the highly toxic degradation product VC (Tiehm and Schmidt, 2011). Aerobes on the other hand have a higher potential to further degrade VC to non-chlorinated, non-toxic compounds through epoxide formation.

It is noticeable that the chlorine isotope fractionation is almost always smaller in magnitude than the carbon isotope fractionation, as opposed to the trends observed for physical processes. Furthermore, different reactive processes lead to remarkably different carbon and chlorine isotope enrichment factors for the same compound. As commonly observed, the exact reaction mechanism dictates whether a large primary or lower secondary chlorine isotope is effect observed or whether dilution and masking have an influence on the magnitude of fractionation.

During aerobic oxidation of DCM, a considerable carbon and chlorine isotope effect was observed by Heraty et al. (1999). The first and rate-limiting step was proposed to involve breaking of a C-Cl bond by nucleophilic substitution (SN_2), leading to CH_2O (formaldehyde) and HCl via the unstable intermediate CH_2ClOH. VC and 1,2-DCA on the other hand undergo a different reaction mechanism during aerobic oxidation, which manifests itself in the smaller enrichment factors (Abe et al., 2009b). Here, the first step does not involve C-Cl bond cleavage but C-C bond cleavage and the formation of an epoxide between the two carbon atoms.

Two distinct aerobic degradation pathways for 1,2-DCA can be distinguished using dual carbon-chlorine isotope plots (Palau et al., 2014a). Only a small carbon isotope effect is measured with C-H bond cleavage as the initial step, which is the case for the reaction with the monooxygenase enzyme used by the *Pseudomonas* sp. strain. The dehalogenase enzyme that is employed by *Xanthobacter autotrophicus* and *Ancylobacter aquaticus* causes a large carbon isotope effect, as here the initial step is C-Cl bond cleavage via nucleophilic SN_2 substitution. For both pathways, Palau et al. (2014a) observed a similar chlorine isotope effect $\varepsilon_{Cl,bulk}$. During anerobic degradation by *Dehalococcoides* and *Dehalogenimonas*, 1,2-DCA exhibits a significant chlorine isotope effect (Palau et al., 2017). The reason is thought to be a dichloroelimination pathway, where not one but two C-Cl bonds are cleaved. This dichloroelimination pathway also appears during anaerobic degradation of 1,1,2-TCA to VC (Rosell et al., 2019). Compared to 1,2-DCA, the chlorine isotope effect is not quite as pronounced due to masking effects caused by differences in the enzyme kinetics.

The anaerobic reductive dechlorination of chlorinated ethenes can either proceed via single-electron transfer or nucleophilic addition as initial step (Aelion, 2010). The former would lead to primary carbon and chlorine isotope effects due to C-Cl bond cleavage, while the latter would attack the C=C double bond resulting in a large primary carbon isotope effect and secondary chlorine isotope effects (Cretnik et al., 2014). For TCE, Kuder et al. (2013) applied triple element hydrogen-carbon-chlorine isotope analysis and observed secondary chlorine isotope effects consistent with the nucleophilic addition pathway.

Masking effects may also have a major impact on the observed isotope fractionation during reductive dechlorination. For PCE, Renpenning et al. (2014) obtained significantly different dual carbon-chlorine isotope slopes for the culture *Sulfurospirillum multivorans* ($\varepsilon_C/\varepsilon_{Cl}$ ranging from 2.2 to 2.8) versus the culture's enzyme corrinoid cofactor ($\varepsilon_C/\varepsilon_{Cl}$ ranging from 4.6 to 7.0), which was attributed to the preceding rate-limiting steps. Interestingly, this masking effect was not observed during reductive dechlorination of TCE.

While considerable work has been done in measuring chlorine isotope fractionation during biotic degradation, the masking effects that are inherent to biotic processes as well as the ambiguity of some dual carbon-chlorine isotope slopes can complicate the attribution of specific mechanisms to observed isotope fractionations. Improvements in analytical methods can assist in lowering the uncertainty of chlorine CSIA data, especially for TCE.

11.3.3 Chlorine isotope fractionation during abiotic processes

Abiotic degradation of chlorinated contaminants in soil is usually slower than biotic degradation, but has the advantage that it is likely to result in completely dechlorinated, generally non-toxic compounds. There are many different constituents of natural soil that can be relevant to abiotic degradation during natural remediation, namely iron-bearing sulfide, oxide, hydroxide and clay minerals (He et al., 2015). Additionally, engineered remediation at polluted sites often involves reactive barriers that employ zero-valent metals or compounds that contain metals at their reduced state.

The isotope effects associated with abiotic degradation in chlorinated hydrocarbons are often very pronounced for carbon isotopes and for chlorine isotopes. Audí-Miro et al. (2013) studied the chlorine isotope effect during degradation of chlorinated ethenes in an engineered reactive barrier that used zero-valent iron (ZVI). The process is based on an electron transfer where the chlorinated ethenes are dechlorinated reductively, while the ZVI is oxidized. Abiotic TCE degradation yielded *cis*-1,2-DCE via a hydrolysis pathway as well as ethene and ethane via dichloroelimination (Figure 11.2). This additional pathway is often observed during abiotic degradation and involves the simultaneous cleavage of two chlorine substituents, explaining the large chlorine isotope effect. Torrentó et al. (2017) explored different abiotic degradation mechanisms for chloroform (CF). The oxidization reaction was proposed to involve C-H bond cleavage as the initial step, hence only a small secondary chlorine isotope effect was observed. Rodríguez-Fernandez et al. (2018) investigated the suitability of different iron minerals in engineered remediation, and observed a pronounced pH dependence of the chlorine isotope enrichment factors, which was suspected to be caused by surface passivation of the iron minerals at alkaline pH.

Most chlorine isotope enrichment data for abiotic processes have so far been obtained with a focus on zero-valent iron in engineered remediation. The question of which pathways are taken during reaction of chlorinated contaminants with natural constituents of soils could be investigated using dual carbon-chlorine isotope analysis.

11.4 Application

11.4.1 Source apportionment

The molecular isotopic signature of environmental contaminants can often be used to trace their sources on local to global scales. On a local scale it is often necessary to allocate a contamination to a specific source in order to allow appropriate means of risk reduction and/or to identify responsible parties in litigation. In particular, work in the latter area has been termed "environmental forensics" in the US (Morrison, 2000). Traditional approaches in environmental forensics use chemical fingerprinting, biomarker analysis, and chemometrics. However, the potential of isotopic signatures of single contaminants in this area has only recently been explored. Frequently, it is possible to allocate sources of a chemical or to trace the time of contaminant releases because isotopic signatures of chemicals show differences between manufacturers depending on the conditions and the pathways used to synthesize the compound.

Chemical fingerprinting of the n-alkane fraction in crude oils and refined products in combination with isotopic characterization of carbon in the individual homologues has been successfully used to allocate sources of sediment contamination (Rogers and Savard, 1999) and bird feather oiling (Mazeas and Budzinski, 2002). Pond et al. (2002) suggests the preferred use of hydrogen isotopic composition of longer chain alkanes (n-C19 to n-C27) for source identification because the isotopic signature of hydrogen in crude oil components varies much more compared with carbon and is hardly changed during weathering and degradation of crude oil. However, no application to source allocation based on hydrogen isotopic data has been reported so far.

On a regional scale, source apportionment of polycyclic aromatic hydrocarbons (PAHs) both in the atmosphere and in sediment records has been studied intensely utilizing $\delta^{13}C$ analysis. Interestingly, $\delta^{2}H$ analysis of individual PAHs has not been reported to date. With a combination of concentration measurements and $\delta^{13}C$ isotopic analysis of individual PAHs in sediments from Lake Erie, it was possible to distinguish three areas of different contamination history. Furthermore, it could be shown that the main emission pathway for PAHs was fluvial input. Many studies show the necessity to combine chemical fingerprinting techniques and compound-specific isotope analysis. Often, neither CSIA nor fingerprinting alone are conclusive for source apportionment but the information gain from isotopic analysis will certainly make CSIA indispensable in future source allocation investigations. Furthermore, several studies have shown that by determining the isotopic composition of two or more elements [e. g. by combining $\delta^{13}C$ and $\delta^{37}Cl$ (Drenzek et al., 2002; Shouakar-Stash et al., 2003) or $\delta^{13}C$ and $\delta^{2}H$ (Shouakar-Stash et al., 2003; Gray et al., 2002)], a much better differentiation can be obtained.

11.4.2 Identification and quantification of biodegradation processes

In the context of natural attenuation, it is essential to estimate the different sinks for organic contaminants such as dilution, sorption, or biodegradation, because the last of these is the only process of contaminant destruction. Degradation and in particular biodegradation is frequently accompanied by a substantial kinetic isotope effect, whereas many other environmental processes such as dispersion, sorption, or volatilization are not, or only to a much lower extent, subject to isotope fractionation. In such cases, stable isotope analysis provides a complementary opportunity to identify degradation processes in situ.

CSIA has been applied successfully as a means to investigate biodegradation in many studies. Most of the studies report substantial isotope fractionation during microbial degradation of the investigated compounds. In such biodegradation studies, only the isotope ratio of the residual substrate is analyzed and the first enzyme reaction in the degradation pathway has been identified as the fractionating step. Other processes such as uptake of the substrate or diffusion through the aqueous phase to the organisms did not significantly influence the isotope fractionation. Another important point is that the first enzyme reaction in the degradation of both methylated aromatic hydrocarbons and chlorinated hydrocarbons is an "irreversible" step that does not allow chemical or isotope equilibrium of the substrate with products produced in subsequent enzyme reactions in the pathway. Thus, even if a subsequent reaction would be rate limiting and produce a pronounced isotope fractionation it would not affect the isotope ratio of the substrate.

In a few studies microbial degradation without isotopic degradation was reported. Morasch et al. (2002) have shown that the extent of fractionation may depend on the enzyme mechanism. In this case, degradation of aromatic hydrocarbons by a ring dioxygenase did not yield a significant fractionation in carbon. Several recent studies utilizing both carbon and hydrogen isotope measurements concluded that the frequently much stronger isotope fractionation in hydrogen is a more powerful tool to provide evidence of biodegradation at contaminated field sites, in particular for small extents of biodegradation (Gray et al., 2002; Morasch et al., 2002; Hunkeler et al., 2001). However, one should be aware that isotope fractionation for a specific element depends very much on the reaction mechanism in the rate-limiting step of the biochemical reaction. As was found for source apportionment studies, the combined use of hydrogen and carbon isotope analysis might improve the assessment of biodegradation.

Isotope fractionation for anaerobic toluene degradation was determined in batch experiments with various terminal electron acceptors. Remarkably constant α values were found under different anaerobic redox conditions with various pure cultures. The isotope fractionation factor α obtained from the sulfate-reducing bacterial culture was later on used to predict toluene degradation in a more complex environment, that is, anoxic column experiments with sediments from a contaminated site and sulfate as electron acceptor (Richnow et al. , 2003). A calculation of the biodegraded toluene fraction was performed with the measured isotope ratios, the initial toluene concentration at the inlet of the column, and the α value from the batch experiments (Meckenstock et al. ,1999). It perfectly matched the measured concentration profile along the column (Meckenstock et al. , 2002). This finding showed that isotope fractionation can be used to quantify biodegradation not only in batch experiments which are closed systems but also in sediment columns.

Under the assumption that a contamination plume in an aquifer behaves similar to such sediment column systems, the quantitative isotope fractionation concept has been successfully applied in a number of field cases so far. One of the first was a tar oil-contaminated site where an almost 1000-m-long hydrocarbon plume was located in an anoxic aquifer (Richnow et al. , 2003). By using carbon isotope fractionation factors for anaerobic degradation of toluene and o-xylene, the observed isotope ratios and the initial substrate concentration at the source, the expected residual substrate concentration C_t along the plume was calculated. The calculated concentrations could describe the steep concentration gradients along the monitoring transect (Richnow et al. , 2003). Similar results were obtained for sites contaminated with benzene, toluene, ethylbenzene and xylenes (BTEX) from a tanker truck accident (Meckenstock et al. , 2002), and in a landfill leachate plume (Richnow et al. , 2003). The calculation of the residual substrate concentration along such monitoring transects relies on R_0, the isotope ratio of the substrate in the source. This parent value (usually the most negative) may also be replaced by the isotope ratio of the substrate in the most upstream monitoring well near the source if the source itself is not available without changing the result of the calculation.

Investigations of larger areas at several contaminated sites revealed that the spatial distribution of the extent of biodegradation can also be described with the help of isotope fractionation data (Richnow et al. , 2003). At these sites, the monitoring wells were not located along a single groundwater flow path. In the case of a former gas work plant, the calculated residual toluene and xylene concentrations perfectly matched the measured contaminant concentrations (Griebler et al. , 2003). However, it is important to note that the determination of the spatial distribution of biodegradation based on isotope data is only feasible if the area of interest is contaminated by only one source with a defined source isotopic composition of the contaminant, R_0. If multiple plumes from different sources commingle at a site, this approach is not possible. Especially in such cases, the isotope data have to be tested by a plot of $\ln(R_t/R_0)$ versus lnf according to Eq. 11. 8. Data points that do not lay on a straight line in this plot might belong to plumes of other sources and may not be taken for a quantitative calculation of biodegradation. Such wells on the contaminated gas work site mentioned above were identified based on the isotope fractionation data. Geochemical parameters such as chloride concentrations were analyzed in parallel and confirmed that the groundwater from the spotted wells was not hydrologically connected to the contaminant source area. Although stable isotope fractionation has obviously a great potential for the assessment of biodegradation, this emphasizes the necessity to put isotope data into the context of geochemical parameters.

A thorough error propagation revealed that the total error of the calculated residual substrate concentration C_t is mainly dependent on the input value of the initial substrate concentration C_0 (Griebler et al., 2003). Input errors of other parameters in the Rayleigh equation are only of minor relevance to the final result. This clearly shows that the substrate concentrations in monitoring wells of the source area have to be accurately determined keeping in mind that only the aqueous concentration of the contaminant of interest is relevant here.

However, even if no concentration data are available, a semiquantitative description of the microbial degradation activity is possible which was termed percentage of biodegradation B. The percentage of biodegradation B is calculated with Eq. 11. 13.

$$B = 100\left(1 - \frac{C_t}{C_0}\right) = \left[1 - \left(\frac{R_t}{R_0}\right)^{\frac{1}{\alpha - 1}}\right]100 \tag{11.13}$$

Thus, B is dimensionless and relies only on the measured isotoperatios (R_0 and R_t) in the monitoring wells and the isotope fractionation factor α. The extent of biodegradation B depicts the extent to which the substrate has been degraded independently from the initial concentration. The contaminant concentration might have been reduced in addition by other sinks such as adsorption or dilution but the remaining substrate in the sample was then degraded to a certain percentage B. The extent of biodegradation B is a useful value to indicate different levels of biodegradation and can easily be converted into a quantitative value if the source concentration for the respective sample is accessible.

11. 4. 3 Case study: Demonstrating a natural origin of chloroform in groundwater using stable carbon isotopes

For groundwater management and remediation, it is important to know to what extent chloroform in groundwater originates from natural sources. In Denmark, a maximum of 1μg/L of volatile halogenated hydrocarbons including chloroform is allowed for groundwater to be used for drinking water, although up to 25μg/L is permitted if chlorination is needed for disinfection. One major water work already had to seek for a dispensation from the strict rules and others may follow as water works tend to place new water wells in forested areas rather than in farmland in order to avoid problems with nitrate and pesticides. In this study, the carbon isotope composition of natural and anthropogenic chloroform was characterized. Natural chloroform present in groundwater can be distinguished from anthropogenic chloroform based on its carbon isotope signature.

11.4.3.1 Field site

Viborg Site. Viborg is located in Jutland, Denmark. The study area is situated in a coniferous plantation that has been in place since mid-1800. The study area is underlain by sandy glacial outwash deposits acting as an aquifer with a water table around 5m below land surface. The soil type is an early podsolic profile. Groundwater flows toward the waterworks of Viborg located 2km down gradient of the study site. At the Viborg waterworks, chloroform was detected since 1995 at a concentration around 1.5μg/L. In this study, groundwater samples were taken in four multilevel groundwater monitoring wells where chloroform concentration above 0.1μg/L has been detected in previous studies. The groundwater wells consist of 63mm diameter HDPE tubes with 1～2m long screens. The soil gas samplers are 4cm long brass screens situated in 5～7m depths from 0.5m bgs to just below the water table. The soil gas screens are situated few meters from the well No. 66.1797, Figure 11.3.

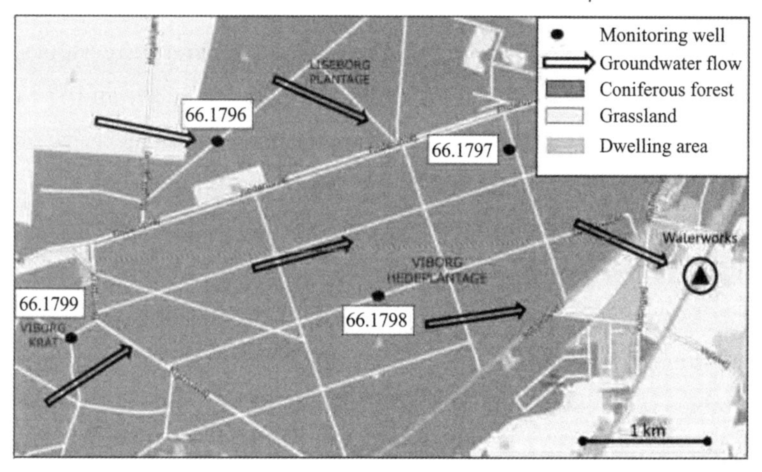

Figure 11.3 Map of Viborg study site with groundwater flow arrows illustrating the location of the coniferous plantation, the multilevel groundwater wells, the soil gas sampler, and the municipal waterworks.

Tisvilde Site. Tisvilde is located at approximately 60km in north of Copenhagen in Zealand, Denmark. The study area is a forest far from industrial activity. This area was forested in the 19th century to prevent soil drift. The coniferous forest consists mainly of

Scots Pine (Pinus sylvestris) and Norway spruce (Picea abies). The topsoil is constituted of an organic horizon mainly composed of partly degraded needles and branches. The aquifer is constituted of sandy glacial outwash deposits with a water level at around 5m below surface. The monitoring well has a screen in 11～14m bgs.

Vellev Site. Vellev is located 27km east of Viborg in an agricultural area with clayey soils and underlain sand gravel aquifer. The investigated well which has a screen between 31.5m and 37.5m bgs and is situated adjacent to an industrial landfill for refrigerator waste. The groundwater contains beside chloroform(～10μg/L) high amounts of tetra chloromethane, trichloroethylene and other solvents. Thisted City Site. The well is situated in the central part of Thisted City in a minor park and is a part of the public waterworks network. The well intake is between 20m and 35m below surface and the water has a CFC-model-age of 5 years. The location is adjacent to several former small industries as a car repair and a dry-cleaning shop.

Tved Site. Tved Plantation is situated 5km from the coast 9km north of Thisted and consists mainly of coniferous forest of spruce and pine planted 110 years ago. The top soil is aeolian sand which is widespread with shifting sand deposits caused by the drifting sand. The sand is underlain by glacial weathered until and the aquifer top about 16m bgs coincides with the Prequaternary surface of Cretaceous white chalk. Two abstraction wells are equipped with screens from 12m to 18m and 24m to 40m bgs, respectively.

11.4.3.2 Results and discussion

The isotopic signature of industrial chloroform was characterized by analyzing chloroform from several suppliers and taking into account literature values (Table 11.4). The $\delta^{13}C$ of industrial chloroform was comprised between −43.2‰ and −63.6‰ ($n=1\sim6$; $1\sigma=0.08‰\sim0.39‰$) consistent with its production from methane (Zwank et al., 2003). The similarities in $\delta^{13}C$ values between some chloroform suppliers suggest that the samples may come from the same carbon feedstock and similar production processes. It is also possible that chloroform from different suppliers is produced by the same manufacturer. The isotopic signature of natural chloroform produced in forest soils was constrained based on soil-air samples from a field site in Viborg, Denmark. At this site, a natural origin of chloroform in soils is well established based on spatial and temporal concentration patterns (Albers et al., 2010a, 2010b). Chloroform concentrations and isotope ratios were measured in a soil-gas multilevel sampler reaching to 4m depth located adjacent to well No. 66.1797, where the highest chloroform concentrations occurred (Figure 11.2). Maximum chloroform concentrations were detected during summer and autumn at 0.5m depth (Figure 11.3) consistent with a biological production of chloroform in the organic-rich soil zone during the warmer periods of the year. The $\delta^{13}C$ of chloroform was within a narrow range from

−22.8‰ to −26.2‰ ($n=2$; $1\sigma=1.08‰\sim1.97‰$; Figure 11.3), which is close to the typical isotope signature of soil organic carbon (Figure 11.4) and strongly enriched in ^{13}C compared to industrial chloroform (Figure 11.3). The somewhat higher standard deviation of these results is a consequence of a change in the P&T-method to use compressed soil air. However, these results confirm that the soil organic matter constitute the main carbon source of natural chloroform.

Table 11.4 ^{13}C Isotope Signature of Different Industrially Produced Chloroform.

$\delta^{13}C$ (‰, V-PDB)	$\pm 1\sigma$	n	Supplier	Purity(%)	References
−43.2		1			Holt et al., 1997
−43.3		1			Holt et al., 1997
−47.1	0.20	5	Fluka	99.5	this study
−47.9	0.08	5	Alfa Aesar	99.0	this study
−48.6	0.30	5	Acros Organics	99.8	this study
−51.5	0.22	3	Fisher Scientifc	>99	this study
−51.7	0.39	2			Jendrzejewski et al., 2001
−63.4	0.22	6	Sigma-Aldrich	99.8	Hunkeler and Aravena, 2000
−63.6	0.14	4	Sigma-Aldrich	99.8	Hunkeler and Aravena, 2000

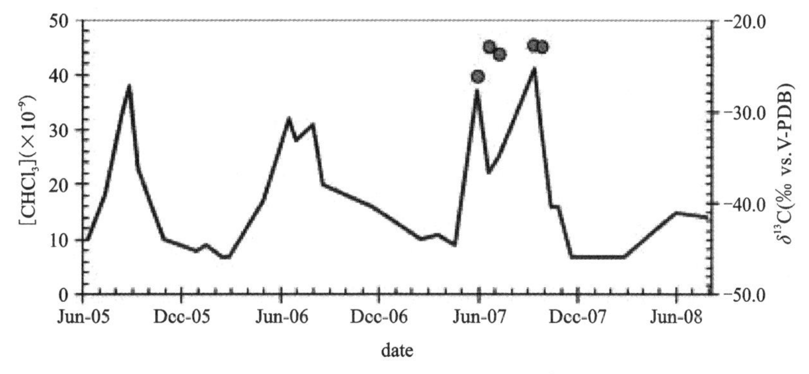

Figure 11.4 Concentration and carbon isotope signature of the natural chloroform source of Viborg. The black line corresponds to the evolution of the chloroform concentration between June 2005 and August 2008 (modified from Albers et al., 2010). The circles represent the isotope $\delta^{13}C$ ratios of chloroform.

To demonstrate that carbon isotope analysis can be used to determine the origin of chloroform in groundwater, the $\delta^{13}C$ of chloroform at five sites was determined using a new method that permits to measure $\delta^{13}C$ at low concentration levels. Groundwater was collected at the Viborg site where chloroform is known to be produced naturally in the soil organic horizon and down gradient of an industrial landfill at Vellev, Denmark, where chloroform is likely of anthropogenic origin. In addition, groundwater was also sampled at the Tisvilde and Tved sites with no identified anthropogenic chloroform sources.

At Viborg, groundwater samples were taken in four multilevel monitoring wells in which chloroform was previously detected (Figure 11.3). The highest chloroform concentrations were detected in No. 66.1797 ranging from 0.9μg/L to 4.1μg/L ($n=24$) (Figure 11.5b). The CFC groundwater-age reached 35 years in the deepest sampling point (Figure 11.5c). The $\delta^{13}C$ of chloroform at the water table (−22.0‰) corresponded well to the $\delta^{13}C$ of soil gas chloroform (−22.8‰ and −26.2‰) demonstrating that chloroform maintains its characteristic isotope signature during transport through the unsaturated zone. The $\delta^{13}C$ values of chloroform in deeper zones varied between −16.7‰ and −26.8‰ ($n=5$) (Figure 11.5a). The range in $\delta^{13}C$ chloroform of uppermost groundwater was found within −17‰ to −29‰ and the nearly horizontal flowing groundwater from five screens reflect the different infiltration areas of chloroform formation in the forest. At the Tved site, the chloroform concentrations in groundwater between 12m and 18m bgs. were 1.54～1.86μg/L ($n=9$), the $\delta^{13}C$ between −24.3‰ and −25.2‰($n=10$), and CFC-groundwater ages approximately 20 years. At the Tisvilde site, the $\delta^{13}C$ of chloroform was −20.1‰ (Figure 11.6).

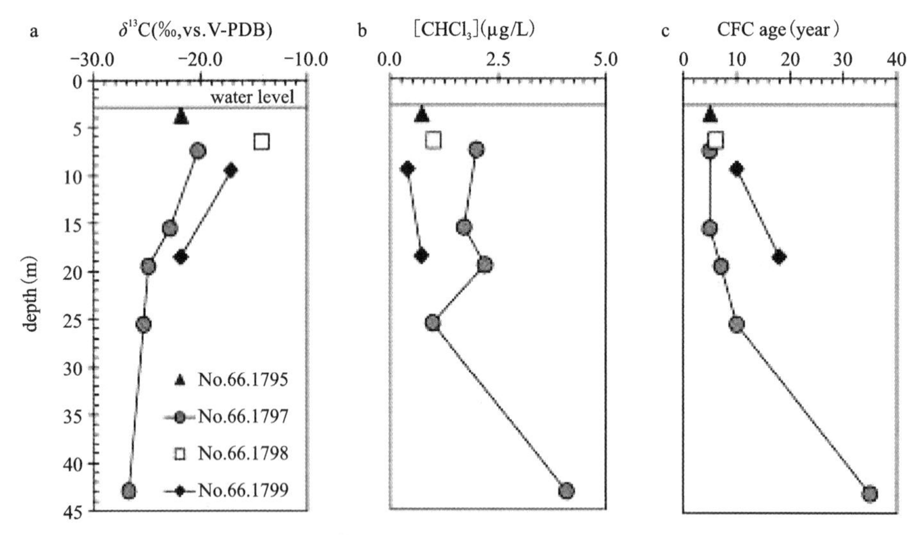

Figure 11.5 Profile of the $\delta^{13}C$ values, concentration and CFC age of groundwater at different depths at the Viborg plantation.

At the Vellev site, the chloroform concentration in groundwater was 10.7μg/L, which is 2 to 5 times higher than the concentration measured at the other sites in forested areas. The $\delta^{13}C$ of chloroform at the Vellev site amounted to −42.1‰. At the Thisted site the chloroform concentration was minor than the Vellev site, <0.2μg/L and the chloroform $\delta^{13}C$ was −47.0‰.

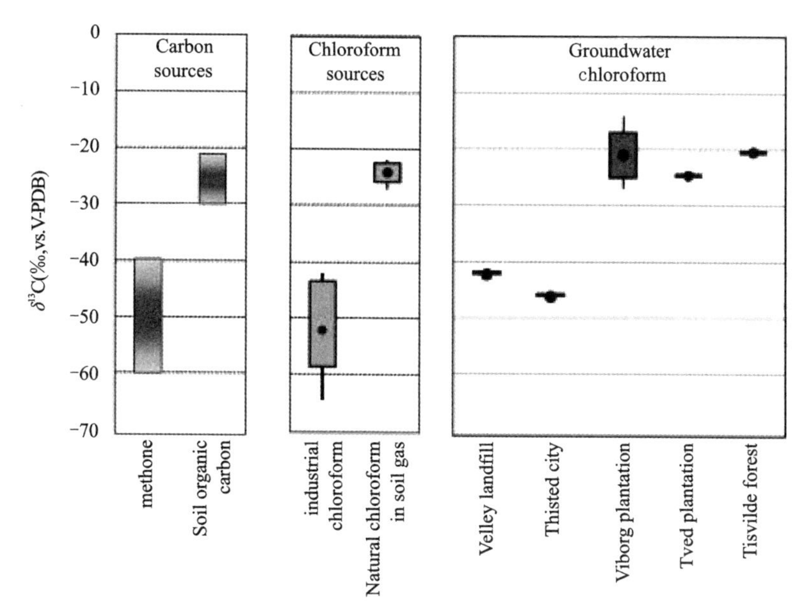

Figure 11. 6 Carbon isotopic signature of carbon and chloroform sources and groundwater chloroform. The gray bars represent the standard deviation around the mean $\delta^{13}C$ value (black dots) and the vertical black lines correspond to the minimum and the maximum $\delta^{13}C$ values.

At three of the study sites (Viborg, Tved and Tisvilde), the $\delta^{13}C$ of groundwater chloroform was close to the values of Viborg soil-gas chloroform indicating a natural origin of chloroform (Figure 11. 6a and 11. 6c). This conclusion is plausible as the three sampling sites were located within spruce and pine plantations, which are associated with soils that are favorable for chloroform production. In contrast, chloroform from the Vellev site had clearly an anthropogenic signature (Figure 11. 6) consistent with is likely origin from an industrial landfill and a concentration higher than at the other three sites. Chloroform was detected in groundwater samples as old as 35 years demonstrating that chloroform can persist over extended periods in oxic groundwater. The persistence of chloroform in groundwater and its high mobility makes it particularly difficult to relate chloroform in groundwater to its source based on concentration data alone as chloroform might be transported over several kilometers. However, the Viborg study indicates that the origin of chloroform can still be determined based on isotope data even if chloroform was transported over an extensive distance. The $\delta^{13}C$ of chloroform at the Viborg water works down gradient of the plantation with an average CFC-groundwater age of 30 years had a $\delta^{13}C$ of $-24.2‰$, which is still within the range of the $\delta^{13}C$ in soil gas at the Viborg plantation.

In summary, the study demonstrates that natural and anthropogenic chloroform have a distinctly different carbon isotope signature that can be related to the carbon source from which chloroform originates, soil organic matter and methane, respectively. The strong difference in $\delta^{13}C$ makes it possible to clearly identify the origin of chloroform in groundwater even if some changes of the isotope ratios occur during transport. The study demonstrates that chloroform can be naturally present in groundwater at the low microgram level and persist over decades. The isotope method opens new possibilities for a comprehensive assessment of the natural occurrence of chloroform in groundwater. It also helps to take appropriate measures when detecting chloroform in groundwater. We have in this study demonstrated that it is possible to distinguish between the naturally formed chloroform and industrial produced chloroform using isotopic signature. This has led to the Danish EPA to change the limit of chloroform in groundwater to be used for drinking water from 1μg/L to 10μg/L when the origin of chloroform is a natural source.

11.4.4 Case study: Use of dual carbon-chlorine isotope analysis to assess the degradation pathways of 1,1,1-trichloroethane in groundwater

As mentioned above, compound-specific isotope analysis (CSIA) is an innovative tool to investigate degradation pathways of organic contaminants because the extent of isotope fractionation (ε_{bulk}) during compound transformation is highly reaction-specific (Vanstone et al., 2008). During the course of a reaction, combined changes in isotope ratios (e.g. $\Delta\delta^{13}C$ vs. $\Delta\delta^{37}Cl$) for a given reactant generally yield a linear trend in a dual element isotope plot (Abe et al., 2009; Cretnik et al., 2013; Palau et al., 2014a). The dual element isotope slope ($\Lambda=\Delta\delta^{13}C/\Delta\delta^{37}Cl\approx\varepsilon C_{bulk}/\varepsilon Cl_{bulk}$) reflects isotope effects of both elements and, thus, different slopes may be expected for distinct transformation mechanisms involving different bonds with distinct elements. Following this approach, dual isotope slopes observed in the field can be compared to the slopes determined in laboratory experiments to identify degradation pathways. A significant advantage of the dual isotope approach is that the Λ value often remains constant, regardless of the occurrence of transport and retardation processes (Thullner et al., 2013).

In this study, dual C-Cl isotope analysis of 1,1,1-TCA in groundwater samples was performed with the purpose of elucidating the fate of 1,1,1-TCA in a contaminated aquifer. In order to evaluate the potential of the multi-isotope analysis and the dual C-Cl isotope slopes to identify degradation pathways of 1,1,1-TCA in the field, the isotope ratios of 1,1,1-TCA ($\delta^{13}C$ and $\delta^{37}Cl$) and 1,1-DCE ($\delta^{13}C$) from field samples, in conjunction with concentration data, were compared to the isotope patterns determined from a previous laboratory experiment of 1,1,1-TCA transformation by DH/HY (Palau et al., 2014b). In addition, the isotopic composition of TCE ($\delta^{13}C$ and $\delta^{37}Cl$) detected in the aquifer was also determined to assess its transformation.

11.4.4.1 Field site

The dual C-Cl isotope approach was evaluated at a site where the subsurface is impacted by a mixture of CAHs. The origin of the contamination was related to an industrial plant where 1,1,1-TCA and TCE were used as solvents for cleaning and degreasing metal parts since the 1960's. In the late 1980's, an environmental survey at the site revealed important subsurface contamination in the north-eastern part of the plant, where the waste disposal and the delivery zones were located (Figure 11.7).

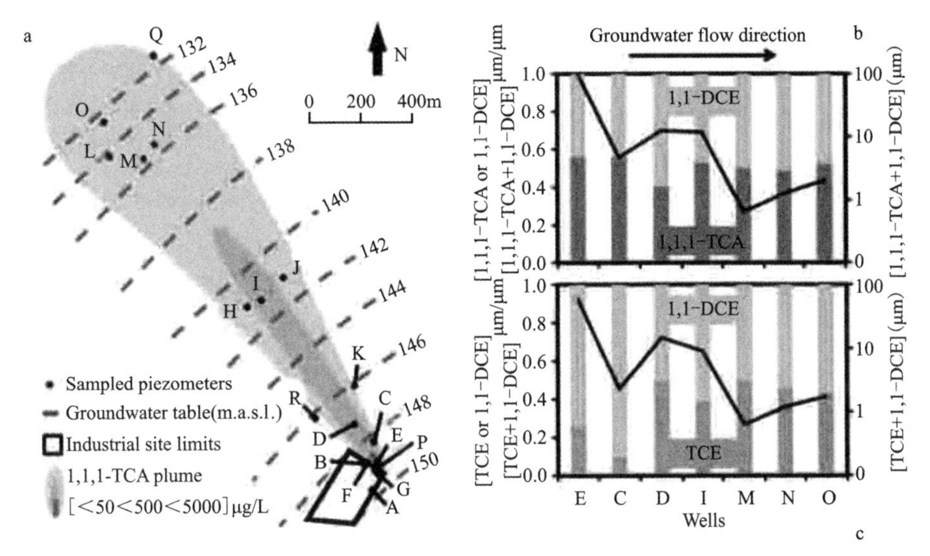

Figure 11.7 a. Site map and groundwater monitoring wells network. Dashed lines correspond to the groundwater surface (in m above sea level) and contour lines depict the 1, 1, 1-TCA concentrations in the aquifer. b. Total concentration (i.e. 1,1,1-TCA+1,1-DCE) (right *y*-axis, line) and concentration of 1,1,1-TCA and 1,1-DCE normalized by the total concentration (left *y*-axis, bars). c. Total concentration (i.e. TCE+1,1-DCE) (right *y*-axis, line) and concentration of TCE and 1,1-DCE normalized by the total concentration (left *y*-axis, bars).

The lithology at the site consists of, from top to bottom, Quaternary loess deposits (from 5m to 18m thick), a layer of flint conglomerate resulting from chalk alteration and dissolution (from 4m to 8m thick), Senonian chalks forming the fractured bedrock aquifer (thickness of ~30m) and Campanian smectite clay corresponding to the low permeability basis of the aquifer. The chalk unit can be considered as a dual porosity aquifer composed of high matrix porosity (up to 45%) and much lower fracture porosity (on the order of 1%~5%). Despite the relatively low fracture porosity, its contribution to the hydraulic conductivity is predominant. In the studied area, the chalk aquifer is unconfined and the groundwater table is found between 16.9m and 28.6m below ground surface, showing an annual fluctuation of up to 2m and inter-annual variations of approximately 5m. Groundwater flows towards north-west and the average hydraulic gradient is ~1% (Figure 11.7). According to the hydraulic conductivity range determined at the site by pumping tests and assuming an effective porosity of 0.01, the groundwater seepage velocity can be estimated to be 0.3~8.6m/d.

11.4.4.2 Sampling and analysis

The field site is equipped with a groundwater monitoring network consisting of 30 wells situated along the CAHs plume. Water samples from selected wells (18 wells, Figure 11.7a) were collected for chemical and isotope analysis in February 2011 (first campaign) and March 2013 (second campaign). In order to evaluate if the observed isotope pattern of primary compounds and potential metabolites is related to reactive processes, measured concentrations are transformed to relative concentrations taking into account reaction equations and related to isotope ratios in analogy to the Rayleigh equation (Eq. 11.1). As several reactive processes might occur simultaneously, the slope of such a plot will not necessarily correspond to a specific laboratory enrichment factor (ε_{bulk}). The substrate remaining fraction (f) at a certain well is estimated according to Eq. 11.13 and Eq. 11.14 for 1,1,1-TCA and TCE, respectively:

$$f_{1,1,1\text{-TCA}}=\frac{[1,1,1\text{-TCA}]}{[1,1,1\text{-TCA}+\text{HAc}+1,1\text{-DCE}]}=\frac{[1,1,1\text{-TCA}]}{[1,1,1\text{-TCA}+3.6\times1,1\text{-DCE}]} \tag{11.13}$$

$$f_{\text{TCE}}=\frac{[\text{TCE}]}{[\text{TCE}+1,1\text{-DCE}]} \tag{11.14}$$

where [1,1,1-TCA] and [TCE] are the aqueous concentration of 1,1,1-TCA and TCE, respectively, and [1,1,1-TCA + HAc + 1,1-DCE] and [TCE + 1,1-DCE] are the total concentration of 1,1,1-TCA, TCE and their respective products for the DH/HY and hydrogenolysis pathways, respectively. Mole fractions are used instead of absolute concentrations as the first take into account the effect of dilution. Regarding the hydrogenolysis products of TCE, *cis*-1,2-DCE is not considered in Eq. 11.14 as its concentration in groundwater (up to 14mg/L) is much smaller than that of 1,1-DCE (up to 10mg/L). For 1,1,1-TCA, HAc produced by hydrolysis was not analyzed in groundwater samples. In the aquifer, HAc is readily biodegraded because it is used as electron donor and carbon source by the microorganisms. Therefore, for $f_{1,1,1\text{-TCA}}$, the expression [1,1,1-TCA]/[1,1,1-TCA+3.6×1,1-DCE] is used, which accounts for the produced HAc. The yield of HAc (hydrolysis product) relative to 1,1-DCE (dehydrohalogenation product) was estimated by first order curve fitting of concentration-time data series obtained in a previous laboratory study (Palau et al., 2014b). The uncertainty of the calculated $f_{1,1,1\text{-TCA}}$ and f_{TCE} in the field, i.e. 39% and 17%, respectively, was estimated by error propagation in Eqs. 11.13 and 11.14, and an uncertainty of 10% was assumed for commercial concentration analysis of volatile organic compounds (Hunkeler et al., 2008).

11.4.4.3 Field geochemical conditions and CAHs concentration

High DO and nitrate concentrations were measured in groundwater, ranging between 2.6mg/L and 8.4mg/L for DO and from 51.6mg/L to 94.4mg/L for nitrate, which indicate the presence of aerobic conditions in the aquifer. Concentrations of dissolved Mn and dissolved Fe are low ($\leqslant$0.01mg/L and $\leqslant$0.07mg/L, respectively), which is in agreement with the presence of oxygen. Aerobic conditions are unfavorable for microbial reductive dechlorination of CAHs.

The main CAHs present in groundwater, i. e. 1,1,1-TCA, TCE and 1,1-DCE, are detected at concentrations $>$1mg/L, reaching a value of up to 20mg/L for 1,1,1-TCA in the source area (well E in February 2011) and forming a CAHs plume spreading northwest (see the 1, 1, 1-TCA plume in Figure 11.7a). Several compounds are detected at lower concentrations, including 1,1,2-TCA (up to 500mg/L), 1,1-DCA (up to 140mg/L), 1,2-dichloroethane (1,2-DCA, up to 270mg/L) and *cis*-1,2-dichloroethene (*cis*-1,2-DCE, up to 14mg/L), and their mole fractions relative to the total concentration of chlorinated ethanes and ethenes are $<$ 7%. The presence of 1, 1, 1-TCA and TCE reductive dechlorination products such as 1,1-DCA and *cis*-1,2-DCE, respectively, could be related to the occurrence of micro-anaerobic environments in the aquifer. The contribution of 1, 1, 2-TCA dehydrohalogenation to 1,1-DCE concentration is probably very small according to the low molar concentration of 1,1,2-TCA relative to 1,1,1-TCA ($<$8%).

Concentrations of 1, 1, 1-TCA, TCE and 1, 1-DCE show a similar distribution in the aquifer and a large concentration range of two orders of magnitude is observed for all of them in the wells situated close to the plume centerline. High concentrations of 1, 1-DCE are already present in the wells located in the source area, up to 10mg/L in Well E in February 2011 (Figure 11.7a). Changes in aqueous CAH concentrations in the plume can be related to transformation processes but also to non-degradative processes such as hydrodynamic dispersion and sorption. To account for dispersion, relative variations in 1,1,1-TCA, TCE and 1,1-DCE concentrations along the plume can be expressed as mole fractions. Increasing mole fractions of 1,1-DCE downgradient from the source would be indicative of 1,1,1-TCA and/or TCE degradation during transport. However, the mole fractions of 1, 1-DCE in several wells located close to the plume centerline show a small variation relative to [1,1,1-TCA+1,1-DCE], from 0.41 to 0.60 (Figure 11.7b), and the fractions of 1,1-DCE relative to [TCE+1,1-DCE] are higher for the wells situated close to the source, i. e. Wells E and C (Figure 11.7c). Therefore, additional data is necessary to confirm the contribution of degradation processes to the observed changes in mole fractions.

11.4.4.4 Isotope patterns of 1,1,1-TCA, TCE and 1,1-DCE in the aquifer

The chlorine isotope composition of 1,1,1-TCA in groundwater range from +2.4‰ to +7.6‰. In previous studies, chlorine isotope ratios of pure phase 1,1,1-TCA from different manufacturers showed values ranging from −3.54‰ to +2.03‰ (Shouakar-Stash et al., 2003). Compared to the manufacturers' range, the higher range of $\delta^{37}Cl$ values in groundwater suggests that 1,1,1-TCA could be affected by degradation processes. Similarly, the carbon isotopic composition of 1,1,1-TCA in groundwater, which ranges from −21.1‰ to −25.1‰ (with the exception of the value of −26.3‰ measured in the Well E in February 2011), is also higher than the manufacturers' range, which varies between −25.5‰ and −31.6‰ (Hunkeler and Aravena, 2010), supporting 1,1,1-TCA transformation in the aquifer. To evaluate in more detail whether the variations of isotope ratios of 1,1,1-TCA in groundwater are due to degradation, $\delta^{37}Cl$ and $\delta^{13}C$ values are related to the concentration data according to the Rayleigh equation (Figure 11.8c and 11.8d). Chlorine and carbon isotope ratios of 1,1,1-TCA exhibit an enrichment in heavy isotopes (i.e. ^{37}Cl and ^{13}C) with decreasing mole fractions, with the exception of data from Wells A, E and G (red markers in Figure 11.8c and 11.8d), confirming that isotope variations of 1,1,1-TCA are related to its degradation. The $\delta^{13}C$ values of 1,1-DCE in groundwater, from −18.5‰ to −25.3‰, are generally depleted in ^{13}C compared to those of 1,1,1-TCA (Figure 11.8d), which is consistent with the abiotic formation of 1,1-DCE from 1,1,1-TCA via dehydrohalogenation. In addition, this isotope pattern also suggests that 1,1-DCE is not further degraded in most of the wells.

In Well A, carbon and chlorine isotopes ratios of 1,1,1-TCA are significantly enriched in both ^{13}C and ^{37}Cl. These higher values could be explained either by a distinct source of 1,1,1-TCA with a heavier isotope composition or by the effect of biodegradation. Relatively low DO values varying from 1.0mg/L to 1.7mg/L were measured in this well between 2005 and 2008, which could indicate that micro-anaerobic environments favorable to microbial reductive dechlorination of 1,1,1-TCA took place at that time and that 1,1,1-TCA affected by biodegradation is still present in the vicinity of Well A. In contrast, for Wells E and G, $\delta^{13}C_{1,1,1\text{-TCA}}$ values are slightly depleted in ^{13}C (up to −26.3‰ in E, February 2011), while $\delta^{37}Cl_{1,1,1\text{-TCA}}$ values are lightly enriched in ^{37}Cl (up to +5.8‰ in G, March 2013). Such behavior could be related to the effect of vaporization and diffusion processes on the residual 1,1,1-TCA contamination in the unsaturated zone (Jeannottat and Hunkeler, 2012). Wells E and G are located in the vicinity of the source area (Figure 11.7a) and previous reports at the site showed that, when the water level rises, it sometimes reaches highly contaminated parts of the unsaturated zone in the source area, leading to a direct input of residual contaminants into the aquifer. For the remaining 15 out of 18 wells investigated (i.e. Wells

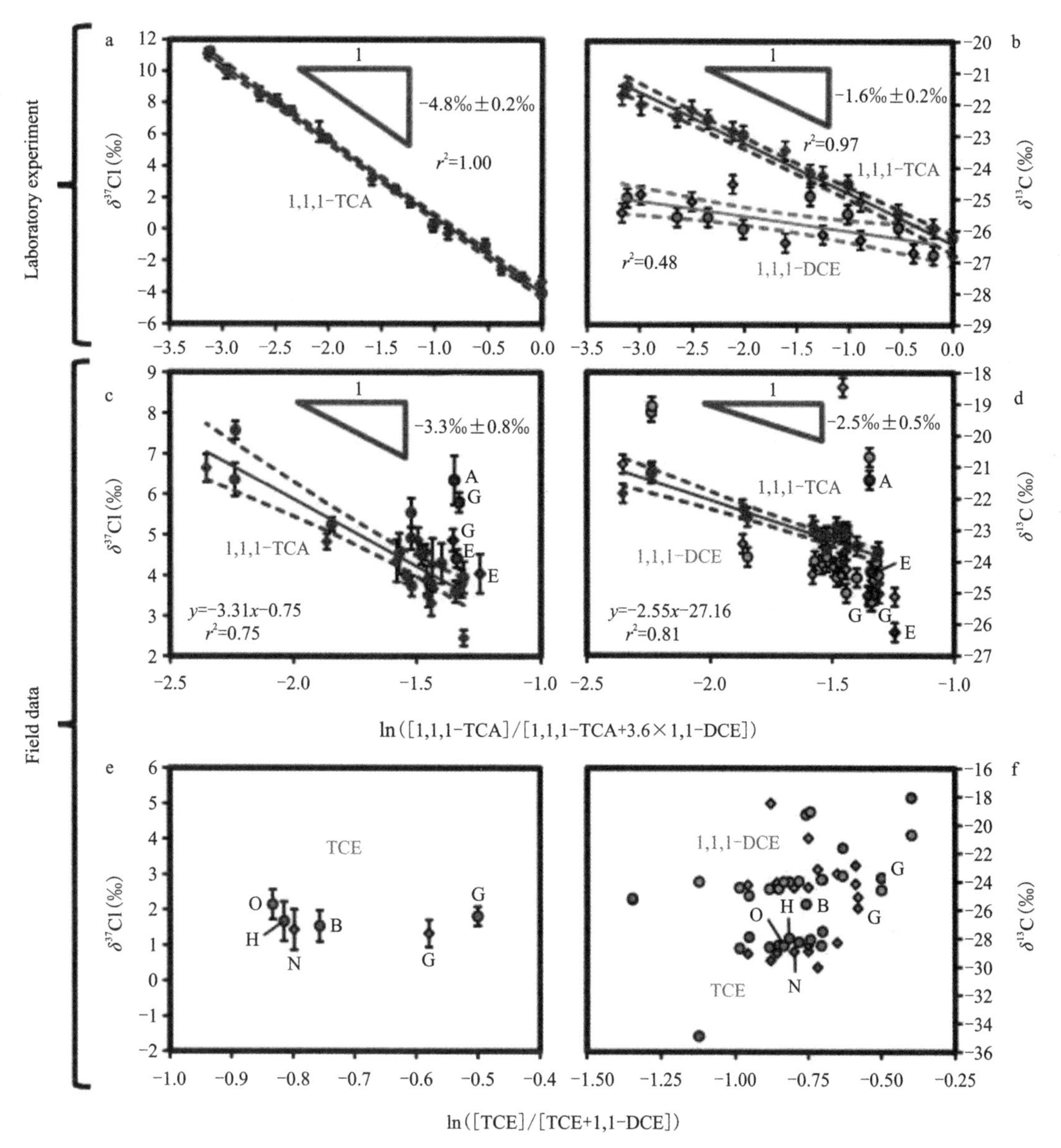

Figure 11. 8 a, b. Isotopic composition of 1, 1, 1-TCA (δ^{37}Cl and δ^{13}C) and 1, 1-DCE (δ^{13}C) during DH/HY of 1, 1, 1-TCA in batch experiments. Data from duplicate experiments are combined (i. e. rhombus and circle symbol marks). c, d. Isotopic composition of 1, 1, 1-TCA (δ^{37}Cl and δ^{13}C) and 1, 1-DCE (δ^{13}C) in groundwater samples. Dashed lines correspond to the 95% confidence intervals of regression parameters. e, f. Isotopic composition of TCE (δ^{37}Cl and δ^{13}C) and 1, 1-DCE (δ^{13}C) in groundwater samples. For field samples, data from both sampling campaigns are combined, i. e. rhombus (February, 2011) and circles (March, 2013).

B~D, F and H~R), observed variations with regard to both Cl and C isotope values are well described by a linear trend ($r^2 \geq 0.75$, Figure 11. 8c and 11. 8d). The intercepts of the correlation lines, i. e. −0. 7‰±1. 9‰ for Cl and −27‰±1‰ for C (the uncertainties were estimated by error propagation in the regression equations for Cl and C isotope data indicated in Figure 11. 8c and 11. 8d), can be considered as an estimate of the initial isotopic composition of 1,1,1-TCA, which agree very well with the ranges reported for pure 1,1,1-TCA from different manufacturers, i. e. between −3. 54‰ and +2. 03‰ for Cl (Shouakar-Stash et al. , 2003) and between −25. 5‰ and −31. 6‰ for C (Hunkeler and Aravena, 2010).

In order to compare the field and laboratory isotope patterns, the isotope data of 1,1,1-TCA and 1,1-DCE measured during 1,1,1-TCA transformation by DH/HY in the laboratory (Palau et al. , 2014b) were reevaluated in this study according to Eqs. 11. 1 and 11. 12(Figure 11. 8a and 11. 8b). In general, the field $\delta^{13}C$ values of 1,1,1-TCA and 1,1-DCE (Figure 11. 8d) exhibit a pattern similar to the one observed in the laboratory batch experiment (Figure 11. 8b), providing further evidence for 1,1,1-TCA dehydrohalogenation in the aquifer. Compared to the laboratory experiment, the correlation lines for field isotope data show a smaller slope for Cl, i. e. −3. 3‰±0. 8‰ (field, Figure 11. 8c) and −4. 8‰±0. 2‰ (laboratory, Figure 11. 8a), and a larger slope for C, i. e. −2. 5‰±0. 5‰ (field, Figure 11. 8d) and −1. 6‰±0. 2‰ (laboratory, Figure 11. 8b). However, when taking their uncertainty into consideration, the slopes for field and laboratory data are relatively similar for both elements. The larger slope obtained from field carbon isotope data compared to the laboratory DH/HY experiment can be associated with the simultaneous occurrence of biodegradation processes of 1,1,1-TCA in addition to DH/HY in the field, which is further investigated using a dual isotope approach. For TCE, several groundwater samples with different $\delta^{13}C$ values (data points labeled in Figure 11. 8f) were selected for chlorine isotope analysis, showing similar $\delta^{37}Cl$ values (from +1. 3‰±0. 4‰ to +2. 1‰±0. 4‰, Figure 11. 8e). The $\delta^{37}Cl$ values of TCE in groundwater fall within the reported range of pure TCE from different manufacturers which varies between −3. 19‰ and +3. 90‰ (Hunkeler and Aravena, 2010), suggesting little transformation of TCE. The carbon isotopic composition of TCE varied from −21. 6‰ to −30. 0‰, with an average of −27‰±2‰ ($\pm 1\sigma$, $n=24$), except for the Wells A (−18. 1‰) and K (−34. 9‰) on March 2013. As observed for chlorine, most of the $\delta^{13}C_{TCE}$ values fall within the range of TCE from different manufacturers, i. e. between −24. 5‰ and −33. 5‰ (Hunkeler and Aravena, 2010), supporting little degradation of TCE in groundwater. Contrary to the isotope patterns of 1,1,1-TCA, $\delta^{37}Cl_{TCE}$ and $\delta^{13}C_{TCE}$ values do not show any enrichment in ^{37}Cl and ^{13}C with decreasing mole fractions of TCE (Figure 11. 8e and 11. 8f), confirming that TCE is not significantly degraded in the aquifer. This result is in agreement with the aerobic conditions determined in the aquifer. In addition, the $\delta^{13}C$ values of 1,1-DCE are generally enriched in ^{13}C compared to TCE (Figure 11. 8f). According to the normal carbon isotope fractionation of TCE during reductive dechlorination (Hunkeler and Morasch, 2010), the $\delta^{13}C$ values of produced 1,1-DCE would be lower than those of TCE. Therefore, for most of the samples, the observed changes in $\delta^{13}C_{TCE}$ can probably be associated with some variability in the carbon isotopic composition of source TCE.

11.4.4.5 Dual C-Cl isotope approach to investigate degradation pathways in the field

Carbon and chlorine d isotope values of 1, 1, 1-TCA in groundwater samples were combined in a dual isotope plot (Figure 11.9). Isotope values from Wells A, E and G are not included because, as indicated above, isotope data from these wells could be affected by processes different than compound transformation. The plotted data show a linear trend ($r^2=0.75$) with a dual isotope slope ($\Delta=\Delta\delta^{13}C/\Delta\delta^{13}Cl\approx\varepsilon C_{bulk}/\varepsilon Cl_{bulk}$) of 0.6 ± 0.2, confirming that transformation of 1,1,1-TCA is an important process in the aquifer. This field Δ value is very different from that determined in a recent laboratory study for oxidation (Figure 11.9), clearly indicating that oxidation cannot be the main process involved (Palau et al., 2014b). In contrast, the field slope is closer to the one determined for 1,1,1-TCA transformation via DH/HY in the laboratory (0.33 ± 0.04, Figure 11.9; Palau et al., 2014b). The significant difference between the dual isotope slopes determined for the field and the DH/HY experiment (ANCOVA, $p=0.000\,3$) suggests that additional degradation processes of 1,1,1-TCA likely occur in the aquifer, as pointed out by the carbon isotope patterns in Figure 11.8. A higher L value (1.5 ± 0.1) associated with the reduction of 1,1,1-TCA by zero-valent iron was previously reported (Palau et al., 2014b), however, significant biotic and/or abiotic reductive dechlorination of 1,1,1-TCA are discarded due to the aerobic conditions in the aquifer. On the other hand, in aerobic conditions, microbial cooxidative degradation of 1,1,1-TCA to 2,2,2-trichloroethanol via CeH bond cleavage in the first reaction step has been reported in several studies (Field and Sierra Alvarez, 2004; Yagi et al., 1999). The occurrence of microbial oxidation of 1, 1, 1-TCA would be consistent with the different slopes determined from $\delta^{13}C_{1,1,1\text{-}TCA}$ data for the field and the DH/HY experiment in Figure 11.8. As observed during abiotic oxidation of 1,1,1-TCA in a recent study (Palau et al., 2014b), a much higher isotope effect associated with C-H bond cleavage is expected for C compared to Cl. This might explain, taking as a reference the slopes determined from the laboratory experiment, the higher slope obtained for C, $-2.5‰\pm0.5‰$(field) and $-1.6‰\pm0.2‰$ (laboratory) (Figure 11.8b, d), compared to the smaller slope observed for Cl, $-3.3‰\pm0.8‰$ (field) and $-4.8‰\pm0.2‰$ (laboratory) (Figure 11.8a, c). Therefore, a combination of DH/HY and microbial oxidation may be taking place.

In this case, oxidation and DH/HY pathway-specific contributions to total 1, 1, 1-TCA degradation may be estimated using the expression derived by van Breukelen (2007), Eq. 11.15:

$$F=\frac{\Delta\cdot\varepsilon_{O}^{Cl}-\varepsilon_{O}^{C}}{(\varepsilon_{DIH}^{C}-\varepsilon_{O}^{C})-(\varepsilon_{DIH}^{Cl}-\varepsilon_{O}^{Cl})} \tag{11.15}$$

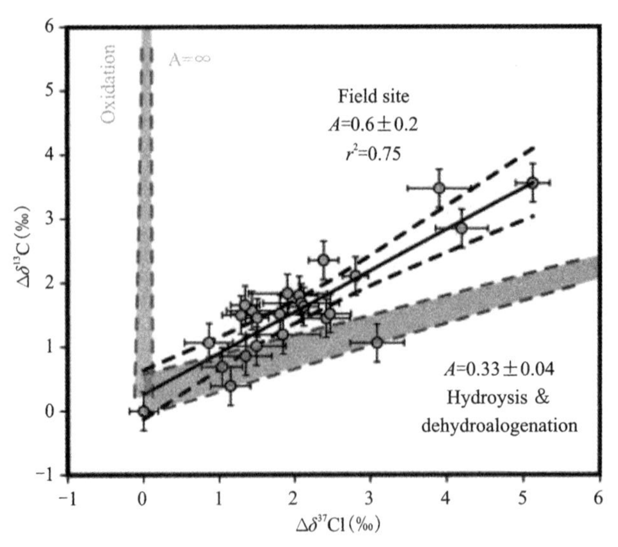

Figure 11. 9 Dual CeCl isotope trends during transformation of 1, 1, 1-TCA in the investigated test site and in two experimental systems. Data from both campaigns are combined (blue circles). L values (±95%C. I.) are given by the slope of the linear regressions and the black dashed lines correspond to the 95%C. I. Shaded areas (95%C. I.) indicate exclusive occurrence of either one of the two pathways.

where F is the distributionof DH/HY and oxidation pathways, $\varepsilon_{\mathrm{DIH}}^{\mathrm{C}}$ and $\varepsilon_{\mathrm{DIH}}^{\mathrm{Cl}}$are the C and Cl isotope fractionation values during DH/HY of 1,1,1-TCA and $\varepsilon_{\mathrm{O}}^{\mathrm{C}}$ and $\varepsilon_{\mathrm{O}}^{\mathrm{Cl}}$ correspond to the C and Cl isotope fractionation values for 1,1,1-TCA oxidation. For this equation, in addition to the $\varepsilon_{\mathrm{bulk}}$ values of 1, 1, 1-TCA for both reactions involved, only the dual isotope slope determined from field data ($\Delta = 0.6 \pm 0.2$) is necessary. The $\varepsilon\mathrm{C}_{\mathrm{bulk}}$ and $\varepsilon\mathrm{Cl}_{\mathrm{bulk}}$ values of 1,1,1-TCA during DH/HY and oxidation reactions were reported in a recent study (Palau et al., 2014b), showing values of $-1.6‰ \pm 0.2‰$ and $-4.7‰ \pm 0.1‰$ (DH/HY), $-4.0‰ \pm 0.2‰$ and no chlorine isotope fractionation (Oxidation). In this previous study, the isotope fractionation values of 1, 1, 1-TCA during oxidative C-H bond cleavage were determined abiotically by reaction with persulfate. Chlorine isotope fractionation values for microbial oxidation of 1, 1, 1-TCA are still not available in the literature, however, isotope fractionation values determined from abiotically mediated oxidation may be used as a rough approximation. In fact, isotope fractionation values from abiotic reactions are often considered closest to the intrinsic isotope effects (Lollar et al., 2010). According to the reported reaction-specific $\varepsilon_{\mathrm{bulk}}$ values, the contribution of DH/HY was of $80\% \pm 10\%$ (the uncertainty was estimated by error propagation in Eq. 11. 15). This result indicates a relatively small contribution of the oxidation pathway, provided that the $\varepsilon_{\mathrm{bulk}}$ values for microbial oxidation of 1,1,1-TCA by indigenous microorganisms at the site are confirmed in future biodegradation studies. Eq. 11. 15 assumes simultaneous activity of both pathways, which is a likely assumption in our case judging by the good linear correlation between $\delta^{37}Cl$ and $\delta^{13}Cl$ values (Figure 11.9).

The expected rate of 1,1,1-TCA degradation by DH/HY at the measured groundwater temperature can be estimated using the Arrhenius equation(Eq. 11.16):

$$k = A \cdot \exp(-E_a/RT) \tag{11.16}$$

where k is the first order rate constant (s^{-1}), A is the frequency factor (s^{-1}), R is the gas constant (8.314×10^{-3} kJ/mol · K), E_a is the activation energy (kJ/mol) and T is the absolute temperature(K). According to the E_a(122.8kJ/mol) and A ($8.7 \times 10^{13} s^{-1}$) values determined by Gauthier and Murphy (2003) from several previous studies and the average groundwater temperature at the site [(284 ± 1)K, $\pm 1\sigma$, $n=34$], the transformation rate is estimated to be $1.95 \times 10^{-4} d^{-1}$ (i. e. half-live of around 10 years). This slow reaction rate contrasts with the relatively fast groundwater seepage velocity in the saturated zone (up to 8.6m/d), suggesting that significant contaminant retardation would be necessary to explain the high concentrations of 1,1-DCE in the source area. In this site, owing to the high chalk matrix porosity (up to 45%), 1,1,1-TCA is probably subject to retardation by diffusion into the matrix pore water. In addition to degradation of 1,1,1-TCA in the saturated zone, dehydrohalogenation of 1,1,1-TCA to 1,1-DCE might also occur in the unsaturated part of the aquifer (up to 28m thick). Here, downward migration for dissolved compounds in groundwater was estimated at ~1m/a by different studies (Orban et al., 2010). Degradation of 1,1,1-TCA in the unsaturated zone is supported by the detection of 1,1-DCE in relatively high concentrations in soil samples from the unsaturated zone analyzed in previous reports.

References

ABE Y, HUNKELER D, 2006. Does the Rayleigh equation apply to evaluate field isotope data in contaminant hydrogeology? [J]. Environ. Sci. Technol., 40: 1588-1596.

ABE Y, ARAVENA R, ZOPFI J, et al., 2009. Carbon and chlorine isotope fractionation during aerobic oxidation and reductive dechlorination of vinyl chloride and cis-1,2-dichloroethene[J]. Environ. Sci. Technol., 43 (1): 101-107.

AELION C M, HOHENER P, HUNKELER D, et al., 2010. Environmental Isotopes in Biodegradation and Bioremediation[M]. Boca Raton: CRC Press: 249-293.

ALBERS C N, HANSEN P E, JACOBSEN O S, 2010. Trichloromethyl compounds - natural background concentrations and fates within and below coniferous forests[J]. Sci. Total Environ., 408 (24): 6223-6234.

ALBERS C N, LAIER T, JACOBSEN O S, 2010. Formation, fate and leaching of chloroform in coniferous forest soils[J]. Appl. Geochem., 25: 1525-1535.

AUDÍ-MIRO C, CRETNIK S, OTERO N, et al., 2013. Cl and C isotope analysis to assess the effectiveness of chlorinated ethene degradation by zero-valent iron: evidence fromdual element and product isotope values[J]. Appl. Geochem., 32:175-183.

BAERTSCHI P, KUHN W, KUHN H, 1953. Fractionation of isotopes by distillation of some organic substances[J]. Nature, 171:1018-1020.

BRADLEY D C, 1954. Fractionation of isotopes by distillation of some organic substances[J]. Nature, 173: 260-261.

BOURG I C, SPOSITO G, 2007. Molecular dynamics simulations of kinetic isotope fractionation during the diffusion of ionic species in liquid water[J]. Geochem. Cosmochim. Acta, 71: 5583-5589.

Brand W A, 2004. Handbook of Stable Isotope Analytical Techniques[J]. Elsevier, Amsterdam, The Netherlands, I:835-856.

CRANE R, SCOTT T, 2012. Nanoscale zero-valent iron: future prospects for an emerging water treatment technology[J]. J. Hazard Mater: 211-212.

CRETNIK S, BERNSTEIN A, SHOUAKAR-STASH O, et al., 2014. Chlorine isotope effects from isotope ratio mass spectrometry suggest intramolecular C-Cl bond competition in trichloroethene(TCE) reductive dehalogenation[J]. Molecules, 19:6450-6473.

CRETNIK S, THORESON K A, BERNSTEIN A, et al., 2013. Reductive dechlorination of TCE by chemical model systems in comparison to dehalogenating Bacteria: Insights from dual element isotope analysis ($^{13}C/^{12}C$, $^{37}Cl/^{35}Cl$)[J]. Environ. Sci. Technol., 47 (13): 6855-6863.

EINSTEIN A, 1905. Über die von der molekularkinetischen theorie der Warme geforderte bewegung von in ruhenden flüssigkeiten suspendierten teilchen[J]. Ann. Phys., 322: 549-560.

GODIN J P, FAY L B, HOPFGARTNER G, 2007. Mass Spectrom[J]. Rev., 26: 751.

HE Y T, WILSON J T, SU C, et al., 2015. Review of abiotic degradation of chlorinated solvents by reactive iron minerals in aquifers[J]. Groundwater Monitoring & Remediation, 35:57-75.

HERATY L J, FULLER M E, HUANG L, et al., 1999. Isotopic fractionation of carbon and chlorine by microbial degradation of dichloromethane[J]. Org. Geochem., 30: 793-799.

HORST A, LACRAMPE-COULOUME G, SHERWOOD LOLLAR B, 2016. Vapor pressure isotope effects in halogenated organic compounds and alcohols dissolved in water [J]. Anal. Chem., 88: 12 066-12 071.

HUNKELER D, MECKENSTOCK R, SHERWOOD LOLLAR B, et al., 2008. A guide for assessing biodegradation and source identification of organic ground water contaminants using compound specific isotope analysis (CSIA)[R]. Technical Report EPA 600/R-08/148, US EPA, Washington D C, USA.

HUNKELER D, ARAVENA R, SHOUAKAR-STASH O, et al., 2011b. Carbon and chlorine isotope ratios of chlorinated ethenes migrating through a thick unsaturated zone of a sandy aquifer[J]. Environ. Sci. Technol., 45: 8247-8253.

JEANNOTTAT S, HUNKELER D, 2012. Chlorine and carbon isotopes fractionation during volatilization and diffusive transport of trichloroethene in the unsaturated zone[J]. Environ. Sci. Technol., 46: 3169-3176.

JOCHMANN M, BLESSING M, HADERLEIN S, et al., 2016. Rapid Commun[J]. Mass Spectrom., 20: 3639.

KRUMMEN M, HILKERT A W, JUCHELKA D, et al., 2004. Rapid Commun[J]. Mass Spectrom., 18: 2260.

KUDER T, VAN BREUKELEN B M, VANDERFORD M, et al., 2013. 3d-CSIA: carbon, chlorine, and hydrogen isotope fractionation in transformation of TCE to ethene by a Dehalococcoides culture[J]. Environ. Sci. Technol., 47: 9668-9677.

LANGVAD T, RICHTZENHAIN H, SALAKIVI V, et al., 1954. Separation of chlorine isotopes by ion-exchange chromatography[J]. Acta Chem. Scand., 8: 526-527.

LOLLAR B S, HIRSCHORN S, MUNDLE S O, et al., 2010. Insights into enzyme kinetics of chloroethane biodegradation using compound specific stable isotopes[J]. Environ. Sci. Technol., 44 (19): 7498-7503.

ORBAN P, BROUYERE S, BATLLE-AGUILAR J, et al., 2010. Regional transport modelling for nitrate trend assessment and forecasting in a chalk aquifer[J]. J. Contam. Hydrol., 118 (1-2): 79-93.

PALAU J, CRETNIK S, SHOUAKAR-STASH O, et al., 2014a. C and Cl isotope fractionation of 1, 2-Dichloroethane displays unique $\delta^{13}C/\delta^{37}Cl$ patterns for pathway identification and reveals surprising C-Cl bond involvement in microbial oxidation[J]. Environ. Sci. Technol., 48 (16): 9430-9437.

PALAU J, SHOUAKAR-STASH O, HUNKELER D, 2014b. Carbon and chlorine isotope analysis to identify abiotic degradation pathways of 1, 1, 1-trichloroethane [J]. Environ. Sci. Technol., 48 (24): 14 400-14 408.

POULSON S R, DREVER J I, 1999. Stable isotope (C, Cl, and H) fractionation during vaporization of trichloroethylene[J]. Environ. Sci. Technol., 33: 3689-3694.

RENPENNING J, KELLER S, CRETNIK S, et al., 2014. Combined C and Cl isotope effects indicate differences between corrinoids and enzyme (sulfurospirillum multivorans PceA) in reductive dehalogenation of tetrachloroethene, but not trichloroethene[J]. Environ. Sci. Technol., 48: 11 837-11 845.

RODRÍGUEZ-FERNÁNDEZ D, HECKEL B, TORRENTO C, et al., 2018. Dual element (CCl) isotope approach to distinguish abiotic reactions of chlorinated methanes by Fe(0) and by Fe^{2+} on iron minerals at neutral and alkaline pH[J]. Chemosphere, 206: 447-456.

ROSELL M, PALAU J, MORTAN S H, et al., 2019. Dual carbon-chlorine isotope fractionation during dichloroelimination of 1,1,2-trichloroethane by an enrichment culture containing *Dehalogenimonas* sp. [J]. Sci. Total Environ., 648: 422-429.

SCHMIDT T C, JOCHMANN M A, 2012. Origin and fate of organic compounds in water: characterization by compound-specific stable isotope analysis[J]. Annu. Rev. Anal. Chem., 5: 133-155.

THULLNER M, FISCHER A, RICHNOW H H, et al., 2013. Influence of mass transfer on stable isotope fractionation[J]. Appl. Microbiol. Biotechnol., 97 (2): 441-452.

TIEHM A, SCHMIDT K R, 2011. Sequential anaerobic/aerobic biodegradation of chloroethenes-aspects of field application[J]. Curr. Opin. Biotechnol., 22: 415-421.

TORRENTO C, PALAU J, RODRÍGUEZ-FERNÁNDEZ D, et al., 2017. Carbon and chlorine isotope fractionation patterns associated with different engineered chloroform transformation reactions[J]. Environ. Sci. Technol., 51: 6174-6184.

NUMATA M, NAKAMURA N, KOSHIKAWA H, et al., 2002a. Chlorine isotope fractionation during reductive dechlorination of chlorinated ethenes by anaerobic bacteria[J]. Environ. Sci. Technol., 36: 4389-4394.

VAKILI F, 2017. Stable Isotope Fractionations of Chlorinated Ethenes Associated with Physical Processes[D]. Ontario, Canada: University of Waterloo.

VAN BREUKELEN B M, 2007. Extending the Rayleigh equation to allow competing isotope fractionating pathways to improve quantification of biodegradation[J]. Environ. Sci. Technol., 41 (11): 4004-4010.

VANSTONE N, ELSNER M, LACRAMPE-COULOUME G, et al., 2008. Potential for identifying abiotic chloroalkane degradation mechanisms using carbon isotopic fractionation[J]. Environ. Sci. Technol., 42 (1): 126-132.

WANNER P, HUNKELER D, 2019. Isotope fractionation due to aqueous phase diffusion-what do diffusion models and experiments tell? —a review[J]. Chemosphere, 219: 1032-1043.

ZWANK L, BERG M, SCHMIDT T C, et al., 2003. Compound-specific carbon isotope analysis of volatile organic compounds in the low-microgram per liter range[J]. Anal. Chem., 20: 5575-5583.

Chapter 12 Groundwater Dating

Of the considerable amount of groundwater recharge that occurs on a daily basis, most circulates rapidly to surface water drainage systems such that storm hydrographs respond closely with precipitation. Alluvial aquifers along drainage channels or shallow bedrock aquifers can respond rapidly to recharge and have groundwater with mean residence times on the order of months to a few years. Only a small fraction recharges to groundwater systems with much longer circulation times. Deeper and confined aquifers host groundwater that may circulate over hundreds to thousands of years. Others may retain groundwater recharged under climate conditions different from current and host fossil groundwater that are not being recharged today.

Determining the mean subsurface age of groundwater is important for several reasons. Groundwater age translates roughly to circulation and renewability, which is critical for water resource planning. Groundwater resources identified as being modern are regularly recharged and can be sustainably developed, whereas exploiting fossil groundwater is essentially mining the resource. The age of groundwater relates also to its vulnerability to contamination. Modern groundwater will have more direct pathways from the surface, facilitating infiltration of contaminants from surface activities in the recharge area. Old groundwater are isolated from the surface and perhaps less vulnerable to anthropogenic contamination.

Groundwater age is defined as the amount of time that has elapsed since a particular water molecule of interest was recharged into the subsurface environment system until this molecule reaches a specific location in the system where it is either sampled physically or studied theoretically for age-dating.

The selection of techniques depends on the anticipated age range of the groundwater flow system, as well as the need for resolution in refining the age calculation. All methods have constraints, whether from a sampling and analytical perspective or from an interpretive perspective. The use of two or more methods provides some measure of assurance when the results converge on an estimate of mean groundwater age. Figure 12.1 shows the typical effective age range for various methods.

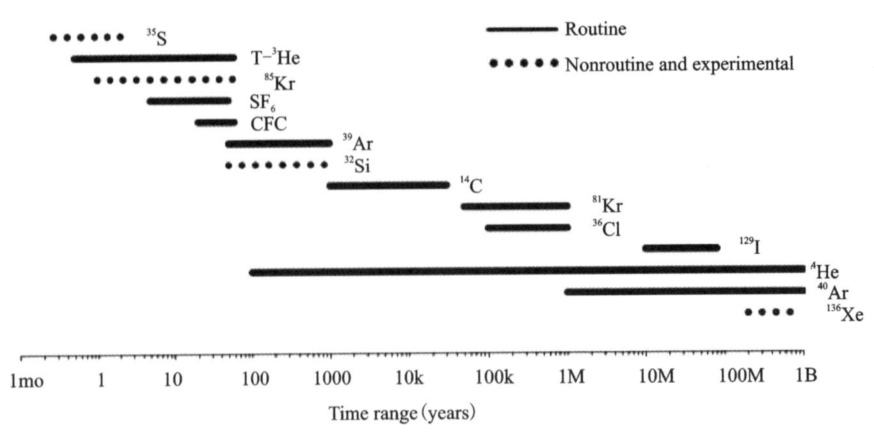

Figure 12. 1 Age range for routine, nonroutine, experimental radioisotopes and anthropogenic gases in groundwater dating.

12. 1 ^{3}H

Hydrogen has three isotopes: ^{1}H (common hydrogen or protium, H); ^{2}H (deuterium, D); and ^{3}H (tritium, T). Deuterium (also referred to as heavy stable isotope of hydrogen) and protium are stable, but tritium is radioactive with a half-life of 12. 43 years (some references give 12. 3 years). If tritium replaces one of the hydrogens in the water molecule, H_2O, the result would be what is called tritiated water, THO. Tritium content of a water sample is expressed as TU, where one TU represents one THO molecule in 1018 H_2O molecules. In simpler terms, one TU means one mg THO in 109 tones H_2O. This is equal to 6. 686×1010 tritium atoms/kg and has an activity of 0. 118 1Bq/kg (3. 193pCi/kg) or 7. 1 disintegration per minute per liter of water. Tritium dating method is the first technique developed to date groundwater.

In nature, tritium can be produced by cosmic-ray bombardment of nitrogen and deuterium in the upper atmosphere:

$$^{14}_{7}N+n \rightarrow {}^{12}_{6}C+{}^{3}_{1}H \tag{12.1}$$

$$^{2}_{1}H+{}^{2}_{1}H \rightarrow {}^{3}_{1}H+{}^{1}_{1}H \tag{12.2}$$

The amount of this production is estimated at 0. 5±0. 3atoms/cm^2/s. Before the onset of atmospheric nuclear weapons tests, the global equilibrium tritium inventory was estimated at about 3. 6kg or 35 mega curies (some references give different figures). This leads to a rainfall tritium concentration of 3～6TU in Europe and North America and 1～3TU in South Australia (Solomon and Cook, 2000). Such estimations, however, are not quite accurate, and various references argue for a range of 0. 5～20TU pre-bomb rainfall. It is albeit clear that because of higher geomagnetic latitude, rainfall tritium content in the Northern Hemisphere is more than that in the Southern Hemisphere.

Recently, tritium is more produced through thermonuclear tests, which started in 1952, and, though banned, still form a significant source of tritium in the atmosphere and other spheres. In all textbooks, graphs are presented that illustrate temporal concentration of tritium in the rainfall as early as 1953 (in Ottawa, Canada, for example), reaching its peak in 1963 and 1964 in the Northern and the Southern Hemispheres, respectively, and declining since then to an almost natural level in the Southern Hemisphere. One such graph is illustrated in Figure 12. 2, to show the concentration of tritium in the precipitation in Vienna, Austria. The first increase in tritium concentration due to bomb testing resulted in a concentration of tritium in precipitation of (66 ± 1)TU at Chicago on November 18, 1952. In the periods March-July 1954 and March-September 1956, tritium in Chicago rainfall exceeded about 100TU and between December 18, 1957, and July 29, 1958, the tritium level in Chicago rain and snow (and presumably in Nevada precipitation) ranged from a minimum of 54. 0TU to a maximum of 2160TU.

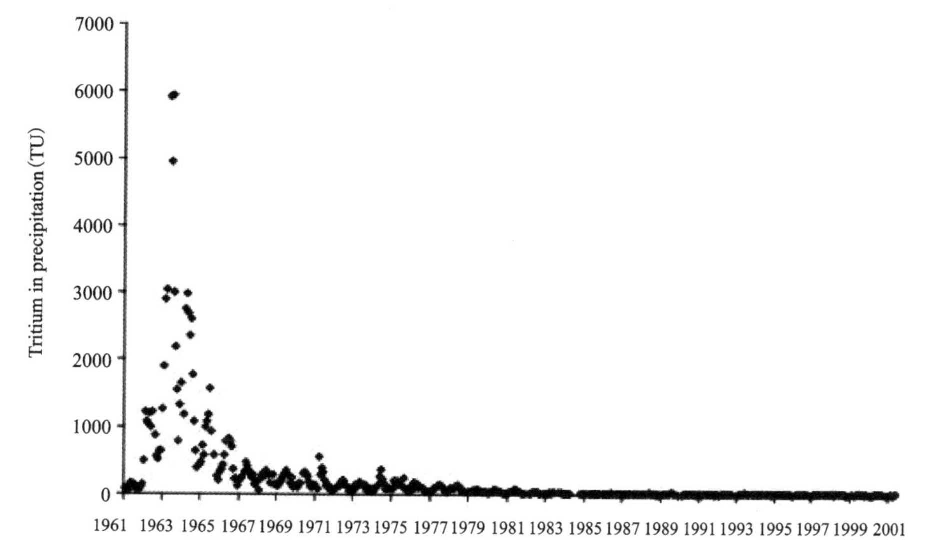

Figure 12. 2 Concentration of tritium in the atmospheric precipitation at Vienna, Austria.

In addition, tritium can also be produced by nuclear reactor operations, watch industry wastes and neutron radiation of lithium in rocks. The amounts of these ways were much less than the thermonuclear tests.

12. 1. 1 Analysis

When sampling, one-liter bottles are preferred to allow repeated measurements if necessary. There is no special precaution except to avoid watches and compasses, which contain some tritium. Further, samples with a high amount of radon-222 should be degassed because ^{222}Rn is also an emitter of beta particles and can cause interference. In some cases, ^{222}Rn is removed by dissolving it in an immiscible organic solvent (which is added to the sample) and separating the water and the solvent afterward. Radon-222 is 10 times more soluble in the solvent than in the water, so all ^{222}Rn are dissolved in the solvent only. Another approach is to store sample for a few days to allow for the complete disintegration of ^{222}Rn atoms because of the short half-life of 3.8 days.

The ^{3}H content of water samples is measured by the Liquid scintillation counting (LSC), or direct counting with a detection limit of about 8～10TU involving distillation of the sample and counting the disintegrations. The sample volume required is at least 20mL. If the water had lower ^{3}H content, then low-level counting and ultra low-level counting can be use with the detection line of 1TU and 0.1TU, respectively, and the sample volume required is 250mL and 500mL, respectively.

The result of tritium analysis of the water samples is usually reported in tritium units. They can be reported in activities as well, e.g. 1Bq/L. Activities are convertible to TU by dividing to 0.118 1.

12.1.2 Groundwater dating by tritium

The tritium artificially produced by thermonuclear tests can be used to evaluate the young groundwater age via ① locate the position of the bomb-peak tritium or ② find high tritium samples that we can confidently relate to the thermonuclear era. For ①, we need to measure the vertical concentration of tritium in a profile such as in a piezometer nest. The age of the groundwater would then be the "date of sampling-1963 (in Northern Hemisphere) or date of sampling-1964 (in Southern Hemisphere)." This approach, which will give us a quantitative age, works well at a flow divide where downward flow prevails; horizontal flow causes the loss of a distinct peak due to dispersion. For ② we have to have water samples whose tritium concentration is over 8TU. Such samples are certainly younger than 70 years (2022－1952＝70), but their precise age cannot be determined. This occurs for the majority of cases. We should point to the fact that although thermonuclear testing facilitated the practice of dating young groundwater until about 1990, but it has prevented using naturally derived tritium as a dating tracer. This unfortunate situation will last for another 3 decades, until the impact of thermonuclear tests on the tritium in the environment is completely erased. By the year 2046, 99% of the 1963 bomb-peak tritium atoms will be decayed. We will be then able to again use naturally derived tritium as a dating tracer because we will have a near-precise initial value, which is the natural level of cosmogenic production. This statement will hold true if further bombs are not tested or exploded and if other significant widespread sources of tritium are not created. In such case, the dating range of the tritium methods is from 1.9 to 41 years if 10% and 90% of the initial atoms disintegrate into helium.

12.2 $^3H/^3He$

The fading of the tritium dating method has led the scientists to revive an old technique, namely $^3H/^3He$, to replace it. The ^3He in the groundwater was produced by atmosphere, the fission of ^6Li by neutrons (nucleogenic helium), tritium decay and mantle release. have their own particular $^3H/^3He$ ratio and lead to specific $^3H/^3He$ ratios in groundwater. This is how heliums of different sources are differentiated. In young (and in most cases shallow) groundwater, the contribution of nucleogenic and mantle helium is insignificant. However, it is still a difficult task to separate the atmospheric helium from the tritiogenic helium if there is considerable excess air component, though excess air can be quantified by measuring dissolved gases in water. Furthermore, if other sources of helium such as nucleogenic and mantle are significantly present, dating by $^3H/^3He$ becomes practically impossible.

12.2.1 Analysis

To avoid contamination with air helium, samples for ^3He analysis are usually collected in 10mL copper tubes sealed with stainless steel pinch-off clamps. Helium is then separated from all other gases dissolved in the sample and is analyzed by a mass spectrometer. The unit to report tritium concentration is TU and the unit to report ^3He concentration is also TU. One TU of ^3He represents one ^3He atom per 1018 hydrogen atom and is approximately equal to 2.487pcm^3 ^3Heper kg of water. In the other words, one TU of ^3H decays to produce 2.487pcm^3 ^3He per kg of water at STP. Often in the literature, the concentration of ^3He in the groundwater samples is reported as $\%\delta^3He$:

$$\%\delta^3He=\frac{R_{sample}-R_{air}}{R_{air}}\times 100 \tag{12.3}$$

where R_{sample} is the $^3H/^3He$ ratio (units in cm^3 STP/g H_2O) of the water sample, R_{air} is the $^3H/^3He$ ratio in the atmosphere, which is 1.384×10^{-6}. Analytical precision for ^3He is about $\pm 1\%$. Note must be made that ^4He is always simultaneously measured with ^3He, because sampling and sample preparation are the same (no further cost) and also because ^4He is needed to precisely determine the tritiogenic ^3He.

12.2.2 Groundwater dating by $^3H/^3He$

The tritium-helium method measures the relative abundance of tritium and ^3He in a groundwater sample. The amount of ^3He from the decay of tritium is measured along with the amount of tritium remaining in the water. That sum is equal to the amount of tritium that was present at the time of recharge, or the initial value. Mathematically we write:

$$^{3}H_{0} = {}^{3}H + {}^{3}He_{tri}$$

$$^{3}H = {}^{3}H_{0} \ln e^{-\lambda t} \tag{12.4}$$

Combining these two equations, we obtain:

$$t = \frac{1}{\lambda} \ln\left(\frac{^{3}He_{tri}}{^{3}H} + 1\right) \tag{12.5}$$

It is seen from the equation that in order to measure the age of a groundwater sample, we simply need to measure its tritium and $^{3}He_{tri}$ simultaneously. There is no need to have the initial value of ^{3}H, i. e. C_0 or $^{3}H_0$.

In real, we need to measure the excess air component of helium and the ^{4}He content of the sample to calculate $^{3}H/^{3}He$ ratio in order to determine the atmospheric component of helium. If mantle and the nucleogenic components of helium. are thought to be significant, they also need to be calculated.

12.3 ^{85}Kr

Krypton is a colorless and inert gas belonging to the 8th group of the Mendeleyev table. Krypton is highly soluble in water, together with xenon. Typical abundances of atmospheric krypton in groundwater range from $(7.61 \sim 12.57) \times 10^{-8}$ cm^3 STP/g and from $(2.26 \sim 3.80) \times 10^{-8}$ in seawater (Ozima and Podosek, 2001). Krypton has the atomic number and atomic weight of 36 and 83.8, respectively. It has 26 isotopes ranging in atomic number from 71 to 95. Six of these are stable: ^{78}Kr (0.35%), ^{80}Kr (2.28%), ^{82}Kr (11.58%), ^{83}Kr (11.49%), ^{84}Kr (57%), and ^{86}Kr (17.3%). All others are radioactive with half-lives of milliseconds to a few hours with the exception of ^{85}Kr and ^{81}Kr, whose half-lives are 10.76 and 229 000$\pm$ 11 000 years, respectively. Krypton-81 has since long been proposed as a groundwater dating tool. Krypton-85 is used for dating young groundwater.

Natural production of ^{85}Kr takes place in small amounts in the atmosphere by spallation and neutron activation of stable ^{84}Kr.

$$^{84}_{36}Kr + n \rightarrow {}^{85}_{36}Kr + \gamma \tag{12.6}$$

Manmade ^{85}Kr is formed when plutonium and uranium undergo fission. The main anthropogenic sources of ^{85}Kr to the atmosphere are, therefore, nuclear weapon testing and nuclear reactors used for both commercial energy production and weapons plutonium production. The nuclear weapons tests in 1945—1963 contributed about 5% of the total ^{85}Kr in the atmosphere. In nuclear reactors, most of the ^{85}Kr produced is retained in the fuel rods. When these fuel rods are reprocessed, the ^{85}Kr is released to the atmosphere. Most ^{85}Kr is produced in the northern hemisphere, specifically in North America, Western Europe, and the former Soviet Union. The present background concentrations of ^{85}Kr in the atmosphere are about 1Bq/m^3 and are doubling every 20 years (WMO, 2001).

12.3.1 Analysis

The main obstacle in preventing extensive use of ^{85}Kr as a dating tool is the extensive sample preparation and analysis. The volume of sample required is large (in excess of 100 liters) and depends on the krypton concentration of the water sample. The full procedure to analyze groundwater samples for ^{85}Kr includes the following various steps:

(1) Degassing: Gases dissolved in the sample must be extracted from the liquid.

(2) Separation of gases: Having done step(1), we will have a mixture of various gases. We have to separate each gas; in this case we have to separate ^{85}Kr from the rest. In a lot of cases, not all gases can be extracted from the water sample, and not whole krypton from the rest. To check the percentage of krypton extracted from the sample, it is compared with the amount of expected krypton in the sample based on the solubilities of krypton. Separation of krypton from other gases is carried by gas chromatography.

(3) Counting: The separated krypton is place in a counter and its activity is measured using low-level gas proportional counting facilities. The counting time is about 2~3 days to one week per sample, and a few samples are counted simultaneously.

Krypton-85 measurements of the groundwater samples are reported in disintegration per minute per cubic centimeter of krypton atstandard pressure and temperature (dpmcm-3 krypton STP).

12.3.2 Groundwater dating by ^{85}Kr

Krypton-85 disintegrates by beta decay to stable ^{85}Rb. In this respect, it can be used as a "clock" or a radioactive tracer.

$$^{85}_{36}Kr \rightarrow ^{85}_{37}Rb+\beta^- \tag{12.7}$$

However, its use in age-dating is more based on its nearly linear increasing concentration in the atmosphere. Atmospheric ^{85}Kr is dissolved in rainwater and is carried out to the unsaturated and then saturated zones. The higher the concentration of ^{85}Kr is, the younger the groundwater age is. If ^{85}Kr is combined with an additional radioactive isotope with a similar half-life (such as ^3H), additional confidence in the results can be gained. Krypton-85 is commonly used with tritium because the two tracers have similar half-lives but completely different input functions. Tritium input has been decreasing since the 1960s, but ^{85}Kr concentration has been on the rise. Its short half-life and increasing concentrations in the atmosphere make ^{85}Kr a potential replacement for ^3H as tritium levels continue to decline. A simpler chart, like Figure 12.3, can also be used to directly relate measured activity of the water sample to its age, but this need upgrading every year, i.e. each year the scale of the horizontal axis needs to change.

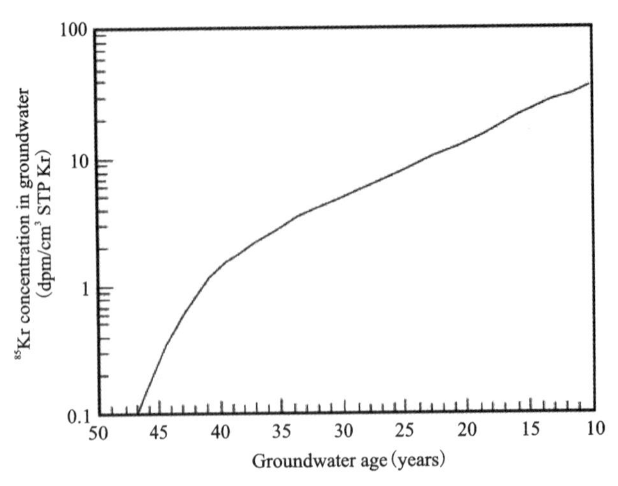

Figure 12. 3 Relationship between groundwater age and ^{85}Kr activity for samples collected in 2004 based on the activity of ^{85}Kr in the atmosphere of the Northern Hemisphere.

12.4 ^{32}Si

Silicon has the relative atomic weight of 28 and 3 stable isotopes: ^{28}Si (92.27%), ^{29}Si (4.68%), and ^{30}Si (3.05%). It also has a number of radioactive isotopes such as ^{26}Si, ^{27}Si, ^{31}Si, and ^{32}Si; but only ^{32}Si is of hydrologic and geoscientific interest. Despite being the second most abundant element (after oxygen) in the Earth's crust, forming 25.7% of the Earth's crust by weight, silicon isotopes are not widely used in geosciences because the nature of Si compounds is such that the δ^{30}Si variations are small and the fractionation processes are limited. The half-life of silicon-32 was and still is under intense investigation and doubt. The most recent estimation, which is the weighted average of a number of earlier measurements, puts it at 140±6 years.

Silicon-32 is produced in the upper atmosphere by cosmic-ray spallation of argon according to the reaction:

$$^{40}_{18}Ar + P \rightarrow ^{32}_{14}Si + P + 2\alpha \qquad (12.8)$$

The production rate of ^{32}Si, like that of other cosmogenic radionuclides, shows latitude and seasonal variations, anticorrelation to solar activity, and highest natural production in Antarctica (Morgenstern et al., 1996). Silicon-32 has also been produced by thermonuclear tests, but this was washed out from the atmosphere quickly within a few years afterwards. So far, there is no evidence to suggest underground or other source of production for ^{32}Si. In the laboratory, silicon-32 is prepared by spallation from chlorine in a beam of protons at 340~420MeV by a reaction of the type:

$$^{37}_{17}Cl + P \rightarrow ^{32}_{14}Si + 2P + \alpha \qquad (12.9)$$

Silicon-32 disintegrates to ^{32}P through a beta emission reaction; the half-life of ^{32}P is 14.3 days:

$$^{32}_{14}Si \rightarrow ^{32}_{15}P + \beta^- + \bar{\nu} \qquad (12.10)$$

12.4.1 Analysis

The activity of ^{32}Si in the water samples is usually low, and hence a sample volume of between $5m^3$ to $20m^3$ is required to achieve satisfactory results. Due to the low activity of ^{32}Si, its measurement in ground-and surface water samples is undertaken by counting its daughter product (^{32}P) through a five-step procedure described follow:

(1) Scavenging of dissolved silicon by a ferric hydroxide precipitation from the water samples (this process is undertaken in the field and only the precipitate and 30 liters of water sample are transported to the lab).

(2) Radio chemically pure silica is then extracted from the ferric hydroxide scavenge slime.

(3) ^{32}Si decays to ^{32}P in the pure silica extract and reaches a secular equilibrium within 1~3 months. A precise amount of carrier phosphate is added two weeks before counting to make the total $Mg_2P_2O_7$ content of the sample 62mg.

(4) Rradio chemically pure sample of ^{32}P is then re-extracted (milked).

(5) Finally, the milked ^{32}P activity is counted over a period of about 40 days.

The results of ^{32}Si analyses have largely been reported as the activity, i. e., dpm/m^3 or $milliBq/m^3$.

12.4.2 Groundwater dating by ^{32}Si

Since 1961, ^{32}Si activity in atmospheric precipitation has been measured at three laboratories including Physical Research Laboratory, Ahmedabad, India, University of Copenhagen, Denmark, and Bergakademie Freiberg, Germany. The activities are generally less than $10milliBq/m^3$ except largely for bomb-produced ^{32}Si during 1963—1965. Atmospheric silicon-32 is dissolved in rain-and snow water in the form of silicic acid. It is then deposited on the soil surface where it either infiltrates into the subsurface environment or is absorbed on the soil particles. About 50% of the deposited ^{32}Si is retained on the soil surface, and the remaining is transported to the subsurface groundwater system. That portion of silicon-32 that reaches groundwater decays to ^{32}P, and this forms the basis of groundwater dating. Using the basic law of isotopic decay, the age of a groundwater sample can be determined by having an initial ^{32}Si value of the recharged water and the given sample ^{32}Si concentration:

$$^{32}Si = ^{32}Si_0 e^{-\lambda t} \qquad (12.11)$$

Theoretically, the silicon-32 dating method can date groundwater whose ages range from 21 to 465 years, if 10% and 90%, respectively, of the original silicon-32 atoms have decayed.

This is, however, an invalid suggestion since the effectiveness of the method is under serious doubt.

12.5 ^{39}Ar

Argon, a noble gas with the relative atomic mass of 39.948g/mol, is inert, colorless, and odorless and makes up about 0.934% of the Earth's atmosphere, the third abundant gas after nitrogen and oxygen. Argon has 3 stable and 12 radioactive isotopes ranging in atomic mass from 32 to 46. All radioactive isotopes, but ^{39}Ar, have short half-lives of milliseconds to days. The most abundant isotopes of argon in the terrestrial atmosphere are ^{40}Ar (99.6%), ^{36}Ar (0.337%), and ^{38}Ar (0.063%). However, argons-37,-39, and -40 are the most extensively studied isotopes of argon for their various applications in the fields of environment, planetary, and earth sciences. Argon-39 has a half-life of 269 years. The decay constant, λ, of argon-39 is 0.002 577 year^{-1}. The solubility of argon in water in equilibrium with the atmosphere at 10℃ and 1 atmosphere is 3.9×10^{-4} cm^3 STP/mL water.

In the atmosphere, ^{39}Ar takes place by nuclear weapon testing at a rate of less than 0.005dpm/L and cosmic-ray activity, primarily with ^{40}Ar with a rate of 0.1dpm/L:

$$^{40}_{18}Ar+n \rightarrow ^{39}_{18}Ar+2n \tag{12.12}$$

Such a reaction does not occur underground because it requires neutrons with energy in excess of 12.8MeV.

In the subsurface environment, ^{39}Ar was produced:

(1) through irradiation of ^{39}K with fast neutrons in a neutron-capture, proton-emission reaction:

$$^{39}_{19}K+n \rightarrow ^{39}_{18}Ar+P \tag{12.13}$$

(2) via a less important reaction of ^{39}K:

$$^{39}_{19}K+\overline{\mu} \rightarrow ^{39}_{18}Ar+v_{\mu} \tag{12.14}$$

(3) by neutron activation of ^{38}Ar:

$$^{39}_{18}Ar+n \rightarrow ^{39}_{18}Ar+\gamma \tag{12.15}$$

(4) via alpha emission by calcium-42.

$$^{42}_{20}Ca+n \rightarrow ^{39}_{18}Ar+\alpha \tag{12.16}$$

12.5.1 Analysis

Sampling groundwater for ^{39}Ar analysis requires a huge volume of water, and this is the major drawback of this technique, especially where groundwater flow rate into boreholes and piezometers is slow. The amount of water required depends on the concentration of argon in the sample, the degassing efficiency, and the amount of the argon required (about 2 liters of argon is usually needed). It is $15m^3$ on average with a minimum of $10m^3$. Initially, all gases present in the water sample are separated (degassing) by pumping water through an evacuated cylinder of acrylic glass from where the gas is extracted with a vacuum pump followed by two high-pressure compressors. Different gases in the extracted gas such as argon, krypton, etc. are separated using distillation, gas chromatography, and chemical procedures. Once argon gas is separated through the lengthy process, its argon-39 activity is measured by low-level proportional radioactive counting.

Argon-39 results of analysis are reported in four ways:

(1) Activity: is the direct measurement of activity of ^{39}Ar in the sample, i. e. 0.15dpm/L of water (or per liter of argon) or 0.002 5bq/L. If the reported activity is the activity in water, its quantity (value) is much less compared to when it is reported in argon.

(2) Atoms in groundwater, i. e. atoms/L or atoms/mL. Activity values can be converted to atoms/mL if we have the concentration of argon in groundwater. The formula for this, which needs careful attention in harmonizing the units if used, is:

$$^{39}Ar(\text{atom/L water}) = \text{Actvity}(\text{dpm/L of argon}) \times Ar(\text{cm}^3 Ar/\text{cm}^3 \text{ water}) \times 204\,020\,779 \quad (12.17)$$

Note that argon-39's half-life in seconds divided by $(\ln 2 \times 60)$ equals 204 020 779 [8 483 184 000/41.58=204 020 779]. The coefficient of 60 is used to convert dpm to Bq.

(3) Percent modern argon (pma): This unit is the ratio of the ^{39}Ar activity in the sample to the argon-39 activity in the atmosphere and it is expressed as percentage, e. g. 76% modern argon. 100pma is equal to $1.78 \times 10^{-6} Bq/cm^3$ of Ar (1.05×10^{-4} dpm), or $1.67 \times 10^{-2} Bq/cm^3$ of air.

(4) $^{39}Ar/^{40}Ar$ ratio. This ratio expresses the number of ^{39}Ar atoms to the number of argon-40 atoms in the sample. In the atmosphere, this ratio is 8.1×10^{-16}, which means the number of argon-40 atoms is almost $10^{15} \sim 10^{16}$ times that of ^{39}Ar.

12.5.2 Groundwater dating by ^{39}Ar

Atmospheric ^{39}Ar is dissolved in the rain-and snow water molecules and reaches the Earth's surface. Upon entering the subsurface environment, the communication between ^{39}Ar content of rainwater (now groundwater) molecules and the atmospheric ^{39}Ar stops. From this point onward, ^{39}Ar decays to ^{39}K by a beta emission as:

$$^{39}_{18}Ar \rightarrow {}^{39}_{19}K + \beta^- + \bar{\nu} \quad (12.18)$$

Therefore, as time goes by, the number of ^{39}Ar atoms in the groundwater decreases, albeit if there is no subsurface addition of +Ar. The age of the groundwater can then be determined using the simple decay equation:

$$^{39}Ar = {}^{39}Ar_0 e^{-\lambda t} \quad (12.19)$$

What we measure by analyzing water samples is ^{39}Ar; $^{39}Ar_0$ is the initial concentration of

^{39}Ar in the recharging rainwater (usually can be taken as 0.107dpm/L because the atmospheric production is well known and constant), λ is the decay constant for ^{39}Ar (0.002 577 $year^{-1}$), and t is the age of the given groundwater sample in year. Theoretically, ^{39}Ar dating method can be used to date groundwater in the age range of 41 to 894 years, if 10% and 90%, respectively, of the original atoms have decayed. This is, however, a rough suggestion and the actual dating range and the effectiveness of the method depend on the accurate estimation of subsurface production of ^{39}Ar as well as on the precision and detection limit of the analytical facilities, which are not available widely. In addition, it is becoming more evident that the application of this method should be limited to sedimentary and carbonate terrains where subsurface production of ^{39}Ar is limited.

12.6 ^{14}C

Carbon has 15 isotopes ranging in mass from 8 to 20, but only three of these occur naturally. There are two stable carbon isotopes: carbon-12, or ^{12}C (98.89%), and carbon-13, or ^{13}C (1.11%), which was introduced in Chapter 5. Carbon-14, or radiocarbon, the naturally occurring radioactive isotope of carbon.

Sources of ^{14}C and how they enter rain-and groundwater molecules are shown in Figure 12.4. As a result of cosmic radiation, a small number of atmospheric nitrogen nuclei are continuously transformed by neutron bombardment into radioactive nuclei of ^{14}C. A neutron knocks a proton out of nitrogen and takes its place.

$$^{14}_{7}N + n \rightarrow ^{14}_{6}C + P \qquad (12.20)$$

The amount of cosmic rays penetrating the Earth's atmosphere varies with the sun's activity and the Earth's transit through magnetic clouds in the Milky Way Galaxy. This governs the amount of ^{14}C produced and the half-life used to date various materials. The average annual natural production rate of ^{14}C, which is temporally and spatially variable, is 7.5kg, and the total mass of ^{14}C on Earth is about 75 tons. The ^{14}C formed is rapidly oxidized to $^{14}CO_2$ and enters the Earth's plant and animal life through photosynthesis and the food chain. The rapidity of the dispersal of ^{14}C into the atmosphere has been demonstrated by measurements of radioactive carbon produced from thermonuclear bomb testing. ^{14}C also enters the Earth's oceans in an atmospheric exchange and as dissolved carbonate.

12.6.1 Analysis

An earlier method of measuring ^{14}C of groundwater samples depended on the radioactive decay counting, which requires a large volume of the water sample. As ^{14}C decays back to ^{14}N, it emits a weak beta particle (b), or electron. For the counting method, which is still in use, at least 2g of carbon should be present in the water sample. The major carbon-

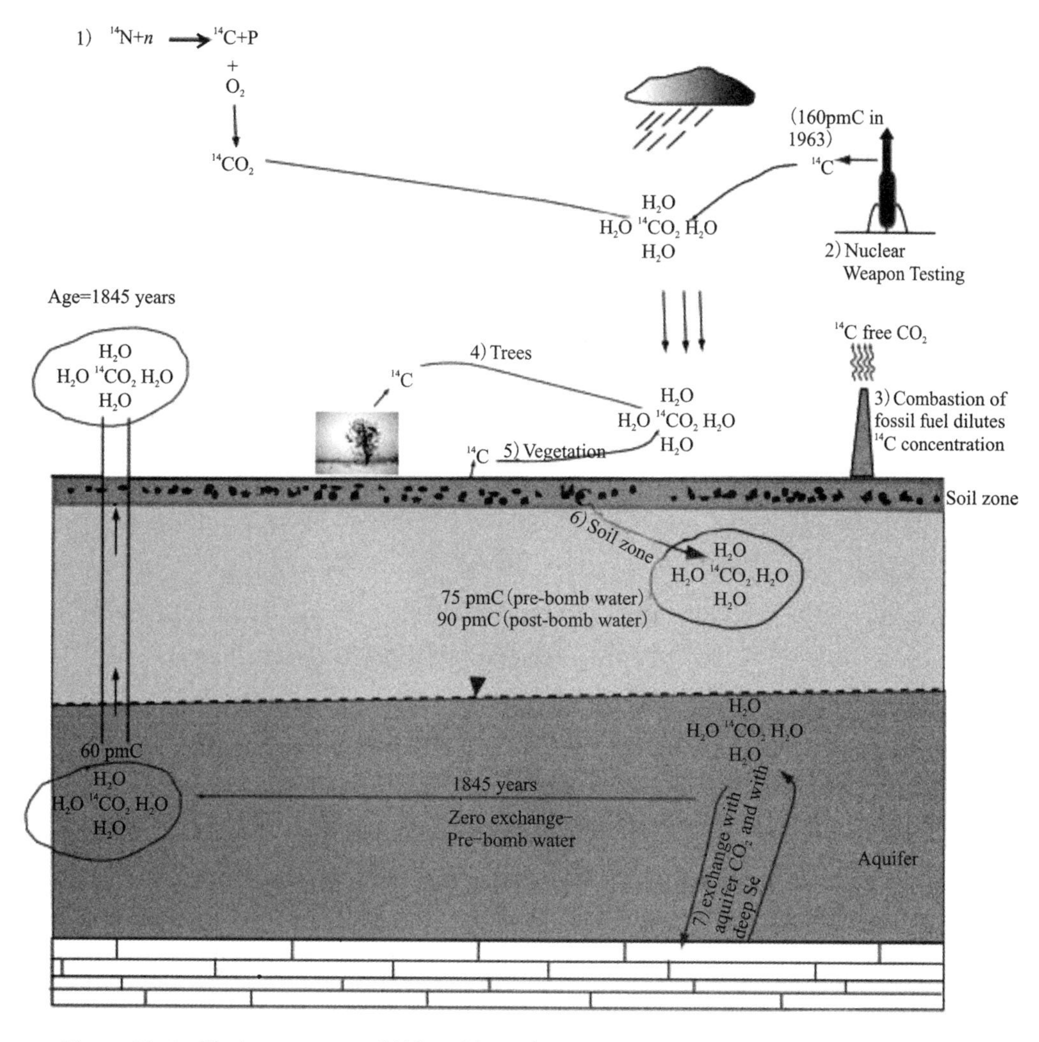

Figure 12.4 Various sources of ^{14}C and how they mix and enter the subsurface environment. An example of dating a pre-bomb groundwater sample is also illustrated.

containing species in the groundwater is bicarbonate. We should, therefore, determine the volume of sample required for ^{14}C analysis by determining the sample's bicarbonate concentration. This is done in the field prior to sampling for ^{14}C. For example, if the bicarbonate concentration of a groundwater is 205mg/L, it contains 40.3mg of carbon in each liter. Therefore, we need to collect a sample volume of 50L to obtain 2g carbon. The carbon content of the groundwater sample can be precipitated as $BaCO_3$ in the field. The problem of a large sample volume was overcome with the development of the accelerator mass spectrometry, which literally extracts and counts the ^{14}C atoms in the sample and at the same time determines the amount of the stable isotopes ^{12}C and ^{13}C. The AMS injects negative carbon ions from the analyte into a nuclear accelerator. The negative ions accelerate toward the positive potential where they pass through a thin carbon film or tube filled with low pressure gas. Nitrogen ions are so unstable that they self-destruct before getting to the accelerator terminal leaving the ^{14}C alone to be counted. The molecules dissociate into their component atoms and the kinetic energy they had accumulated is distributed among the separate atoms all different from the ^{14}C. Accelerating the ions to high energy also has the advantage of allowing well-established nuclear physics techniques to detect the individual ^{14}C ions. As a consequence, a measurement that may last 12h and require several grams of sample using decay counting may take only 30min and consume only a few milligrams using AMS. Carbon-14 analyses are reported in percent modern Carbon, pmC.

12.6.2 Groundwater dating by ^{14}C

Atmospheric ^{14}C is dissolved in the percolating rainwater, as shown in Figure 12.3 and reaches the groundwater table. In groundwater, ^{14}C starts decaying to nitrogen:

$$^{14}_{6}C \rightarrow {}^{14}_{7}N+\beta^{-} \tag{12.21}$$

If no further ^{14}C exchange occurs, measurement of the remaining ^{14}C atoms can be used to date groundwater following the first-order kinetic rate law for decay:

$$C=C_0 e^{-\lambda t} \tag{12.22}$$

where C_0 is the activity assuming no decay occurs (initial activity or activity at $t=0$), and C is the observed or measured activity of the sample. The groundwater datable by ^{14}C ranges in age from 870 to 19 000 years if 10% and 90%, respectively, of the original atoms are assumed to have decayed. However, ages up to 40 000 years or longer have been reported by this method. This can be achieved if the right conditions are met. AMS technique has also helped to increase the dating range of ^{14}C method by enabling lower detection limits.

The ^{14}C content of the rainwater is modified through a number of processes numbered from 1 to 6 in Figure 12.4 and listed below. This makes dating by ^{14}C a delicate exercise. The final outcome of these processes together is an unknown and difficult to calculate initial value (C_0), which could vary from less than 75pmC to over 200pmC. The ideal situation for determining initial value, C_0 is to be able to measure the ^{14}C concentration of a young but tritium-free pre-bomb water in the upgradient area of the portion of the aquifer we want to date. The classical approach to determine the initial value is by tree rings for almost 7000 years, with no way to accurately determine it prior to 7000 years ago.

In addition to the complex processes above, which affect the initial value of ^{14}C and make the interpretation of ^{14}C measurements a challenge, a number of geochemical processes also affect the dating results. A series of geochemical processes in groundwater that can alter the ^{14}C quantities in groundwater in ways not fully reliant on radioactive decay. These include: ①The congruent dissolution of carbonate minerals, which adds carbon without ^{14}C activity to the groundwater, which results in a lower ^{14}C ratio for the sample; ② The dissolution of carbonate or other calcium-containing minerals accompanied by the precipitation of calcite, which could remove ^{14}C; ③ The addition of dead carbon from other sources such as the oxidation of old organic matter, sulfate reduction, and methanogenesis can reduce the ^{14}C activity of the sample; ④ Possible isotopic exchange involving CO_2 and carbonate minerals could lower the ^{14}C activity, though this process is negligible at normal groundwater temperatures.

To obtain true ages of groundwater samples, two adjustments must be done on the ages obtained. The first is to correct the half-life used for dating from 5730 years to Libby's 5568 years. This causes insignificant modification; for example, a 17 000-year-old age is reduced to 16 519 years ($17\,000 \times 5568/5730 = 16\,519$ years). The second adjustment is to take into account the secular variations in ^{14}C production and to calibrate the ages to years before 1950 (a standard calibration curve is needed to do this). For younger groundwater, these two adjustments do not result in substantial changes, but for older groundwater changes become appreciable.

One can interpret ages from a single sample or use ion and isotopic data from many samples using mass balance modeling to sort out the major inputs and outputs of carbon. The simplest way is to account only for the most important process affecting ^{14}C activity, which is the congruent dissolution of calcite.

$$CO_2 + H_2O + CaCO_3(s) \rightarrow Ca^{2+} + 2HCO_3^- \tag{12.23}$$

At equilibrium according to this reaction, half the bicarbonate would be generated from a source containing ^{14}C (CO_2), and the other half would be generated from a dead source like calcite. A number of complex computer codes have been developed to account for various mineral dissolution and isotopic exchange reactions. Some of the preceding models assume an open system, in which groundwater ^{14}C interacts and exchanges with the surrounding environment, and some assume a closed system.

12.7 ^{81}Kr

As described above, ^{85}Kr is described as a tracer to date young groundwater, and ^{81}Kr has since long been proposed as an old groundwater dating tool. Krypton-81 is produced in the upper atmosphere by cosmic-ray-induced spallation of five heavier Kr isotopes, e. g.

$$^{83}_{36}Kr + P \rightarrow ^{81}_{36}Kr + P + \alpha \tag{12.24}$$

and through neutron capture by ^{80}Kr:

$$^{80}_{36}Kr + n \rightarrow ^{81}_{36}Kr + \gamma \tag{12.25}$$

So far, all research suggests that there is neither significant subsurface production nor an appreciable anthropogenic source for ^{81}Kr. It can, therefore, be viewed as a solely atmospheric tracer and an ideal one because of its long half-life, largeness of the atmospheric reservoir, low direct yield from spontaneous fission of ^{238}U (i. e. insignificant subsurface production), and shielding from other decay products by ^{81}Br.

12.7.1 Analysis

In the field, field gas extraction is required as relatively large amounts of water needed to be sampled to satisfy the requirements of dating. According to Henry's law, dissolved gas will be released from water when exposed to a vacuum environment. Some characteristics of such systems must be satisfied: ①Be well sealed to avoid ambient air contamination. ②High extraction yield and speed to minimize the water volume and extraction time. ③Be sufficiently robust and convenient for field work. There are two methods to realize the extraction, vacuum extraction chambers and membrane contactors. The vacuum extraction system was introduced herein.

The diagram of the vacuum extraction system is shown in Figure 12.5. A diaphragm pump is first used to prepare the vacuum before field working. Sand in the water pumped from the wells are filtered out by cotton filters. The water pressure is increased by a booster pump to satisfy the atomization. A pressure gauge and a flow meter record the water pressure and flux. Water is sprayed through 4 atomizer nozzles into a transparent cylinder made of plexiglass. At the bottom of the cylinder, water is continuously pumped out by magnetic drive pumps which can work at negative pressure. The water flux is controlled to match the pump rate and to reach a dynamic equilibrium of water levels. Gas at the top of the cylinder is continuously transferred to a pre-vacuumed sample container (typically 4L) through a compressor.

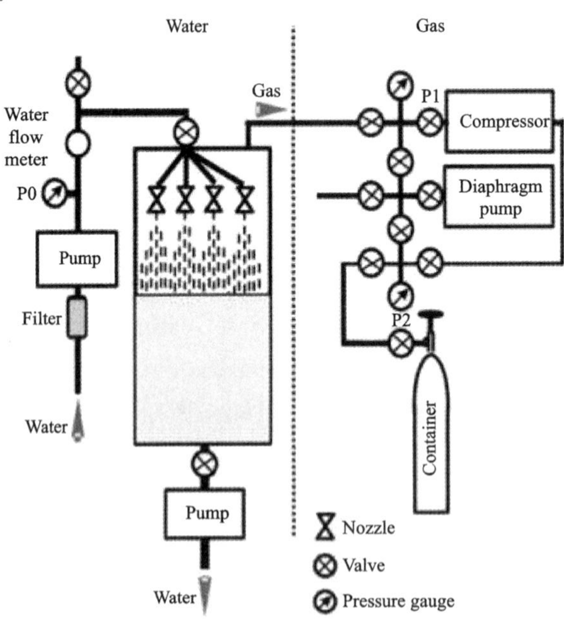

Figure 12.5 Diagram of vacuum extraction system.

An extraction yields higher than 75% is reached, according to the oxygen concentration before and after extraction measured by a dissolved oxygen meter. The flow rate is about 5～8L/min, mainly limited by the drainage pump. The leak rate is about 5Pa • L/min, which represents less than 0.1% contamination from modern air during 2h operation.

The separated gas needs to be purified before the analysis. the two steps of the purification process, cryogenic distillation and gas chromatography (GC) separation, are applied to separate ppm-level krypton from a large quantity of gas which have been collected from environmental samples. Accelerator mass spectroscopy (AMS) has been used to date ^{81}Kr using 500μL of krypton extracted from 16 tons of groundwater (Collon et al., 2000). The low efficiency and isobar interference make it difficult for ^{81}Kr dating. Resonance ionization mass spectrometry (RIMS) was applied to measuring the ^{81}Kr in the Milk River aquifer, Canada. With the problems of poor isotope selectivity and pre-enrichment unsolved, RIMS has so far not worked successfully at multisamples (Lehmann et al., 1991). Atom trap trace analysis (ATTA) is a laser-based method and mainly used to ^{85}Kr and ^{81}Kr, counting particular rare isotope atoms in a magneto-optic trap (MOT) with micro-liters (STP) of sample gas. Only atoms resonating with the laser frequency will be trapped and detected, without interference from any other species. The results of ^{81}Kr analyses are reported relative to the modern atmospheric value:

$$\frac{R}{R_{\text{air}}}=\frac{[^{81}\text{Kr}/\text{Kr}]_{\text{sample}}}{[^{81}\text{Kr}/\text{Kr}]_{\text{air}}} \tag{12.26}$$

12.7.2 Groundwater dating by ^{81}Kr

For a present-day sample, $R_{81}=1$. The apparent ^{81}Kr-age of the sample can be calculated as follows:

$$t=-t_{1/2}\ln(R_{81})/\ln 2 \tag{12.27}$$

where $t_{1/2}=229\pm11$ka is the half-life of ^{81}Kr.

For cases of $R_{81}<1$, the sample is too old for any original ^{85}Kr to be present, thus the ^{85}Kr decay activity (denoted R_{85}) is expected to be zero. Any presence of ^{85}Kr indicates a mixture of young or modern gas sources, and a correction must be made in the age calculation. The contamination can come from different sources: ① Mixing with young groundwater (age<50 years). Additional information about the source of mixture is needed to obtain the ages of different end members. ②Mixing with air during the gas extraction and purification process. Air can be introduced into the groundwater and contaminate the gas sample in the well head or in the pumping system. In addition, leaks can occur in the degassing system or the sample bottle.

For case②, effects of air contamination can be corrected. The relationship between the corrected ^{81}Kr abundance (R_{81_corr}) and the measured ^{81}Kr abundance (R_{81_meas}), both relative to the modern air value, is:

$$R_{81\_meas} = R_{81\_corr} \times (1-\eta) + \eta \tag{12.28}$$

where η is the fraction of the krypton coming from modern contaminants. The contamination fraction η can be extracted using the measured ^{85}Kr activity (R_{85_meas}) and the ^{85}Kr activity of the local air at the sampling site (R_{85_air}):

$$R_{85\_meas} = R_{85\_air} \times \eta \tag{12.29}$$

By combining Eq. 12. 28 and Eq. 12. 29, the corrected ^{81}Kr abundance (R_{81_corr}) can be calculated.

12.8 ^{36}Cl

Chlorine has 16 isotopes whose mass numbers range from 31 to 46, but only three of these occur naturally. The remaining 13 have half-lives shorter than 1h and this is why they do not occur in nature. The three abundant isotopes of chlorine include two stable isotopes, chlorine-35 and chlorine-37, with 75. 53% and 24. 47% abundances, respectively, and one radioactive isotope, ^{36}Cl, with a half-life of 301 000±2000 years. Chlorine-36 is produced through the following processes:

In the atmosphere, by cosmic-ray splitting of ^{40}Ar and neutron activation of ^{36}Ar:

$$^{40}_{18}Ar + P \rightarrow {}^{36}_{17}Cl + n + \alpha \tag{12.30}$$

$$^{36}_{18}Ar + n \rightarrow {}^{36}_{17}Cl + P \tag{12.31}$$

The amount of ^{36}Cl produced through these processes depends mainly on the geographic latitude and is greater in middle latitudes. It is estimated that splitting of ^{40}Ar and activation of ^{36}Ar are responsible for 67% and 33% of total natural atmospheric production of ^{36}Cl, respectively.

From nuclear power and nuclear fuel re-processing facilities. Chlorine-36 together with iodine-129 are the only two elements that can escape from the nuclear fuel waste disposal repositories and enter the surrounding environment. Almost all other radionuclides will remain trapped, either in the fuel itself or within a very short distance of it (Wiles, 2002). It has also been recently suggested that the operation of two nuclear reactors in the United States has led to the production of ^{36}Cl (Davis et al. , 2003).

By neutron activation of stable ^{35}Cl at the Earth's surface or in the shallow subsurface and in the deep subsurface:

$$^{35}_{17}Cl + n \rightarrow {}^{36}_{17}Cl + \gamma \tag{12.32}$$

The rate of production of ^{36}Cl through this process varies depending on the type of rocks, minerals, and solutions, and the availability of neutron sources. In general, it is higher in rocks with higher uranium and thorium contents such as uranium ore deposits and is lower in lithologies like basalt and sandstone.

12.8.1 Analysis

The collected groundwater should contain a minimum of 1～2mg of chloride, but up to 25mg may be required. If samples are too low in chloride, preconcentration must be undertaken. Collected samples are then prepared for target loading. This means preparation of about 8～10mg of pure AgCl and pressing it into the copper sample holders. The samples are prepared for target loading through addition of $AgNO_3$ solution to the samples to precipitate AgCl. The precipitated (AgCl) is then purified by dissolving it with NH_4OH. Next, $Ba(NO_3)_2$ is added to precipitate unwanted sulfur as $BaSO_4$. HNO_3 is added next to neutralize the solution and in the last stage, the AgCl is reprecipitated with $AgNO_3$ for target loading. If the Cl^- concentration of the sample is very low, a ^{36}Cl free carrier must be added to it. The carrier may be AgBr of very low sulfur and chlorine content. The prepared water samples are then analyzed for ^{36}Cl content using an accelerator mass spectrometer (AMS). Usually standard and blank samples are tested to identify the reproducibility and accuracy of the measurements.

The measured chloride-36 is reported in two ways:

(1) As a ratio of $^{36}Cl/Cl \times 10^{15}$ (the number of ^{36}Cl atoms to the total number of chlorine atoms). The multiplication factor (10^{15}) is used because the actual concentration of ^{36}Cl is very small.

(2) As atoms of ^{36}Cl per liter of water using the following equation:

$$^{36}\mathrm{Cl}(\text{in atoms per liter}) = 1.699 \times 10^4 \times B \times R \tag{12.33}$$

where B is the concentration of chlorine in mg/L and R is the (^{36}Cl/total Cl) $\times 10^{15}$ ratio. Therefore, to determine the concentration of ^{36}Cl, the concentration of Cl^- in water samples must be measured with conventional methods such as silver nitrate titration, ion selective electrode, and ion chromatography.

12.8.2 Groundwater dating by ^{36}Cl

The principle of this method is simple. It is based on the radioactive decay of ^{36}Cl in the subsurface groundwater system. Above the Earth's surface, ^{36}Cl (atmospheric ^{36}Cl) with an initial value, $^{36}Cl_0$, enters groundwater by rainwater infiltration. After time t, it decays to ^{36}S and ^{36}Ar:

$$^{36}_{17}\mathrm{Cl} \rightarrow {}^{36}_{18}\mathrm{Ar} + \beta^- + \bar{\nu} \tag{12.34}$$

$$^{36}_{17}\mathrm{Cl} \rightarrow {}^{36}_{18}\mathrm{S} + \beta^- + \bar{\nu} \tag{12.35}$$

and reaches a new concentration, ^{36}Cl, according to the decay equation:

$$C = C_0 e^{-\lambda t} \tag{12.36}$$

If one knows the initial concentration, C_0, and the present concentration, C, then one can calculate the length of time that the ^{36}Cl has resided in the subsurface groundwater system. Chlorine-36 dating method is capable of dating groundwater with an age range of 46 000 to 1 000 000 years if we assume that 10% and 90%, respectively, of the original ^{36}Cl atoms are disintegrated.

However, many inaccuracies and problems surround C_0 and C values, the initial and the present concentration of ^{36}Cl, respectively. We do not accurately know what the concentration of ^{36}Cl in rainwater was when it entered the subsurface system. This is the biggest obstacle for the ^{36}Cl dating method (Love et al. ,2000). Davis et al. (2001) explains 6 ways to estimate the C_0 value and argue that all these ways contain weaknesses and inaccuracies. These approaches include:

(1) Calculation of theoretical cosmogenic production and fallout.

(2) Measurement of ^{36}Cl in the present-day atmospheric precipitation and use it as C_0.

(3) Assuming that shallow groundwater contains a record of C_0.

(4) Extraction of ^{36}Cl from vertical depth profiles in desert soils.

(5) Recovering ^{36}Cl from cores of glacial age.

(6) Calculation of subsurface production of ^{36}Cl for water that has been isolated from the atmosphere for more than one million years.

As with regard to the measured C, the situation is not satisfying, too. The assumption inusing the decay equation is that the ^{36}Cl atoms that enter the subsurface groundwater system behave as an isolated packet when they migrate through the flow system (piston flow theory). As explained above, ^{36}Cl is not only produced in the atmosphere, but it is also produced in the subsurface. In addition, subsurface mixing (mixing of low and high ^{36}Cl water), cross-formational flow, diffusion between aquitard and aquifer, and dilution and evaporation processes complicate the task of finding which C should be used in the decay equation: the C measured in the laboratory or the C obtained when the contributions/effects of all the above factors have been eliminated.

12.9 ^{4}He

There are eight known isotopes of helium, ranging in mass from 3 to 10, but only ^{3}He and ^{4}He are stable. The unstable ones have half-lives of less than milliseconds. ^{4}He is an unusually stable nucleus because its nucleons are arranged into complete shells. Helium is unusual in that its isotopic abundance varies greatly depending on its origin. Rocks from the Earth's crust have isotope ratios varying by as much as a factor of ten; this is used in geology to study the origin of such rocks. Heliums-3 and -4 are both used in age-dating groundwater as shown in 12. 2.

There are four different sources of ^4He in groundwater:

(1) Atmospheric helium: Like all other noble gases, a chief reservoir for ^4He is the atmosphere. Atmospheric helium is dissolved in rainwater and is carried into the groundwater. The ratio $^3He/^4He$ in the air is 1.384×10^{-6}, but due to small fractionation, it reduces to 1.36×10^{-6} when helium dissolves in precipitation. Hence, the concentration of ^4He in recharging groundwater is about $48\mu cm^3$(STP)/kg. A portion of atmospheric ^4He in groundwater is the result of excess air entrainment during recharge. This leads to higher ^4He concentration in recharging water. To date groundwater, we must separate ^4He resulting from solubility and ^4He from excess air. This is done by measuring neon concentration in groundwater and comparing helium to neon ratios in air (0. 288 2) with that of those in the groundwater sample.

(2) Radiogenic helium or crustal helium. Helium-4 is produced from aquifer matrix and from the sediment grains by alpha decay of uranium and thorium; the alpha particles that emerge are fully ionized ^4He nuclei:

$$^{238}_{92}U \rightarrow {}^{206}_{82}Pb + 8^4_2He \tag{12.37}$$

$$^{238}_{92}U \rightarrow {}^{234}_{90}Th + {}^4_2He \tag{12.38}$$

$$^{235}_{92}U \rightarrow {}^{207}_{82}Pb + 7^4_2He \tag{12.39}$$

$$^{232}_{92}U \rightarrow {}^{208}_{82}Pb + 6^4_2He \tag{12.40}$$

Helium produced within the Earth's crust through the above reactions enters groundwater system, too, the so-called crustal flux of helium.

(3) Mantle or terrigenic helium. There has recently been strong evidence to suggest that ^4He produced deep in the Earth's mantle can find its way to enter relatively shallow groundwater.

(4) Helium-4 atoms that were entrapped in the crystal lattices of sediments or rock strata in the course of deposition may enter groundwater through solid-state diffusion process. Ancient ^4He is the name for this type of helium.

12. 9. 1 Analysis

Like all noble gases, the main point to observe when sampling groundwater for helium is to avoid mixing it with air. Air helium could diffuse into the sample. Also, correct sample pressure has to be maintained to avoid helium degassing from the water samples. The main method used for sampling for ^4He is that using the copper tube welded at both ends with clamps. A copper tube is impermeable to helium, and samples can be stored, without contamination and degassing, for a year. Further, the malleability of copper allows convenient sample opening for degassing it in an extraction line. Use of the copper tube is an almost standard method for sampling those substances that are present in the atmosphere in high quantity and in the groundwater in small quantity.

Sampled groundwater is degassed in a vessel under high vacuum and the helium is purified from reactive gases (N_2, CO_2, water vapor, hydrocarbons) in ultrahigh vacuum extraction and purification lines. At the end, separated helium is analyzed using quadrupole mass spectrometry or noble gas mass spectrometry. The unit to report ^4He concentration is the cm^3 STP/g of water or the cm^3 STP/L of water.

12.9.2 Groundwater dating by ^4He

Helium accumulates in the groundwater from the in situ radioactive decay of the uranium and thorium in the aquifer matrix, as well as from any flux into the groundwater from the underlying crust and mantle. Therefore, the concentration of helium in the aquifer increases with time. If the rate at which the in situ production and crustal flux have supplied helium to the flowing groundwater is known, then it is possible to calculate the length of time that the groundwater of known helium content has resided in the aquifer. The equation for such a purpose is:

$$t=\frac{^4He_{rad}}{J_{He}} \tag{12.41}$$

where $^4He_{rad}$ and J_{He} are the measured concentration and the production rate (accumulation rate) of ^4He in groundwater, respectively. It is relatively simple to measure the concentration of ^4He in groundwater. But it takes a considerable amount of efforts to calculate the accumulation rate of helium, i. e. J_{He}. To calculate the accumulation rate, the following procedure must be undertaken:

(1) The sources of helium must be identified, whether there is only one in situ, crust, or mantle source or two or all of these sources.

(2) In situ production rate must be calculated by:

$$J_{He}=\frac{\rho_r}{\rho_w}(C_U\times P_U+C_{Th}\times P_{Th})\times\left(\frac{1-\theta}{\theta}\right) \tag{12.42}$$

where ρ_r and ρ_w are the densities of the aquifer material and the water, C_U and C_{Th} are uranium and thorium concentration in rocks ($\mu g/g$), and θ is the porosity of aquifer. Production rates from uranium and thorium decay are $P_U=1.19\times10^{-13}\,cm^3$ STP per/μg_U · yr and $P_{Th}=2.88\times10^{-14}\,cm^3$ STP per/μg_U · yr. Equation 12.42 is based on 100% transfer of produced ^4He into groundwater, which comes true after an equilibrium time between rock and fluid in the order of thousands of years. A number of other equations have been developed to calculate the in situ production rate of ^4He.

(3) Crustal and mantle sources of helium need to be calculated as well, and this is the main obstacle for quantitatively dating groundwater by ^4He.

12.10 ^{129}I

Iodine, a halogen with the atomic number of 53 and atomic weight of 126.9. Iodine has 37 isotopes, whose atomic weight ranges from 108 to 144, but only iodine-127 is stable, forming almost 100% of the mass. Short-lived isotopes of ^{125}I, ^{131}I, and ^{133}I, released from human nuclear-related activities, represent some threats to human health, with ^{133}I posing the greatest risk. It has been estimated that in the Chernobyl accident, 35MCi of ^{131}I were injected into the atmosphere. The most useful radioactive isotope of iodine is ^{129}I. It has a half-life of 15.7±0.04 million years and it is one of 34 known radioisotopes with half-lives of greater than one million years. In radioactive waste disposal business, ^{129}I is recognized as a bad radionuclide because its very long half-life makes it persist and because it is anionic and is therefore not delayed by any of the mineral barriers.

Iodine-129 is produced in seven different ways:

(1) By spallation of stable Xe isotopes in the stratosphere, e.g.

$$^{132}_{54}Xe + P \rightarrow ^{129}_{53}I + P + \alpha \tag{12.43}$$

(2) By cosmic-ray spallation of Ce, Ba, and Te isotopes and importing the produce through cosmic dust and meteorites.

(3) By volcanic emission of ^{129}I which was produced as a result of the fission of uranium isotopes in the Earth's interior.

(4) In the lithosphere, from spontaneous fission of uranium-238 and neutron-induced fission of uranium-235, the former reaction being the main contributor.

(5) From aboveground nuclear testing during 1960s and early 1970s; peak $^{129}I/I$ during this period reached about 10^{-7}.

(6) From the Chernobyl accident in 1986 (about 1.3kg).

(7) From nuclear power and nuclear fuel re-processing facilities. Iodine-129, together with ^{36}Cl, are the only elements that can escape from the nuclear fuel waste disposal repositories and enter the surrounding environment. Almost all radionuclides will remain trapped, either in the fuel itself or within a very short distance of it. It is estimated that about 8% of ^{129}I produced in the radioactive waste may be released to the groundwater.

12.10.1 Analysis

Normal filtering of the samples is necessary to avoid particulate matters. Sample volume required depends on the iodine concentration; about 1~2mg of iodine is needed, but higher amounts are preferred to handle them confidently and to be able to perform replicate analysis. However, because iodine concentration in groundwater is usually low, sample volumes are generally not small. Water samples to be analyzed for ^{129}I have to be prepared for target loading by AMS. The following steps are to be followed to extract the iodine from the water sample for AMS loading: ①Pre-concentration of the sample by rotary vacuum distillation; ②Addition of a 2～4mg low ^{129}I carrier; ③Acidification of samples by adding HNO_3; ④Addition of 5～15mL of CCl_4; ⑤Oxidization of I to I_2 as well as the dissolved organic matter with about 5mL H_2O_2; ⑥Addition of 20～30mL of 1mol NH_4OH-HCl (hydroxylamine hydrochloride) to reduce any IO_3^-; ⑦Back-extracting sample iodine by using a 0.1M $NaHSO_3/H_2SO_4$ acid solution; ⑧Addition of 2mg of Cl^- and co-precipitation of AgI and AgCl using $AgNO_3$; ⑨Dissolution of AgCl using NH_4OH (ammonium hydroxide); ⑩Centrifuging the remaining AgI, rinsing it with deionized water and drying it.

About 80% of the iodine is extracted from the water sample using the above procedure. The final prepared sample is an AgI pellet weighing 2～5mg, which can be loaded to AMS for ^{129}I analysis. In the hydrological studies, ^{129}I concentrations are usually reported as the ratio of ^{129}I to the total I, i.e. $^{129}I/I$ (note that I is virtually all iodine-127). Due to the smallness of this ratio, it is multiplied by a coefficient, i.e. $^{129}I/I \times 10x$. However, the coefficient is not well established as yet, whether it is 10^{12} or 10^{14}. Iodine-129 analysis results can be reported in atoms/liter as well. $^{129}I/I$ value is converted to atoms/L using follow equation:

$$^{129}I(\text{in atoms per liter}) = 4700 \times B \times R \tag{12.44}$$

where B is the concentration of iodine in the sample in $\mu g/L$ and R is $^{129}I/I \times 10^{12}$ or 10^{14}.

12.10.2 Groundwater dating by ^{129}I

Iodine-129 decays by beta emission to ^{129}Xe:

$$^{129}_{53}I \rightarrow ^{129}_{54}Xe + \beta^- + \bar{\nu} \tag{12.45}$$

After entering the groundwater system, ^{129}I starts disintegrating into ^{129}Xe. If there is neither subsurface production nor leaching of ^{129}I from the surrounding formations, as time goes by, the concentration of ^{129}I in the groundwater decreases following the decay principles. To calculate the age of groundwater, we need to have the concentration of ^{129}I in the recharging water (initial value) and in the sample and substitute these two in the simple decay equation:

$$^{129}I = ^{129}I_0 e^{-\lambda t} \tag{12.46}$$

For dating groundwater, the initial $^{129}I/I$ value of $(1.1 \pm 0.4) \times 10^{-12}$, corresponding to 2×10^4 atoms/L, is adapted. For dating pore water, the initial $^{129}I/I$ value of 1.5×10^{-12}, which is that of the recent sediments below the zone of bioturbation, is used.

12.11 Application

12.11.1 Application of combined ^{81}Kr and ^{4}He chronometers to the dating of old groundwater in a tectonically active region of the North China Plain(Matsumoto et al., 2018)

The isotopic tracers were used to a deep aquifer in a tectonically active area of the North China Plain (NCP)-the largest alluvial plain in eastern Asia. The NCP overlies a thick Cenozoic sedimentary basin covering ~150 000km^2, and consists of the piedmont pluvial plain, the central alluvial and flood plain, and the coastal plain (Figure 12.6). It is one of the most densely populated areas of the world and is of great agricultural importance for China. The regional aquifer system consists of thick Neogene and Quaternary deposits. These sediments are dominated by alluvial and lacustrine deposits with interbedded marine deposits in the littoral plain. The aquifer system of the Quaternary-Pliocene formations consists of five aquifers (Table 12.2 and Figure 12.6). Aquifer I, is a phreatic aquifer around 10~20m thick, consisting of fine-grained sand in the littoral plain. Depth to the groundwater table is 2~3m, and the specific yield is 1~2.5m^3/hm. Groundwater is a Na-Cl type with TDS > 5g/L (maximum value of 9g/L). Aquifer Ⅱ, depth to ~150m consists of fine sand and silt. Groundwater is Na-Cl type with TDS 7~8g/L. Aquifer Ⅲ and aquifer Ⅳ are hydraulically connected and are the target of present resource exploitation. Aquifer Ⅲ, depth to ~350m and 25~60m thick, is a confined aquifer consisting of fine sand. The specific yield is about 5~10m^3/hm. Groundwater is Na-Cl-HCO_3 type with TDS<1g/L. Fluoride concentrations range up to 5.5~6.9mg/L. Aquifer IV, depth to ~550m, with a thickness of 20~50m, consists of fine sand and silt. The specific yield is less than 2.5m^3/hm. Groundwater is Na-Cl-HCO_3 type with TDS<1g/L. The fluoride concentration is up to 2.5~3.5mg/L. Aquifer V (subdivided into V1 to V3), depth to ~1100m, is a Pliocene confined aquifer consisting of fine sand and silts in the upper part of this group, and mudstone, sandstone and conglomerate sandstone in the lower part of this group. The deposits of subgroup aquifer V1, depth at above 680m, consist of 8~12 layers of alluvial and lacustrine fine sands with a total thickness of ~70m. Groundwater is HCO_3-Na type with TDS 1.15~1.48g/L. The fluoride concentration is about 2mg/L. The deposit of subgroup aquifer V2, depth at above 820m, consists of sandstone with a total thickness of ~50m. Groundwater is Na-HCO_3 type with TDS ~1.5mg/L.

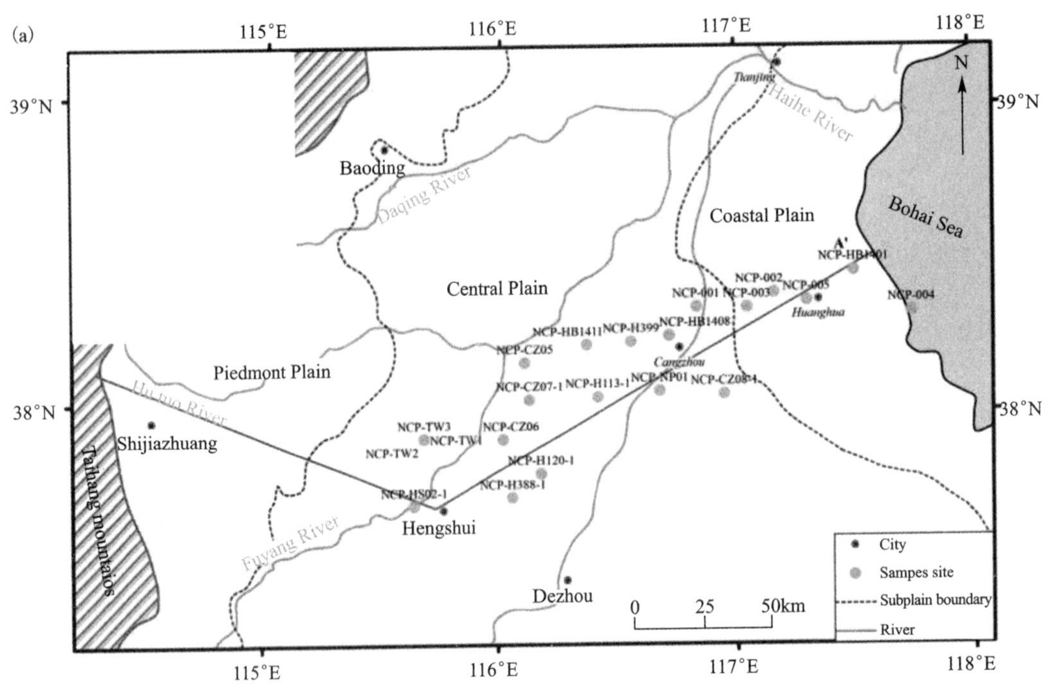

Figure 12. 6 The study area and sample sites in the North China Plain.

12. 11. 2 Helium concentrations and $^3He/^4He$ ratios

Table 12. 1 presents results of noble gas analysis by sampling site. The groundwater samples from the NCP have a range of helium concentrations, spanning from 6×10^{-8} cm^3 STP/g to 7×10^{-5} cm^3 STP/g. As shown in Figure 12. 7a, these concentrations increase towards the coastal plain area, and the new data extend the trends defined by a previous study that covered the area closer to the recharge area (Kreuzer et al. , 2009). Isotope ratios of helium ($^3He/^4He$) from this study and those by Kreuzer et al. (2009) show a monotonic decrease from the recharge area to the middle of the central plain area, and minimum of about 1.1×10^{-7} in well H1201 (Figure 12. 7b). Thereafter, $^3He/^4He$ ratios increase towards the coastal plain. Plotting $^3He/^4He$ ratios versus Ne/He ratios reveals the involvement of three isotopically and elementally distinctive components in the NCP groundwater samples (Figure 12. 7c). Except for data with tritiogenic ^3He, which is evident from $^3He/^4He$ ratios greater than the atmospheric ratios, samples from the Piedmont area and the western part of the Central Plain (i. e. $^3He/^4He>2\times10^{-7}$) nearly all plot along a two-component mixing line between atmospheric and crustal radiogenic sources. The remaining samples show a systematic departure from this binary mixing trend to a source component with elevated $^3He/^4He$ and low Ne/He ratios, revealing additional mantle components in those groundwater samples.

Table 12.1 Results of noble gas isotope analysis and resultant groundwater ages from ^{81}Kr and ^{4}He

Sample	Sample date	Longitude	Latitude	Depth (m) (Aquifer)	Altitude (m. a. s. l.)	Mass spectrometer results			ATTA results	Groundwater ages (10^5 years)	
						He (10^{-6} cm^3 STP/g)	Ne (10^{-7} cm^3 STP/g)	$^3He/^4He$ ($\times10^{-6}$)	$^{81}Kr/Kr$ (R_{sample}/R_{air})[a]	^{81}Kr ages	^4He model age[b]
NCP-001	2013-11-18	116.50	38.20	420(Ⅳ)	10	33.8 (±0.8)	2.40 (±0.06)	0.419 (±0.005)	0.200 (+0.032/−0.028)	5.3 (+0.4/−0.4)	6.8
NCP-002	2013-11-20	117.09	38.23	530(Ⅵ)	4	373 (±0.9)	2.09 (±0.08)	0.340 (±0.004)	0.114 (+0.021/−0.018)	7.2 (+0.5/−0.5)	6
NCP-003	2013-11-20	117.02	38.20	460(Ⅳ)	8	273 (±0.6)	2.49 (±0.10)	0.443 (±0.007)	0.206 (+0.024/−0.024)	5.2 (+0.4/−0.3)	5.5
NCP-004	2013-11-21	117.44	38.19	580(Ⅵ)	4	67.0 (±1.6)	2.23 (±0.06)	0.477 (±0.004)	0.050 (+0.016/−0.013)	9.9 (+0.9/−0.9)	10.4
NCP-005	2013-11-21	117.18	38.22	560(Ⅵ)	4	30.2 (±0.7)	2.28 (±0.06)	0.236 (±0.002)	0.098 (+0.019/−0.016)	7.7 (+0.5/−0.5)	5.9
NCP-CZ05	2016-05-24	116.11	38.16	400(Ⅲ)	10	0.60 (±0.01)	2.07 (±0.02)	0.162 (±0.008)			0.81
NCP-C206	2016-05-24	116.02	37.91	280(Ⅲ)	10	0.47 (±0.01)	2.28 (±0.03)	0.182 (±0.004)			0.6
NCP-CZ07-1	2016-05-24	116.13	38.04	280(Ⅲ)	10	0.73 (±0.02)	2.29 (±0.03)	0.160 (±0.008)			1
NCP-CZ08-1	2016-05-24	116.95	38.06	300(Ⅲ)	10	0.33 (±0.00)	2.20 (±0.02)	0.251 (+0.011)			0.38
NCP-H113-1	2016-05-25	116.42	38.05	320(Ⅲ)	10	2.06 (±0.03)	2.35 (±0.03)	0.159 (±0.013)			2
NCP-H120-1	2016-05-25	116.18	37.80	300(Ⅲ)	10	1.24 (±0.02)	2.36 (±0.03)	0.108 (±0.006)			1.6
NCP-H388-1	2016-05-25	116.06	32.72	350(Ⅲ)	10	1.47 (±0.02)	2.41 (±0.03)	0.135 (+0.007)			1.6
NCP-H399	2016-05-26	116.56	38.23	280(Ⅲ)	10	1.89 (±0.03)	2.35 (±0.03)	0.123 (±0.014)			2.1
NCP-HB1401	2016-05-26	117.50	38.46	700(V2)	10	47.3 (±0.7)	2.60 (±0.03)	0.620 (±0.012)			4.8
NCP-HB1408	2016-05-25	116.72	38.25	400(Ⅳ)	10	1.28 (±0.02)	Not determined	0.348 (±0.007)			1.4
NCP-HB1411	2016-05-26	116.37	38.22	300(Ⅲ)	10	0.99 (±0.01)	2.36 (±0.03)	0.124 (±0.004)			1.3
NCP-HS02-1	2016-05-24	115.64	37.69	350(Ⅳ)	10	0.34 (±0.00)	2.17 (±0.02)	0.240 (±0.006)			0.4
NCP-NP01	2016-05-25	116.68	38.07	300(Ⅲ)	10	8.1 (±0.1)	2.31 (±0.03)	0.221 (±0.008)			3.8
NCP-TW1	2016-05-23	115.68	37.91	600(V1)	10	0.107 (±0.00)	3.19 (±0.04)	1.031 (±0.007)			
NCP-TW3	2016-05-23	115.68	37.91	305(Ⅲ)	10	0.065 (±0.00)	2.70 (±0.03)	1.194 (±0.012)			

a. ^{81}Kr is expressed in terms of the air-normalized ratio, $R_{sample}/R_{air}=[^{81}Kr/K]_{sample}/[^{81}Kr/K]_{air}$. where Rair is the modern atmospheric ratio, $[^{81}Kr/K]_{air}=1.10(\pm0.05)\times10^{-12}$, measured by ATTA(Du et al. 2003).

b. ^4He model ages are alculated based on the ffective ^4He flux and vertical helium diffusion rate determined by using ^{81}Kr and ^4He results as input parameters of the model (see text).

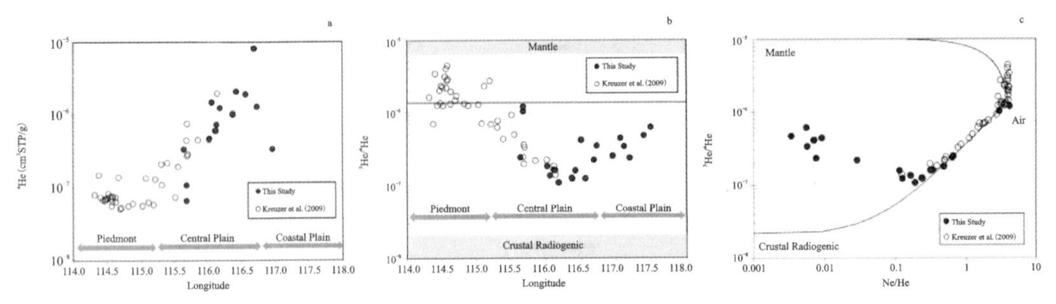

Figure 12.7 ^4He (a) and $^3He/^4He$ (b) ratios of groundwater samples from the North China Plain plotted against longitude of the sampling points. Note that the aquifer flows the eastward, so that flow is from low to high longitude. c. $^3He/^4He$ ratios versus Ne/He ratios, with mixing lines connecting air (Air-equilibrated water), mantle and air and a crustal component. Note that higher $^3He/^4He$ ratios near the atmospheric endmember are due to additional tritiogenic ^3He, as these are very young groundwater (Kreuzer et al., 2009).

12.11.3 ^{81}Kr ages

Previous studies on groundwater dating with $^3H/^3He$ and ^{14}C methods on the western side of the NCP reveal groundwater ages that are progressively older from the piedmont area of the Taihang Mountains to the east. Kreuzer et al. (2009) also showed ^4He concentrations correlated with $^3H/^3He$ and ^{14}C ages, suggesting that ^4He accumulates with increasing groundwater residence time (Figure 12.8). However, it also appears that the correlation becomes uncertain around the 20 000～30 000 years range due to the detection limit of the ^{14}C method. Thus, ^{14}C is not expected to provide meaningful age information for our samples because they are from sites further east to the Central Plain area. To obtain residence time for much older samples, we collected dissolved gas samples from five separate sites from the Coastal plain area for age determination by ^{81}Kr.

Table 12.2 shows the results of ^{81}Kr analyses of these samples. $^{81}Kr/Kr$ ratios in the samples ($=R_{sample}$) range from 5% to 20% of the modern atmospheric ratio [$R_{Air}=(^{81}Kr/Kr)_{air}=1.1\times10^{-12}$; Du et al., 2003]. With the ^{81}Kr decay constant ($\lambda_{Kr}=3.03\times10^{-6}$ yr^{-1}), the age (t_{Kr}) was calculated using:

$$t_{Kr}=-\frac{1}{\lambda_{Kr}}\ln\left(\frac{R_{sample}}{R_{air}}\right) \tag{12.47}$$

The range of ages estimated for these samples was 0.5Ma to 1.0Ma and these ^{81}Kr ages correlate with their radiogenic ^4He contents (Figure 12.8). The oldest among the five samples (NCP-004) is from the easternmost site; the youngest one (NCP-001) is from the westernmost and is 70km away from the NCP-004 site.

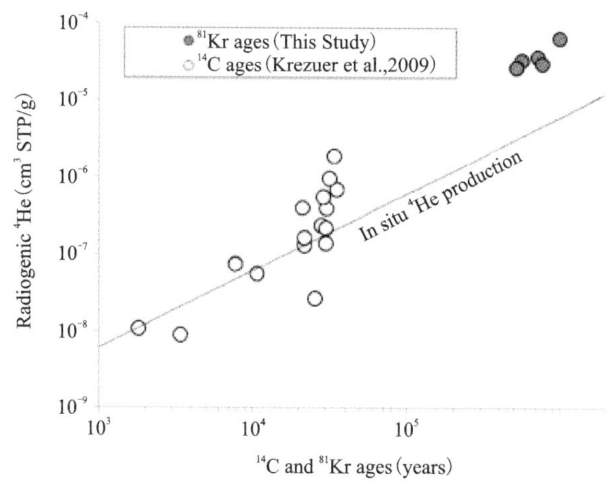

Figure 12.8 ^{14}C (Krezuer et al., 2009) and ^{81}Kr ages determined for the NCP groundwater samples plotted against the radiogenic ^4He. Radiogenic ^4He produced within the aquifer matrix and transferred to groundwater is shown with a reported in situ production rate of 6×10^{-12} cm^3 STP/g/yr (Wei et al., 2015).

12.11.4 Calibration of effective ^4He flux using ^{81}Kr ages

We have shown above that there is an increase in ^4He concentrations along the flow path and that ^4He concentrations correlate with ^{81}Kr ages spanning a much older age range than is accessible using the ^{14}C method (Figure 12.4). These observations lead to an expectation that ^4He concentrations provide a robust age tracer, if the rate at which radiogenic ^4He accumulation in the aquifer samples can be constrained.

Radiogenic ^4He is produced within aquifers by α-decay of U and Th, and its accumulation rate can be estimated with a knowledge of U and Th elemental contents of aquifer rock matrix, along with the assumption of a complete transfer of radiogenic ^4He to groundwater. The simplest approach to estimate groundwater residence times is to assume in situ production as the sole source for the observed amount of crustal ^4He and to use the production rate as rate constant (e.g., Castro et al., 2000; Kipfer et al., 2002; Wei et al., 2015). However, this often results in overestimation of groundwater ages when compared to residence times determined by other techniques (Aggarwal et al., 2015). In the case of the NCP, the in situ production rate of ^4He was estimated to be about 6×10^{-12} cm^3 STP/g/yr based on U and Th measurements from sedimentary core samples (Wei et al., 2015). As discussed in Wei et al. (2015), in situ ^4He can account for observed amounts of radiogenic ^4He in samples with ^{14}C ages younger than about $(2\sim3)\times10^4$ years (Figure 12.9). Geographically, these younger samples are from the Piedmont to the middle of the Central Plain (longitude $<$ ca. 115.5° E). However, for samples with longer residence times, especially those from the Coastal Plain with ^{81}Kr ages of $>5\times10^5$ years, the in situ production is insufficient to account for the observed amount of radiogenic ^4He (Figure 12.9). This discrepancy suggests that the in situ component is responsible for only a part of total radiogenic ^4He in those samples with relatively larger radiogenic ^4He ($>2\times10^{-7}cm^3$ STP/g), and that an external basal ^4He flux into the aquifer is required to control the amount of helium in the samples (e. g. Torgersen and Ivey, 1985). Distribution of radiogenic ^4He in an aquifer with two sources (an in situ component and an external basal ^4He flux) can be modeled as (Torgersen and Ivey, 1985):

$$[^4He]_{x,z} = \left(\frac{P}{U}\right)x + \left[\frac{F(^4He)/\varphi}{U}\right]$$

$$\left[\frac{x}{h} + \left(\frac{hU}{D_{He}}\right)\times\left\{\frac{3z^2-h^2}{6h^2} - \frac{2}{\pi^2}\sum_{n-1}^{\infty}\frac{(-1)^n}{n^2}e^{-\frac{D_{He}n^2\pi^2x^2}{h^2U}}\cos\frac{n\pi z}{h}\right\}\right] \tag{12.48}$$

where P=production rate of ^4He by in situ decay, U=horizontal flow velocity, φ=porosity of the aquifer, x=distance from the recharge zone, h is thickness of the aquifer, z is a depth of sampling from the aquifer top, and D_{He} is an effective (vertical) helium diffusion coefficient. A basal flux $F(^4He)$ enters the aquifer across its bottom. U denotes a horizontal flow rate of groundwater and is written as x/t with a residence time (t). Some of the parameters, such as an in situ ^4He production rate (P), can be estimated from U and Th contents of aquifer matrix rock. In the case of the NCP aquifer system, as noted above, previous work reports $P=6\times10^{-12}cm^3$ STP/g/yr (Wei et al., 2015). Aquifer geometry is also relatively simple for the NCP with its confined layers (layer Ⅲ-V3) consisting of a thickness of about 950m and a relatively flat basement at 1050m b. s. l. Depths of well screens are also known, and porosity is reported to be 0.2. These leaves the effective ^4He flux $F(^4He)$ and the vertical diffusion coefficient D_{He}, as two variables in the equation 12.48, that control the model distribution of radiogenic ^4He within the NCP aquifer system.

It is also possible to calculate the time for the groundwater at the sampling depth to obtain the observed amount of radiogenic ^4He (this required accumulation time will hereafter be called "^4He model age"). The ^4He model age strongly depends on $F(^4He)$ and D_{He}, and differs significantly depending on the depth of samples (namely the z/h ratios). Optimization of $F(^4He)$ and D_{He} is possible by minimizing the differences between the ^4He model ages and the observed ^{81}Kr ages (Aggarwal et al., 2015). A set of five ^4He model ages (NPC-001 to 005) were calculated for a pair of given $F(^4He)$ and D_{He} from $F(^4He)=10^{-7}\sim10^{-5}cm^3$ STP/cm^2/yr and $D_{He}=10^{-11}\sim10^{-7}m^2/s$ and compared with the ^{81}Kr ages. As shown in Figure 12.9, we find the residual sum of squares between the ^4He model ages and the observed ^{81}Kr ages shows a minimum (i. e. best agreement) at $F(^4He)=1.8\times10^{-6}cm^3$ STP/cm^2/yr and $D_{He}=8.5\times10^{-9}m^2/s$. With these fluxes and vertical diffusion, the ^4He model ages agree with the observed ^{81}Kr ages within about 30% (Figure 12.9). Figure 12.9b displays the degree of concordance between the ^4He model ages and ^{81}Kr ages with the optimized values of $F(^4He)$ and D_{He}. Considering the large uncertainties in assumed or assigned parameters (e. g. , depth of sampling, porosity and aquifer geometry) as well as analytical uncertainty in ^{81}Kr, a robust assessment of the uncertainties in these obtained ^4He model ages is not feasible. For now, we assign a 30% error to cover the differences between ages by ^4He and ^{81}Kr.

The method described above yields reasonable agreement between ^{81}Kr and ^4He chronometers in two separate aquifers-the NCP and Guarani aquifers, demonstrating that groundwater dating by ^4He concentrations can be more quantitative when thc model parameters are calibrated by an independent age tracer.

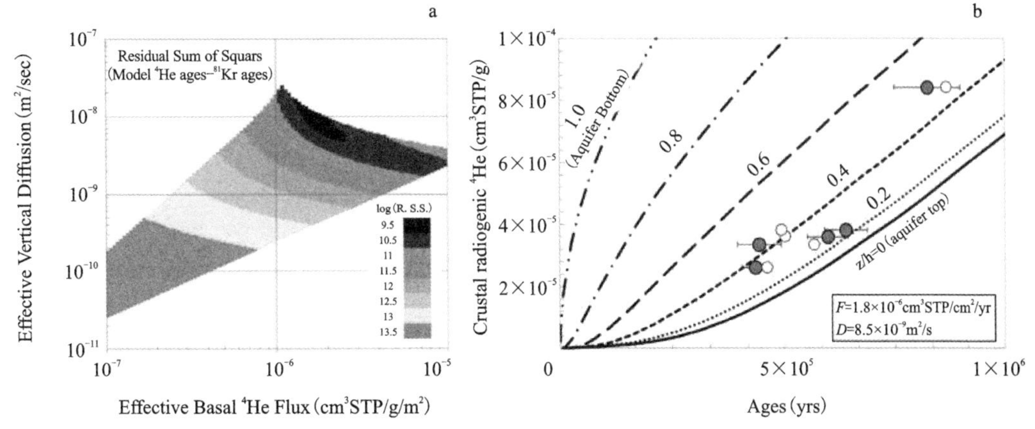

Figure 12. 9 a. Residual sum of squares of the model ^4He ages and the observed ^{81}Kr ages obtained over ranges of basal effective ^4He flux and the effective vertical helium diffusion coefficients. b. Concentration of crustal radiogenic ^4He in the NCP plotted against ^{81}Kr ages (filled circles) and ^4He model ages. Curves represent the model concentrations of ^4He at given depths and ages in the aquifer (expressed as z/h ratios) determined based on the model of Aggarwal et al. (2015), with the effective ^4He flux and diffusion coefficients optimized by using ^{81}Kr ages.

12. 11. 5 Insights from multiple isotopic tracers (^4He, ^{14}C and ^{81}Kr)

In order to see the distribution of groundwater ages across the flow path of the NCP, we compiled ages estimated by ^{81}Kr (this study), ^{14}C (Kreuzer et al. , 2009) and ^4He(^4He model ages based on ^4He concentrations reported here and in Wei et al. , 2015). The model ages were estimated for samples from the eastern part of the NCP, as the application of the ^4He flux optimized based on ^4He and ^{81}Kr from the eastern part should lead underestimation of ages for samples from the western part of the NCP.

As noted earlier, groundwater residence times are at the ^{14}C detection limit in the middle of the Central Plain, and there is a gap between the ages determined by ^{81}Kr and those reliably dated by ^{14}C. These ^4He model ages cover the gap by extending the trends previously defined by the youngest ages (^{14}C ages with$<10^4$ years) to the older age range determined by the ^{81}Kr method ($>5\times10^5$ years). This suggests that there is continuity of groundwater flow of the NCP from the recharge zone to the coastal plain area, at least in the deeper confined sections. This finding is in good agreement with a recent conceptual model for the groundwater in the Quaternary and Neogene aquifers in the NCP (Cao et al., 2016).

The apparent correlation of ages and flow distance in the semilogarithmic diagram means that groundwater ages increase exponentially to the east. This further reveals an exponential decrease in the flow rates to the east. In an age range below 10 000 years in the eastern end of the Piedmont area (longitude of 115.0°E to 115.5°E), ^{14}C ages indicate an eastward component of the flow rate of about 5m/yr (Kreuzer et al., 2009). In the middle of the Central Plain the flow rate suggested by the ^4He model ages is 0.8m/yr to 1m/yr. Further east in the Coastal Plain, the ^{81}Kr ages define a flow rate of 0.2m/yr. This dramatic drop in flow rates towards the coastal area should be reflected in a decreased hydraulic gradient and/or continuous drop of permeability, likely reflecting finer deposits in the Neogene aquifers from the central to coastal plains.

References

AGGARWAL P, MATSUMOTO T, STURCHIO N C, et al., 2015. Continental degassing of ^4He by surficial discharge of deep groundwater[J]. Nat. Geosci., 8: 35-39.

CAO G, HAN D, CURRELL M J, et al., 2016. Revised conceptualization of the North China Basin groundwater flow system: groundwater age, heat and flow simulations[J]. J. Asian Earth Sci., 127: 119-136.

CASTRO M C, STUTE M, SCHLOSSER P, 2000. Comparison of ^4He ages and ^{14}C ages in simple aquifer systems: implications for groundwater flow and chronologies[J]. Appl. Geochem., 15: 1137-1167.

COLLON P, KUTSCHERA W, LOOSLI H H, et al., 2000. ^{81}Kr in the Great Artesian Basin, Australia: A new method for dating very old groundwater[J]. Earth Planet Sci. Lett., 182:103-113.

DAVIS S N, MOYSEY S, CECIL L D, et al., 2003. Chlorine-36 in groundwater of the united states: Empirical data[J]. Hydrogeology Journal, 11:217-227.

DAVIS S N, CECIL L D, ZREDA M, et al., 2001. Chlorine-36, bromide, and the origin of spring water[J]. Chemical. Geology, 179:3-16.

KIPFER R, AESCHBACH-HERTIG W, PEETERS F, et al., 2002. Noble gases in lakes and ground water[J]. Rev. Mineral. Geochem., 47: 615-700.

KREUZER A M, VON ROHDEN C, FRIEDRICH R, et al., 2009. A record of temperature and monsoon intensity over the past 40kyr from groundwater in the North China Plain. Chem. Geol., 259: 168-180.

LEHMANN B E, LOOSLI H H, RAUBER D, et al., 1991. ^{81}Kr and ^{85}Kr in groundwater, Milk River Aquifer, Alberta, Canada[J]. Applied Geochemistry, 6:419-423.

LOVE A J, HERCZEG A L, SAMPSON L, et al., 2000. Sources of chlorine and implications for ^{36}Cl dating of old groundwater, southwestern Great Artesian Basin[J]. WRR, 36:1561-1574.

MATSUMOTO T, CHEN Z, WEI W, et al., 2018. Application of combined ^{81}Kr and ^{4}He chronometers to the dating of old groundwater in a tectonically active region of the North China Plain[J]. Earth Planet. Sci. Lett., 493: 208-217.

MORGENSTERN U, TAYLOR C B, PARRAT Y, et al., 1996. ^{32}Si in precipitation: Evolution of temporal and spatial variation and as a dating tool for glacial ice[J]. Earth Planet Sci. Lett., 144:289-296.

OZIMA M, PODOSEK F, 2001. Noble Gas Geochemistry (second edition) [M]. Cambridge:Cambridge University Press:286.

SOLOMON D K, COOK P G, 2000. ^{3}H and ^{3}He[J]//COOK P G, HERCZEG A L, Environmental Tracers in Subsurface Hydrology[M]. Boston: Kluwer Academic Publishers: 397-424.

TORGERSEN T, 1989. Terrestrial helium degassing fluxes and the atmospheric helium budget: implications with respect to the degassing processes of continental crust[J]. Chem. Geol., Isot. Geosci. Sect., 79: 1-14.

WEI W, AESCHBACH-HERTIG W, CHEN Z, 2015. Identification of He sources and estimation of He ages in groundwater of the North China Plain[J]. Appl. Geochem., 63: 182-189.

WILES D R, 2002. The Chemistry of Nuclear Fuel Waste Disposal[M]. Montreal, Canada:Polytechnic International Press.

WMO, 2001. World meteorological organization global atmosphere watch, No. 143 [R]. Global atmosphere watch measurement guide, WMO TD,(1073):83.